ENERGY IN AGROECOSYSTEM

ENERGY IN AGROECOSYSTEM

Edited by:

Quan Cui

ARCLER
PRESS

www.arclerpress.com

Energy in Agroecosystem

Quan Cui

Arcler Press
2010 Winston Park Drive,
2nd Floor
Oakville, ON L6H 5R7
Canada
www.arclerpress.com
Tel: 001-289-291-7705
001-905-616-2116
Fax: 001-289-291-7601
Email: orders@arclereducation.com

ISBN: 978-1-77407-224-0 (Hardcover)

DECLARATION

ABOUT THE EDITOR

Quan Cui obtained her PhD from Beijing Normal University in 2015. She worked as environmental impact assessment engineer for years. Aiming at linking science to policy, her study interest spans from ecosystem modelling to water resources management. She published peer-reviewed articles about using machine learning models in inflow forecasting, CO_2 management and watershed ecosystem health assessment.

TABLE OF CONTENTS

LIST OF CONTRIBUTORS

Mohamed Ghali
Ecole Supérieure d'Agricultures Angers-Loire (ESA), Social Sciences Laboratory, LARESS, Angers 49007, France

Laure Latruffe
Structures and Markets in Agriculture, Resources and Territories (SMART), French National Institute for Agricultural Research (INRA), Rennes 35000, France

Karine Daniel
Ecole Supérieure d'Agricultures Angers-Loire (ESA), Social Sciences Laboratory, LARESS, Angers 49007, France

Si-Ya Wang
College of Water Resources & Civil Engineering, China Agricultural University; Beijing 100083, China

Jun Qiu
State Key Laboratory of Hydroscience & Engineering, Tsinghua University, Beijing 100084, China
State Key Laboratory of Plateau Ecology and Agriculture, Qinghai University, Xining 810016, China

Fang-Fang Li
College of Water Resources & Civil Engineering, China Agricultural University; Beijing 100083, China
State Key Laboratory of Plateau Ecology and Agriculture, Qinghai University, Xining 810016, China

Ioan Aschilean
National Research and Development Institute for Cryogenic and Isotopic Technologies—ICSI Rm. Valcea, Uzinei Street, No. 4, 240050 Ramnicu Valcea, Romania

SC ACI Cluj SA, Avenue Dorobantilor, No. 70, Cluj-Napoca 400609, Romania

Gabriel Rasoi
National Research and Development Institute for Cryogenic and Isotopic Technologies—ICSI Rm. Valcea, Uzinei Street, No. 4, 240050 Ramnicu Valcea, Romania

Maria Simona Raboaca
National Research and Development Institute for Cryogenic and Isotopic Technologies—ICSI Rm. Valcea, Uzinei Street, No. 4, 240050 Ramnicu Valcea, Romania

Constantin Filote
Stefan cel Mare University of Suceava, Faculty of Electrical Engineering and Computer Science, 720229 Suceava, Romania
MANSiD Integrated Center, PROTHILSYS Lab, Stefan cel Mare University, 720229 Suceava, Romania

Mihai Culcer
National Research and Development Institute for Cryogenic and Isotopic Technologies—ICSI Rm. Valcea, Uzinei Street, No. 4, 240050 Ramnicu Valcea, Romania

Zhihai Yang
College of Economics and Management, Huazhong Agricultural University, No. 1 Shizishan Street, Wuhan 430070, China

Dong Wang
UWA School of Agriculture and Environment, The University of Western Australia, M089, 35 Stirling Highway, Crawley, WA 6009, Australia
Australia-China Sustainable Research and Development Center, Perth, WA 6009, Australia
Inner Mongolia Honghe Energy and Environment Consultancy, Huhhot 010020, China

Tianyi Du
School of Business and Management, Murdoch University, 90 South Street, Murdoch, WA 6150, Australia
Australia-China Sustainable Research and Development Center, Perth, WA 6009, Australia

Anlu Zhang
College of Land Management, Huazhong Agricultural University, No. 1 Shizishan Street, Wuhan 430070, China

Yixiao Zhou
School of Economics and Finance, Faculty of Business and Law, Curtin University, Rm 3016, Bldg 408, Bentley, Perth, WA 6004, Australia

George Vlontzos
Department of Agriculture Crop Production and Rural Environment, University of Thessaly, Fytoko, 38446 Volos, Greece

Spyros Niavis
Department of Planning and Regional Development, University of Thessaly, Pedion Areos, 38334 Volos, Greece

Panos Pardalos
Department of Industrial and Systems Engineering, University of Florida, 401 Weil Hall, Gainesville, FL 32611-6595, USA

Tek Maraseni
Institute for Agriculture and the Environment, University of Southern Queensland, Toowoomba, Queensland 4350, Australia

Guangnan Chen
Institute for Agriculture and the Environment, University of Southern Queensland, Toowoomba, Queensland 4350, Australia

Thomas Banhazi
Institute for Agriculture and the Environment, University of Southern Queensland, Toowoomba, Queensland 4350, Australia

Jochen Bundschuh
Institute for Agriculture and the Environment, University of Southern Queensland, Toowoomba, Queensland 4350, Australia

Talal Yusaf
Institute for Agriculture and the Environment, University of Southern Queensland, Toowoomba, Queensland 4350, Australia

Nan Li
Key Laboratory of Ocean Energy Utilization and Energy Conservation of Ministry of Education, Dalian University of Technology, Dalian 116024, China

Hailin Mu
Key Laboratory of Ocean Energy Utilization and Energy Conservation of Ministry of Education, Dalian University of Technology, Dalian 116024, China

Huanan Li
Key Laboratory of Ocean Energy Utilization and Energy Conservation of Ministry of Education, Dalian University of Technology, Dalian 116024, China

Shusen Gui
Key Laboratory of Ocean Energy Utilization and Energy Conservation of Ministry of Education, Dalian University of Technology, Dalian 116024, China

Yuzhu Wang
College of Information and Electrical Engineering China Agricultural University, Beijing 100083, China

Huanna Niu
College of Information and Electrical Engineering China Agricultural University, Beijing 100083, China

Lu Yang
College of Information and Electrical Engineering China Agricultural University, Beijing 100083, China

Weizhou Wang
State Grid Gansu Electric Power Research Institute, Lanzhou 730000, China

Fuchao Liu
State Grid Gansu Electric Power Research Institute, Lanzhou 730000, China

Ricardo Manso
Marine, Environment and Technology Centre—MARETEC, Department of Mechanical Engineering, Instituto Superior Técnico, Universidade de Lisboa, Avenida Rovisco Pais, 1, 1049-001 Lisbon, Portugal

Tânia Sousa
Marine, Environment and Technology Centre—MARETEC, Department of Mechanical Engineering, Instituto Superior Técnico, Universidade de Lisboa, Avenida Rovisco Pais, 1, 1049-001 Lisbon, Portugal

Tiago Domingos
Marine, Environment and Technology Centre—MARETEC, Department of Mechanical Engineering, Instituto Superior Técnico, Universidade de Lisboa, Avenida Rovisco Pais, 1, 1049-001 Lisbon, Portugal

David Pimentel
5126 Comstock Hall, Cornell University, Ithaca, NY 14853, USA

Alison Marklein
5126 Comstock Hall, Cornell University, Ithaca, NY 14853, USA

Megan A. Toth
5126 Comstock Hall, Cornell University, Ithaca, NY 14853, USA

Marissa Karpoff
5126 Comstock Hall, Cornell University, Ithaca, NY 14853, USA

Gillian S. Paul
5126 Comstock Hall, Cornell University, Ithaca, NY 14853, USA

Robert McCormack
5126 Comstock Hall, Cornell University, Ithaca, NY 14853, USA

Joanna Kyriazis
5126 Comstock Hall, Cornell University, Ithaca, NY 14853, USA

Tim Krueger
5126 Comstock Hall, Cornell University, Ithaca, NY 14853, USA

LeilaFarajian
Department of Agricultural Economics, College of Agriculture and Natural Resources, Science and Research Branch, Islamic Azad University, Tehran, Iran

RezaMoghaddasi
Department of Agricultural Economics, College of Agriculture and Natural

Resources, Science and Research Branch, Islamic Azad University, Tehran, Iran

SafdarHosseini
Department of Agricultural Economics, University of Tehran, Iran

Kurt B.Waldman
Department of Geography, Indiana University, 513 N. Park Ave., Bloomington, IN 47408, USA

Robert B.Richardson
Department of Community Sustainability, Michigan State University, 480 Wilson Rd., East Lansing, MI 48824, USA

Christiane Peter
Leibniz Centre for Agricultural Landscape Research, Institute of Landscape Systems Analysis, Eberswalder Straße 84, 15374 Müncheberg, Germany
Justus Liebig University Gießen, Institute of Farm and Agribusiness Management, Senckenbergstraße 3, 35390 Gießen, Germany

Katharina Helming
Leibniz Centre for Agricultural Landscape Research, Institute of Land Use Systems, Eberswalder Straße 84, 15374 Müncheberg, Germany
Eberswalde University for Sustainable Development, Faculty of Landscape Management and Nature Conservation, Schicklerstraße 5, 16225 Eberswalde, Germany

Claas Nendel
Leibniz Centre for Agricultural Landscape Research, Institute of Landscape Systems Analysis, Eberswalder Straße 84, 15374 Müncheberg, Germany

LIST OF ABBREVIATIONS

AAV	Agricultural added value
AFF	Agriculture, Forestry and Fisheries
AWU	Annual working units
ANN	Artificial neural network
APEC	Asia-Pacific Economic Cooperation
ACCUs	Australian Carbon Credit Units
ACF	Autocorrelation function
ARIMA	Auto Regressive Integrated Moving Average
CFIs	Carbon Farming Initiatives
CSC	China Scholarship Council
CV	Coefficient of variation
CIA	Coleambally Irrigation Areas
CRS	Constant returns to scale
CExC	Cumulative exergy consumption methodology
DEA	Data Envelopment Analysis
DMU	Decision-making unit
EAHE	Earth-air heat exchanger
ERF	Emission Reduction Fund
EMD	Empirical mode decomposition
EREA	Energy Resources Exergy Accounting
EEMD	Ensemble Empirical Mode Decomposition
EEA	Extended Exergy Accounting
FAMEN	Facility agriculture micro energy network
FADN	Farm Accountancy Data Network
FAO	Food and Agricultural Organization
FAOSTAT	Food and Agriculture Organization of the United Nations
FT	Fourier Transform

GDP	Gross domestic product
GLM	Generalized linear models
GA	Genetic Algorithm
GHG	Greenhouse gases
GVA	Gross value added
HDI	Human Development Index
ICSI	Institute for Cryogenic and Isotopic Technologies
ICRISAT	International Crops Research Institute for the Semi-Arid Tropics
IMF	Intrinsic mode functions
kNNs	k-nearest neighbors
LED	Light Emitting Diode
LHV	Lower heating value
LTH	Low temperature heat
LPG	(Liquefied Petroleum Gas)
MPPT	Maximum Power Point Tracking
MAPE	Mean absolute percentage error
MD	Mechanical drive
MTH	Medium temperature heat
MLPNN	Multi-layer perceptive neural network
MNLR	Multiple Nonlinear Regression
MIA	Murrumbidgee Irrigation Area
NREA	Natural resources' exergy accounting
NREA	Natural Resources' Exergy Accounting
NWP	Numerical Weather Prediction
OLS	Ordinary least squares
OECD	Organization for Economic Cooperation and Development
PACF	Partial autocorrelation function
PV	Photovoltaic
PEMFC	Polymer electrolyte membrane fuel cells
PEM	Proton excenge membrane
PTE	Pure technical efficiency

RFS	Renewable Fuel Standard
RMSE	Root mean square error
SECA	Simulation and Analysis of Energy Consumption for Agriculture
SNE	Specific Nutritional Energy
SFA	Stochastic frontier analysis
SVM	Support Vector Machines
TE	Technical efficiency
TFEE	Total-factor energy efficiency
UNFCCC	United Nations Framework Convention on Climate Change
UAA	Utilized agricultural area
VRS	Variable returns to scale
WA	Wavelet Analysis
WTP	Willingness-to-pay

PREFACE

Agroecosystem is the unit of agricultural ecosystem study, including all the elements inside this unit. Its unique is as the name defined, human activities are the core of this ecosystem, like other agricultural activities. Relying on which, agroecosystem has simpler species and backward energy flow into them comparing with natural ecosystems.

Since energy is the sources of agriculture products, the analysis and prediction of energy has always been a research focus. This book covers topics about energy demand modeling, energy efficiency analysis, energy use assessment of agroecosystems. Usually researches follow biomass production and fuel trade-offs to analyze energy efficiency of agroecosystems. However, other measurement concepts as exergy have been more and more popular. While analyses of energy efficiency in agroecosystems' biomass production offer important insights in the context of sustainable land management and biomass production, much of them focused on the energy efficiency of either food or energy provision. Comprehensive analyses at the total agroecosystem level that study long-term change in agroecosystem energetics in the course of the socio-ecological transition have been addressed in this book. On the other hand, basis topics for agroecosystem research such as land-use change, fertilizer/pesticides risk assessment, adaptation to climate change have been kept evolving with novel methodologies. These methodologies and topics were addressed with detailed case studies in this book.

CHAPTER 1

EFFICIENT USE OF ENERGY RESOURCES ON FRENCH FARMS: AN ANALYSIS THROUGH TECHNICAL EFFICIENCY

Mohamed Ghali [1], Laure Latruffe [2], and Karine Daniel [1]

[1]Ecole Supérieure d'Agricultures Angers-Loire (ESA), Social Sciences Laboratory, LARESS, Angers 49007, France

[2]Structures and Markets in Agriculture, Resources and Territories (SMART), French National Institute for Agricultural Research (INRA), Rennes 35000, France

ABSTRACT

Integrating natural resources and ecological services in the production process is crucial to implement sustainable agriculture. However, the measurement of natural resource efficiency remains difficult. This paper aims at contributing to this issue, by investigating French farms' use and excess (slack) of energy resources through Data Envelopment Analyses (DEA). Results show that disentangling energy resources from the rest of intermediate consumption highlights energy use excess which is masked when considering intermediate consumption as a whole. The analysis of the determinants of

Citation: Ghali, M.; Latruffe, L.; Daniel, K. Efficient Use of Energy Resources on French Farms: An Analysis through Technical Efficiency. Energies 2016, 9, 601.

energy use excess and of intermediate consumption shows a discrepancy in results, which policy-makers should take into account when designing energy policies. In addition, results show that large and highly capital intensive farms perform better in terms of energy use excess, while the dependence on public subsidies is a constraint.

Keywords
technical efficiency, energy resources, farms, Data Envelopment Analysis, France

INTRODUCTION

Challenging a productivist agricultural model raises many thoughts and initiatives concerning different ways to design and implement sustainable agriculture capable of preserving natural resources while minimizing the use of external inputs to the ecosystem. In this perspective, Pretty [1], Balmford et al. [2], and Gliessman [3] stress the importance of integrating natural resources and ecological services in the production process. This integration, however, raises a number of limitations including economic assessment [2,4]. Indeed, it is difficult to measure the effectiveness of ecological services because of their difficult physical quantification and monetary valuation [5]. Despite attempts to integrate natural resources in the production function, initiated by Solow [6] and Stiglitz [7], only resources exhibiting economic characteristics of a production factor can be integrated into the production function. Thus, the measurement of natural resource efficiency remains difficult. However some functional approaches exist, at the farm level, which assess whether farmers make efficient use of natural resources to achieve their economic goals.

This is the concept linking economic and ecological efficiencies known as eco-efficiency [8]. This concept was also used by the Organization for Economic Cooperation and Development (OECD) [9], which defines it as "the efficiency with which ecological resources are used to meet human needs", and has been adopted and popularized by the World Business Council for Sustainable Development [10,11] as a way to encourage companies to achieve higher levels of competitiveness and environmental responsibility at the same time. In practical terms, eco-efficiency is the ability to obtain economic performance by making minimal use of natural resources and causing minimal degradation to the environment [12]. Eco-efficiency can be measured using ratios between the economic value of goods or services produced by any single entity (farm, company, sector...), and the environmental pressures generated by the production process [13]. This eco-efficiency improves when environmental impacts diminish and the economic value of production is maintained or increases [12]. On this basis, De Koeijer et al. [14] discussed the measurement of eco-efficiency by using environmental data collected from Dutch sugar beet farms. Picazo-Tadeo et al. [13] assessed the eco-efficiency of Spanish farms in Castilla y León. Basset-Mens et al. [15] analyzed the eco-efficiency of New Zealand dairy farms and the implications of intensification on their eco-efficiency. The eco-efficiency concept was also adapted for the analysis of policy strategies and their possible macroeconomic outcomes [16]. It is used to study efficiency of economic branches, or to connect individual companies to the macroeconomic level [17,18]. These authors consider eco-efficiency as the ratio between added value and environmental pressure, the latter comprising five variables: the specialization of farms, nitrogen and phosphorus levels, the pesticide risk and the energy ratio. Zentner et al. [19] studied the effect of input management and crop diversity on the use of non-renewable energy in Canadian prairie crop systems. Mousavi-Avval et al. [20] evaluated the technical efficiency of energy use on barberry farms in Iran by investigating the relationship between energy waste and farm size.

In the majority of studies dealing with eco-efficiency of farms, businesses, industries, regions or countries, the concept is studied through the relationship between the economic value of goods and services and the impacts of production processes on the environment. Thus, regardless of the method used-parametric, non-parametric, stochastic or econometric-eco-efficiency is measured by considering environmental degradation as an additional input or output. Based on the strict definition of eco-efficiency, which refers to the effective use of environmental resources, the originality of our article is to separate the natural resource input from other intermediate consumption inputs, and to study the technical efficiency of farms in terms of the use of these resources for the case study of France. In our article, we focus on inputs linked to energy resources, including all expenses relating to fertilizers, soil amendments (compost, manure, sulphur...) and fuel, which directly account for the majority of fossil fuel used on the farm. The aim is to demonstrate how, through further analysis of production factors, a more operational diagnostic of farm performance could be drawn. This can shed light on French farms' room for maneuver in terms of improving the efficiency of energy resource use, and can help design environmental policies targeted towards the efficient use of natural resources.

In the second section we describe the methodology with an overview of the efficiency concepts underlying the analysis, and a description of the method employed to compute excess energy factors (slacks). In the third section we present the data used. In the fourth section we discuss the main results in terms of levels of technical efficiency, slack of energy resources and their determinants. Finally, we conclude in Section 5.

METHODOLOGY

In an economy with limited resources, the concept of technical efficiency is useful. It gives an indication on the ability of decision-making units (e.g., firms) to use the existing technology in the most appropriate way. It consists of scale efficiency and pure technical efficiency. Scale efficiency relates to the optimal level of farm size while pure technical efficiency reflects the ability of a decision-making unit to achieve maximum production for a set of inputs regardless of the price of factors and goods. It provides information about managerial practices and the organization of the production unit.

The literature proposes two major approaches to construct a production frontier and measure technical efficiency: parametric [21,22,23] and non-parametric approaches [24,25]. The parametric approach, the stochastic frontier approach, requires a functional form to specify the production frontier and uses econometric tools to estimate it. Deviation from the production frontier can be separated between inefficiency of the decision-making unit and random noise. The non-parametric approach, the Data Envelopment Analysis (DEA) [24], has its origins in the work of Farrell [26] and Farrell and Fieldhouse [27], and offers a relative measure of the efficient production function constructed from observations of inputs and outputs of the sample farms. This approach is not related to a predetermined functional form and prevents the occurrence of misspecification. It has been widely used to study environmental and economic assessments in the agricultural sector (e.g., [28,29,30]) and has gained popularity in energy efficiency analysis [31].

DEA consists of comparing the performance of each farm with the best farms within the current sample, or with a hypothetical farm which uses the same inputs proportionally [32]. The concept of efficiency relates to the distance of the farm considered to the production frontier: a large distance indicates low efficiency. DEA allows calculate both pure technical efficiency and scale efficiency, both of them taken together providing total technical efficiency, and allows obtain input slacks. This, in addition to the fact that DEA avoids misspecification errors due to its nonparametric framework, is why we use DEA here. Input slacks show which inputs are used in excess. This excess input corresponds to the potential additional reduction of an input in addition to the proportional reduction of all inputs as identified by the technical efficiency score. To illustrate this concept, Figure 1 shows a production frontier FF′ made up by the best performing decision-making units, let's say farms, C and D. The frontier shows the minimum use of both inputs considered (X_1 and X_2) to produce one unit of output (Q). Farms A and B are inefficient because they are not on the frontier. As defined by Farrell [26], technical efficiency of A, respectively of B, is given by the ratio OA′/OA, respectively OB′/OB. Hypothetical farms A′ and B′ are considered as peers for farms A and B respectively. Taking A as an example, this farm could reduce both inputs proportionally, that is to say along the ray OA, and produce as much as A′; this is the technical efficiency concept. However, A′ is not as efficient as C, since A′ could further reduce the use of factor $X_2$by the quantity CA′ and continue to produce the same amount of output. Thus, to be efficient in the Koopmans' [33] meaning (Koopmans (1951) gives a more stringent definition of technical efficiency which suggests that the decision-making unit is technically efficient if it operates on the frontier and all the slacks associated with the inputs are zero [32]), farm A must reduce its two inputs proportionally (more precisely by a percentage equal to 1 minus the score of technical efficiency) and further reduce X_2 by CA′. This quantity CA′ is called a slack for input X_2.

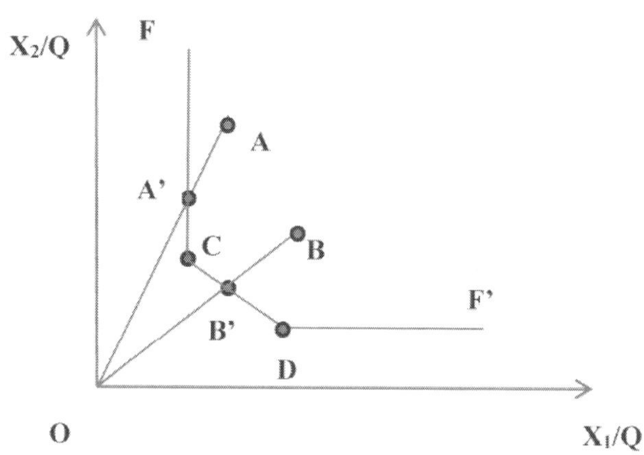

Figure 1: Representation of input slacks: case of two inputs, X_1 and X_2. Source: the authors, adapted from Coelli et al. [32].

Our objective is to measure slacks with respect to energy resource use, and to identify their determinants for French farms. For this, slacks calculated by DEA in a first stage are regressed on potential determinants in a second stage. Tobit models are used in this second stage due to the censored character at 0 of the dependent variables (some farms have no slacks for some inputs).

MODEL SPECIFICATION AND DATA

Data Envelopment Analyses (DEA) Model Specification

Three types of farms are considered here, based on their production specialization, and a separate frontier is constructed for each sample. The computation of technical efficiency and of input slacks is carried out in two ways, depending on the specification of the technology.

Firstly, we use a reference technology (technology T1) which includes three outputs and four inputs, the latter being the inputs conventionally used for measuring technical efficiency of farms: land, labor, capital and intermediate consumption.

Secondly, we separate the energy resources from intermediate consumption, so that the new technology (technology T2) retains the three outputs as in technology T1, but includes five inputs: land, labor and capital as in technology T1; energy resources and the rest of intermediate consumption.

Table 1 summarizes the outputs and inputs selected for the two technologies. The three outputs considered are output from crop activity, output from livestock activity and output from other activities, all expressed in Euros. As for the inputs, land is measured by the utilized agricultural area (UAA) in hectares (ha); labor is measured in annual working units (AWU), that is to say in full-time equivalents; farm capital is measured in Euros; intermediate consumption includes fertilizers, pesticides, seeds, feed, soil amendments, fuel, water, electricity and maintenance costs, and is expressed in Euros; energy resources, which are a part of intermediate consumption, are also expressed in Euros and include direct consumption of energy in the form of fuel, as well as indirect consumption through fertilizers and soil amendments.

Table 1: Description of output and input variables used to calculate technical efficiency and slacks

Name of the Variable	**Description of the Variable**	**Included in Technology T1**	**Included in Technology T2**
Inputs			
UAA	Utilized agricultural area (ha)	Yes	Yes
LAB	Labor (AWU)	Yes	Yes
CAP	Farm capital (Euros)	Yes	Yes
IC	Intermediate consumption (Euros)	Yes	-
ER	Energy resources including fertilizers, soil amendments and fuel (Euros)	-	Yes
R_IC	Remaining intermediate consumption, that is to say intermediate consumption without energy resources (Euros)	-	Yes
Outputs			
OUTPC	Crop output (Euros)	Yes	Yes
OUTPL	Livestock output (Euros)	Yes	Yes
OUTPO	Other farm outputs (Euros)	Yes	Yes

Specification of the Model for the Determinants of Input Slacks

In a second step we seek to identify the determinants of non-proportional excessive use of inputs (the slacks). For each slack we perform three Tobit regressions. In the first regression, we use as the dependent variable the intermediate consumption slacks calculated under technology T1. In the second regression, the dependent variable is the energy resource slacks identified under technology T2. In the third regression, the dependent variable is the slack of the remaining intermediate consumption calculated under technology T2. Variables used as determinants of input slacks are presented in Table 2. The selection of these variables was based on determinants identified in the farm technical efficiency literature (e.g., see the review by Latruffe [34]). In addition, only uncorrelated variables were selected.

Table 2: Variables used to explain input slacks

Name of Variable	Description of Variable
UAA	Utilized agricultural area (ha)
CAP/UAA	Ratio of capital to UAA (Euros per ha)
Debt	Indebtedness ratio defined as debts to capital
Non-Paid/Lab	Ratio of non-paid labor to total labor
S_COP	Share of cereals, oilseed crops and proteinseed crops in the farm's UAA (in %)
S_LEG	Share of legumes in the farm's UAA (in %)
SUB/OUTP	Ratio of subsidies to total output
Age	Age of the farmer
ENVI	ENVI is a dummy variable. ENVI = 0 if the farm is not in Natura 2000 area; ENVI = 1 if the farm is in Natura 2000 area
S_TEXT	Percentage of the clay-sand soil texture in the area of the farm (in %)

Within the determinants, the farm structure is represented by:

- The size of the farm in terms of UAA, which has a mixed effect on technical efficiency in the literature:
- The technology intensity described by the ratio of capital to land, which is expected to increase farm performance;
- The indebtedness ratio, which also has been found to have a mixed effect in the literature;
- The extent of non-paid labor compared to hired labor in total labor, whose effect may also be positive or negative;
- The share of cereals, oilseed crops and protein seed crops, and the share of legumes in the farm's UAA. These shares can give indications on the farm room of maneuver to reduce the use of fertilizers. Indeed they give indications on the nature of agricultural practices used in the different types of farms considered here, such as crop rotation.

Among other variables, the dependence on subsidies is considered and proxied by the ratio of subsidies received by farms related to their total output produced. The subsidies considered here are those received by farms in the frame of the European Union's Common Agricultural Policy and include decoupled subsidies, direct payments made on the basis of hectares of specific crops or heads of specific livestock, payments provided for environmental friendly practices, and lump-sum payments for farms located in disadvantaged areas. Human capital is proxied by farmer's age. Two location variables are used. The first one captures the presence of environmental constraints: it is a dummy indicating whether the farm is located in an environmental zoning area Natura 2000 (Natura 2000 is a European Union's network of nature protection areas established under the 1992 Habitats Directive. It includes Special Areas of Conservation designated by European Union's Member States under the Habitats Directive, and Special Protection Areas delimited under the 1979 Birds Directive [35]). The second one informs on the soil conditions for production: it is the percentage of the clay-sand texture of the soil in the area of the farm considered. Soils with such texture have good water holding capacity but their heavy and compact structure makes it difficult to work them in the rainy season [36].

Data

The database used is the French Farm Accountancy Data Network (FADN) in 2010. This database consists of bookkeeping data for a sample of statistically representative French commercial farms. The analysis is performed for three types of farms based on their production specialization: cereals, oilseed crops and protein seed crop farms; dairy farms; and mixed farms producing crop and livestock outputs. These three types of farms account for 48% of French farms and 62% of the total agricultural area of the country [37]. The samples used here consist of 1,108 farms which specialize in cereals, oilseed crops and protein crops, 1097 dairy farms and 867 mixed farms.

Table 3 and Table 4 present some descriptive statistics of the variables used to calculate technical efficiency and slacks, and of the potential determinants of slacks. Table 3 shows that cereals, oilseed crops and protein seed crops farms are on average larger in terms of land (UAA average of 148 ha) than dairy farms (96 ha) and mixed farms (137 ha). The levels of capital and labor are higher for farms with livestock activity. In terms of the use of energy resources, cereals, oilseed crops and protein seed crops farms use more per hectare on average (213 Euros) than mixed farms (195 Euros) and dairy farms (151 Euros). The share of energy resource expenses in total intermediate consumption is also greater for cereals, oilseed crops and protein seed crops farms (average share of 32.5%, compared to 19.3% for mixed farms and 13.5% for dairy farms). More precisely, the cost of fertilizers alone represents on average 23% of all intermediate consumption expenses for cereals, oilseed crops and protein seed crops farms, compared to 12% and 7.5% for mixed farms and dairy farms respectively. As for the percentage of fuel cost in all intermediate consumption, it is respectively 7.9%, 6.3%, 5.5% for cereals, oilseed crops and protein seed crops farms, mixed farms and dairy farms on average.

Table 3: Descriptive statistics of the variables used to measure technical efficiency and slacks (averages)

Variables	All Farms	Cereals, Oilseed Crops and Protein-seed Crops Farms	Dairy Farms	Mixed Farms
Number of farms	3072	1108	1097	867
Inputs				
UAA (ha)	126	148	96	137
LAB (AWU)	1.9	1.5	1.9	2.3
CAP (Euros/ha)	1658	939	2587	1820
IC (Euros/ha)	892	660	1117	1010
ER (Euros/ha)	190	213	151	195
R_IC (Euros/ha)	701	446	966	815
Fuel (Euros/ha)	58	52	61	63
Fertilizers (Euros/ha)	126	156	83	121
Outputs				
OUTPC (Euros/ha)	778	1128	239	773
OUTPL (Euros/ha)	674	36	1516	802
OUTPO (Euros/ha)	81	60	67	123

Table 4: Descriptive statistics of the determinants of slacks (averages)

Variables	All Farms	Cereals, Oilseed Crops and Protein Seed Crops Farms	Dairy Farms	Mixed Farms
Number of farms	3072	1108	1097	867
CAP/UAA (Euros/ha)	1969	932	2655	2427
Debt	0.422	0.413	0.417	0.441
Non-Paid/Lab	0.90	0.90	0.93	0.85
S_COP (%)	52	84	22	51
S_LEG (%)	2	4	0.3	2
SUB/OUTP	0.27	0.30	0.24	0.28

Age (years)	48.3	50.1	46.8	47.7
ENVI (dummy)	0.04	0.03	0.05	0.04
S_Text (%)	52.7	50.7	55.2	51.9

Table 4 indicates that livestock farms are more capital intensive than cereals, oilseed crops and protein seed crops farms, but that indebtedness level is similar. The subsidy to output ratio is on average 0.27 for all farms, indicating that on average farms in the whole sample receive 0.27 Euro of subsidies for every Euro of output produced.

RESULTS

Technical Efficiency Scores

Table 5 reports average levels of total technical efficiency and pure technical efficiency obtained for the three types of farms considering technology T1, that is to say using the classic inputs (capital, land, labor and intermediate consumption). The results show that farms specializing in dairy are, on average, the most efficient in terms of total technical efficiency: their average efficiency level is 0.74 compared to 0.65 for cereals, oilseed crops and protein seed crops farms and 0.67 for mixed farms. These results indicate that the three samples considered could reduce, on average, all inputs proportionally by 26%, 35% and 33% respectively. This indicates that the average total technical efficiency of crop farms is significantly lower than that of livestock farms, a result which is consistent with the literature (see the meta-analysis of Bravo-Ureta et al. [38]): for example Latruffe et al. [29] found an average of 0.572 for cereals, oilseed crops and protein seed crops French farms between 2001 and 2007, and 0.737 for dairy farms. A lower average technical efficiency for crop farms relative to livestock farms reflects greater heterogeneity in the use of inputs among arable farms than among livestock farms. This can be explained by the fact that climatic and soil conditions that differ across farms influence more the process of production of crop farms than of livestock farms. Table 5 shows that, in terms of average levels of pure technical efficiency, dairy farms are the most efficient with an average of 0.84, significantly different than the average of 0.81 for cereals, oilseed crops and protein seed crops farms and mixed farms. This shows that in terms of production management practices, livestock farms are more efficient than crop farms.

Table 5: Total technical efficiency (TE) and pure technical efficiency (PTE) of the three types of farms calculated under technology T1

Variables	**Cereals, Oilseed Crops and Protein Seed Crops Farms**		**Dairy Farms**		**Mixed Farms**	
Number of farms	1108		1097		867	
Total and pure technical efficiencies	TE	PTE	TE	PTE	TE	PTE
Average	0.65	0.81	0.74	0.84	0.67	0.81
Standard deviation	0.15	0.13	0.13	0.14	0.17	0.17
Minimum	0.13	0.37	0.26	0.41	0.17	0.39
Maximum	1	1	1	1	1	1
Percentage of efficient farms (%)	3.8	7.6	5.3	32.2	6.7	33.5

The efficiency scores estimated using technology T2, that is to say dissociating the energy resources from the rest of intermediate consumption, are presented in Table 6. They show the same trends as those calculated with technology T1 in Table 5. Regarding total technical efficiency, farms specializing in dairy are the most efficient on average, with an average efficiency of 0.75 compared to 0.67 for cereals, oilseed crops and protein seed crops farms and 0.70 for mixed farms respectively. Average levels of pure technical efficiency show that dairy farms are the most efficient with an average of 0.86.

Table 6: Total technical efficiency (TE) and pure technical efficiency (PTE) of the three types of farms calculated under technology T2

Variables	**Cereals, Oilseed Crops and Protein Seed Crops Farms**		**Dairy Farms**		**Mixed Farms**	
Number of farms	1108		1097		867	
Total and pure technical efficiencies	TE	PTE	TE	PTE	TE	PTE
Average	0.67	0.82	0.75	0.86	0.70	0.83
Standard deviation	0.15	0.13	0.14	0.14	0.17	0.16

Minimum	0.15	0.37	0.26	0.41	0.25	0.39
Maximum	1	1	1	1	1	1
Number of efficient farms	56	108	73	368	77	310
Percentage of efficient farms (%)	5.1	9.7	6.7	33.5	8.9	35.8

Input Excess

Table 7 shows, for the three types of farms and under technology T1, the potential reduction of all inputs. Firstly, the proportional potential reduction of all inputs is shown, and it is calculated as 1 minus the total technical efficiency score. Secondly, the average additional non-proportional reduction of each input is given based on slack results and calculated as follows: slack of input *i* obtained from the DEA model divided by the level of the *i*-th input used. Finally, the levels of the total potential reduction of each input (calculated as the sum of non-proportional reduction and proportional reduction) is provided. Table 8 shows similar calculations but when considering technology T2.

Table 7: Percentage of potential reduction of inputs under technology T1 (averages)

Variables	Cereals, Oilseed Crops and Protein Seed Crops Farms	Dairy Farms	Mixed Farms
Proportional reduction of all inputs (1-TE) in % (1)	35.3	26.5	32.6
Additional input slack as a % of input level used (2)			
UAA	6.2	5.6	9.2
LAB	0.9	0.6	0.5
CAP	6.2	3.5	5.4
IC	0.8	1.0	1.0
Total reduction of inputs (1) + (2) in %			

UAA	41.6	32.0	41.8
LAB	36.2	27.1	33.1
CAP	41.5	30.0	38.1
IC	36.1	27.5	33.6

Table 8: Percentage of potential reduction of inputs under technology T2 (averages)

Variables	Cereals, Oilseed Crops and Protein Seed Crops Farms	Dairy Farms	Mixed Farms
Proportional reduction of all inputs (1-TE) in % (1)	33.1	24.5	29.8
Additional input slack as a % of input level used (2)			
UAA	5.6	6.1	9.7
LAB	1.5	0.8	0.6
CAP	5.9	3.7	6.6
ER	5.3	10.6	9.1
R_IC	1.2	1.1	1.7
Total reduction of inputs (1) + (2) in %			
UAA	38.7	30.6	39.4
LAB	34.6	25.3	30.3
CAP	39.1	28.3	36.4
ER	38.5	35.2	38.9
R_IC	34.3	25.7	31.5

Under technology T1 (Table 7) results show large slacks (i.e., non-proportional potential reductions) for land and capital inputs. Land slacks are on average highest for mixed farms, while capital slacks are highest on average for cereals, oilseed crops and protein seed crops farms. In terms of capital, results indicate 6.2% excess capital for cereals, oilseed crops and protein seed crops farms and 5.4% for mixed farms. In total, these farms could reduce their respective use of capital by 41.5% and 38.1%. This illustrates heavy capitalization on farms using arable land. Regarding the intermediate consumption, results show only low values of excess, not exceeding 1% of the inputs used in the three groups of farms.

The results of this model could lead us to conclude that the leeway for farms resides mainly in the improvement of their land and capital factors. In other words, we would say that farmers would benefit from adjusting their capital and land investment strategies without the concern of intermediate consumption expenses, which only show a small percentage of slack. However, intermediate consumption includes various components that may hide important differences. Efficiency computations under technology T2, where energy resources are separated from the rest of intermediate consumption, can help in this way. The results, presented in Table 8, also show that the three groups of farms are too land and too capital intensive. However, the results also indicate that slacks for energy resources are not low, and of a similar level as slacks for land and capital: 5.3% for cereals, oilseed crops and protein seed crops farms, 10.6% for dairy farms and 9.1% for mixed farms. This gives support to our idea that the slack of intermediate consumption, observed in the first model (Table 7), does not reflect truly the use of energy resources.

The difference in excess use in energy resources across the three types of farms shown by Table 8, reflects differences in the heterogeneity of agricultural practices within each type of farms. A low average of energy resource slack for cereals, oilseed crops and protein seed crops farms indicate that farms in this particular sample are clustered towards the sample's mean of slack, while the higher averages for dairy farms and mixed farms indicate a larger dispersion around the mean of slack in the respective samples. This suggests that in the livestock samples farms are more heterogeneous in terms of farming practices with respect to energy use. One explanation may be that some livestock farms use the organic fertilizers from livestock manure and thus reduce their external fertilizers' use, while other livestock farms do not.

Distribution of Slacks According to the Level of Pure Technical Efficiency

The results regarding input excess show some heterogeneity between the three types of farms, especially in the use of energy resources. To further the analysis, we classify each type of farms according to its level of pure technical efficiency, which represents the extent of how management practices could be changed to improve total technical efficiency. Three groups of farms were identified: a group with farms with low efficiency, where the farms' pure technical efficiency score is less than 0.60; a group with farms with medium efficiency, where farms' score is strictly greater than 0.60 but less than 0.85; and a group with farms with high efficiency, where farms' score is strictly greater than 0.85.

Table 9 shows some statistics regarding the slacks of energy resources (obtained under technology T2) for these three groups of farms. Results indicate that the highest slacks on average are observed in the group with highest efficiency scores, and the lowest slacks for lowest efficiency farms. For example, for cereals, oilseed crops and protein seed crops farms, the average slacks for low efficiency farms, medium efficiency farms and high efficiency farms are respectively 0.3%, 3.9% and 7.3%.

Table 9: Slacks of energy resource depending on farms' pure technical efficiency (PTE)

Farm Type	**Variables**	**For Low Efficiency Farms (with PTE ≤ 0.60)**	**For Medium Efficiency Farms (with 0.60 < PTE ≤ 0.85)**	**For High Efficiency Farms (with 0.85 < PTE ≤ 1)**
Cereals, oilseed crops and protein seed crops farms	Average slack in each group (%)	0.3	3.9	7.3
	Maximum slack in each group (%)	12	43	74
	% of farms in each group	7.3	43.1	49.6
Dairy farms	Average slack in each group (%)	6.2	8.5	12.7
	Maximum slack in each group (%)	23	53	69
	% of farms in each group	3.0	44.8	52.2
Mixed farms	Average slack in each group (%)	5.7	9.5	9.5
	Maximum slack in each group (%)	34	55	74
	% of farms in each group	9.9	39.5	50.6

Determinants of Slacks of Intermediate Consumption and Energy Resources

Table 10 (for cereals, oilseed crops and protein seed crops farms), Table 11 (for dairy farms) and Table 12 (for mixed farms) show the results from the Tobit models regressing input slacks on the set of determinants listed in Table 2. The input slacks considered here in three separate models are the slacks for intermediate consumption calculated under technology T1 (IC), the slacks of energy resources calculated under technology T2 (ER) and the slacks of the remaining intermediate consumption also calculated under technology T2 (R_IC).

Table 10: Determinants of input slacks for cereals, oilseed crops and protein seed crops farms

Variables	**Technology T1**	**Technology T2**	
Dependent variable	Slack of IC	Slack of ER	Slack of R_IC
Determinants	Coefficient and significance	Coefficient and significance	Coefficient and significance
UAA	−0.022	−0.051 ***	−0.032 **
CAP/UAA	0.001	−0.005 ***	0.003 **
Debt	−0.016	−0.054 **	0.049 *
Non-Paid/Lab	4.054	22.144 ***	−7.622
S_COP	−22.323 ***	4.236	−25.004 ***
S_LEG	−50.165 **	−14.786	−62.718 ***
SUB/PROEX	−13.502	4.403	5.625
Age	−0.176	−0.227 ***	−0.151
ENVI	−1.409	1.325	3.509
S_Text	0.022	−0.167 ***	0.075
Log-Likelihood	−584.81	−2,045.83	−730.48

Note: *, **, ***: significant at 10%, 5%, 1%.

Table 11: Determinants of input slacks for dairy farms

Variables	**Technology T1**	**Technology T2**	
Dependent variable	Slack of IC	Slack of ER	Slack of R_IC
Determinants	Coefficient and significance	Coefficient and significance	Coefficient and significance

UAA	−0.112 ***	−0.018	−0.167 ***
CAP/UAA	−0.003 ***	−0.002 ***	−0.003 ***
Debt	0.038	−0.188 ***	0.178 **
Non-Paid/Lab	−11.497 *	4.497	−0.353
S_COP	18.744 **	17.711 ***	−2.747
S_LEG	54.695	−15.837	99.549
SUB/PROEX	−9.076	38.676 ***	−10.692
Age	−0.187	−0.138	−0.204
ENVI	−0.806	−2.283	−3.907
S_Text	0.027	0.063	−0.157 **
R^2	0.93	0.99	0.93
Log-Likelihood	−650.78	−2866.46	−634.89

Note: *, **, ***: significant at 10%, 5%, 1%.

Table 12: Determinants of input slacks for mixed farms

Variables	Technology T1	Technology T2	
Dependent variable	Slack of IC	Slack of ER	Slack of R_IC
Determinants	Coefficient and significance	Coefficient and significance	Coefficient and significance
UAA	−0.371 **	−0.197 **	−0.23
CAP/UAA	−0.038	−0.214 ***	0.027
Debt	−14.177 **	7.881 **	−11.993
Non-Paid/Lab	−3.966	−2.842	5.563
S_COP	−0.002 *	−0.002 ***	−0.003 **
S_LEG	−0.078	−0.015	−0.206 **
SUB/PROEX	23.793 ***	9.775 **	18.448 **
Age	−0.121 ***	0.053 ***	−0.187 ***
ENVI	−24.657	−2.571	−28.574
S_Text	16.255 **	17.637 ***	34.986 ***
Log-Likelihood	−454.44	−2,139.37	−593.95

Note: *, **, ***: significant at 10%, 5%, 1%.

The regression results for the slacks of intermediate consumption under technology T1 indicate that only two variables have a significant influence for cereals, oilseed crops and protein seed crops farms (Table 10): the share of cereals, oilseed crops and protein seed crops and the share of legumes in the total farm's UAA. More precisely, *ceteris paribus*, the increase of these shares reduces the slacks of intermediate consumption. This finding is confirmed for the slacks of intermediate consumption without energy resources in technology T2 (last column of Table 10). However, these two shares are not significant determinants of slacks of energy resources. These slacks are influenced in a negative way by the farm UAA, the ratio of capital to land, indebtedness, age and soil texture, and in a positive way by the share of non-paid labor. In other words, larger farms in terms of UAA, farms with a more capital intensive technology, farms with a higher indebtedness ratio, farms with an older manager, farms with a higher share of clay-sand soil texture, and farms with a lower resort to non-paid labor, use less excess of energy resource. In the case of dairy farms (Table 11), slacks of intermediate consumption in technology T1 are negatively influenced by UAA, capital to land ratio, non-paid labor to total labor, and positively influenced by the share of cereals, oilseed crops and protein seed crops in total UAA. Two results are confirmed when energy resources are considered separately (under technology T2): the capital to land ratio decreases the energy resource slacks, and the share of cereals, oilseed crops and protein seed crops in total UAA increases these slacks. In addition, results show that higher indebtedness decreases these slacks and subsidy dependence increases them.

Regarding mixed farms (Table 12), slacks of intermediate consumption under technology T1 and slacks of energy resources under technology T2 are both negatively influenced by the UAA and the share of cereals, oilseed crops and protein seed crops in total UAA, and positively influenced by the dependence to subsidy and the soil texture area share. By contrast, indebtedness and age negatively influence the slacks of intermediate consumption under technology T1 but positively influence the slacks of energy resources under technology T2.

Table 10, Table 11 and Table 12 overall show several findings as regard the determinants of the slacks of energy resources, recalling that the latter include fertilizers, soil amendments and fuel. The first finding is that the capital to land ratio significantly decreases these slacks for all three types of farms, and UAA decreases them for cereals, oilseed crops and protein seed crops farms and for mixed farms. This suggests that highly capitalized farms and large crop farms are able to generate less waste of energy resources. The relationship between farm size and technical efficiency is a debated issue in the literature and has been illustrated by various empirical results (see the review in Latruffe [34]). In our case farm size is a positive determinant of efficient use of energy resources. Our finding regarding the positive effect of capital intensity conforms to the literature on the determinants of technical efficiency. It suggests that high capitalization enables saving on transportation operations and thus on fuel.

The second finding is the mixed effect of indebtedness on energy resource slacks depending on the type of farm: the effect is significantly negative for cereals, oilseed crops and protein seed crops farms and for dairy farms, while it is significantly positive for mixed farms. The literature recognizes that farmers are credit constrained (e.g., [39,40]). In this context, indebtedness can help them make the necessary adjustments to improve their efficiency. Davidova and Latruffe [41] extensively investigated the relationship between indebtedness and technical efficiency for the case of Czech farms, and, as it is the case for our results regarding the relationship between indebtedness and energy resource slacks, they did not find clear-cut conclusions across different types of Czech farms.

The third finding is that the subsidy proxy has a positive effect on energy resource slacks in dairy and mixed farms. This indicates that highly subsidized farms have an inefficient behavior in the sense that they use energy resources more in excess than farms that are low subsidized. The literature on the effect of subsidies on farms' technical efficiency is relatively vast, e.g., see Zhu et al. [42] and a recent meta-analysis by Minviel and Latruffe [43]. Existing studies generally find that subsidies negatively impact farms' technical efficiency. No studies have been concerned with the impact of subsidies on input slacks. Our results are the first on this issue and they are in line with the existing literature on the subsidy-efficiency relationship. One explanation can be found in Serra et al.'s [44] model. The authors explain that subsidies modify farmers' attitudes towards risk, and as a consequence farmers may use more of a risk-increasing input. This is however not found for our sample of cereals, oilseed crops and protein seed crops farms.

The fourth finding regards the crop rotation. Surprisingly, the share of legumes in the farm's UAA does not impact significantly the energy resource slacks, although it could be expected that the use of legumes would imply lower resort to fertilizers. The share of cereals, oilseed crops and protein seed crops in total UAA impacts differently the farms: it increases the energy resource slack for dairy farms, it decreases it for mixed farms, and it has no significant impact for cereals, oilseed crops and protein seed crops farms. The latter result may come from the fact that cereals, oilseed crops and protein seed crops make most of the UAA of cereals, oilseed crops and protein seed crops farms, although Meul et al. [45] report that the share of cereals in arable farms' area in Belgium influences the energy efficiency. By contrast, in the case of dairy farms, this share captures the balance between pasture land and crops used for forage. Our results thus indicate that, for these farms, producing their own forage is a source of inefficient use of energy resources.

The fifth finding arising from Table 10, Table 11 and Table 12 is that the dummy variable indicating whether the farm is located in an environmental zoning area Natura 2000 (ENVI) has no significant effect on all considered slacks for the three types of farms. This indicates that location in Natura 2000 area does not help farms reduce their intermediate consumption slacks nor their energy slacks. This is, in a way, not problematic as Natura 2000 zoning does not aim at such reduction. It could, in opposite, be expected that farmers located in Natura 2000 area actually use more input excess than farmers located outside the zoning, as Natura 2000 zoning constrains farmers' choices and practices which may result in inefficient productive behavior. Our results show that this is not the case, since the dummy ENVI has no significant effect on the slacks.

CONCLUSIONS AND POLICY IMPLICATIONS

In this paper we have investigated the extent of energy resource excess use (slacks) and its determinants, with the help of a DEA model where energy resources are disentangled from the rest of intermediate consumption. The application was to three types of French farms in 2010.

A first conclusion is that our results show for all three types of farms that farms highly technically efficient have a higher slack of energy resources. This finding, contrary to the one in Mousavi-Avval et al. [20] for barberry farms in Iran, may suggest a trade-off for French farms: attaining high technical efficiency means sacrificing the energy efficiency. The management implication is that technical efficiency alone is not a sufficient measure of farms' performance. Accounting for input slacks in performance measures may give incentives to farmers to proceed to the necessary adjustments (such as enterprise reorganization, investments…) to improve further their performance. Another implication relates to policy recommendations. Energy policies based only on the level of pure technical efficiency would consist in assisting lowest efficiency farms, but they should instead be designed in a way to help the highest efficiency farms to reduce their energy use. This confirms Blancard and Martin's [46] result that energy policies need to target the most energy inefficient farms.

In terms of farm specialization, our results indicate that while cereals, oilseed crops and protein seed crops farms are the least technically efficiency on average, they used the lowest excess of energy resource, compared to dairy farms and mixed farms. Hence, despite soil and climate conditions which constrain more their technical efficiency than that of livestock farms, cereals, oilseed crops and protein seed crops farms perform well in terms of energy resource use. Although there exists some literature regarding technical efficiency differentials across various farm production types, the investigation of slacks across farm types is rare. Thus, our study would benefit from replications on various contexts to see if conclusions are robust.

A second important conclusion is that there are differences in the results when considering intermediate consumption as a whole (under technology T1) and when separating the energy resources from it (under technology T2). This is shown when computing the level of slacks (Table 7 and Table 8): taken as a whole, intermediate consumption is not used on excess on average, while when energy resources are disentangled, the latter are used on excess as much as capital and land on average. Similar discrepancies can be highlighted from the analysis of the determinants of slacks. For example, in the case of mixed farms indebtedness significantly decreases the level of slacks of intermediate consumption under technology T1 (Table 12). However, when investigating the two components of intermediate consumption separately (under technology T2), we observe a significant effect on energy resources only, in the way that indebtedness significantly increases the level of slacks of energy resources. In the case of dairy farms the finding is even more problematic since under technology T2 the effect of indebtedness on the two components of intermediate consumption is opposite (Table 11): indebtedness significantly decreases the slacks of energy resources but significantly positively impacts the slacks of the rest of intermediate consumption. Such findings show that when investigating technical efficiency, considering intermediate consumption as a single input may provide misleading recommendations to policy-makers depending on their objective. Considering again the case of mixed farms (Table 12), based on the classic case of studying intermediate consumption as a whole (i.e., under technology T1), it would be recommended that policy-makers design schemes that favor indebtedness so that slacks of intermediate consumption are reduced. However, this may have a positive impact on the excess use of energy resources, and policy-makers should be aware of this side-effect. Also, considering again the case of dairy farms (Table 11), it is evident that policy-makers would need to make a choice: designing debt-favoring policy schemes, which would help reduce slacks of energy resources but increase slacks of other intermediate consumption components; or designing debt-preventing policy schemes, which would have the inverse effects. Hence, if policy-makers are interested in energy consumption by farms, we recommend disentangling energy resources from intermediate consumption. However, this should not be made at the expense of the precision from DEA computation: indeed, increasing the number of inputs in a DEA model may increase efficiency scores due to the "curse of dimensionality".

The literature shows that the degree of dependence on subsidies is often a significant barrier to a farm's technical efficiency. We further show that it constrains French dairy and mixed farms' capacity to reduce the use of energy resources. This may however depends on the type of subsidies. The subsidy variable used in our paper encompasses several types of subsidies which may not have the same impact on farmers' use of resources. This may be particularly the case for payments for environmental friendly practices. Within the European Union's Common Agricultural Policy, these payments relate to agri-environmental schemes which farmers can voluntarily contract. Such schemes may help farmers reduce their fuel use. Thus, we recommend that further research concentrates on this.

Our paper does not aim at providing a novel theoretical nor methodological contribution within the DEA literature. It simply underlines that care should be taken when choosing the inputs for the DEA model, as conclusions may differ depending on the level of disaggregation of inputs. For the particular case of energy resources in French farms in 2010, our results show that policy recommendations may not be the same. The main limitation of our research is that robustness tests have not been conducted. Firstly, robust procedures developed for DEA could be used to confirm our findings. Secondly, other indicators of energy use performance could be computed.

ACKNOWLEDGMENTS

This work has been supported by the cooperative TERRENA and the Regional Council of the Pays de la Loire. Costs to publish in open access are covered by the Social Sciences Laboratory LARESS of the High school of Agriculture of Angers-Loire (ESA). The views defended here are the authors' views and not necessary those of these organizations. The authors would like to thank the two anonymous reviewers and the editor for their helpful

AUTHOR CONTRIBUTIONS

Mohamed Ghali performed the research and wrote the paper. Laure Latruffe contributed by the verification of the model simulations and results and reviewed the paper. Karine Daniel contributed with valuable discussions and supported the research.

REFERENCES

1. Pretty, J. Agricultural sustainability: Concepts, principles and evidence. *Philos. Trans. R. Soc. Lond. B Biol. Sci.* 2008, *363*, 447–465.
2. Balmford, A.; Rodrigues, A.S.L.; Walpole, M.; Ten Brink, P.; Kettunen, M.; Braat, L.; De Groot, R. *The Economics of Biodiversity and Ecosystems: Scoping the Science*; European Commission: Cambridge, UK, 2008; p. 297.
3. Gliessman, S.R. Integrating Agroecological Processes into Cropping Systems Research. *J. Crop. Improv.*2004, *11*, 61–80.
4. Fisher, B.; Turner, R.K.; Morling, P. Defining and classifying ecosystem services for decision making. *Ecol. Econ.* 2009, *68*, 643–653.
5. Ghali, M.; Daniel, K.; Colson, F. Ecologically intensive agriculture: An economic approach. *Econ. Rural.* 2014, *341*, 83–99. (In French)
6. Solow, R.M. The economics of resources or the resources of economics. *Am. Econ. Rev.* 1974, *64*, 1–14.
7. Stiglitz, J.E. A neoclassical analysis of the economics of natural resources. In *Scarcity and Growth Reconsidered*; Smith, V.K., Ed.; RFF Press: New York, NY, USA; Johns Hopkins University Press: London, UK, 1979; pp. 36–66.
8. Schaltegger, S; Müller, K.; Hindrichsen, H. *Corporate Environmental Accounting*; Wiley: Chichester, UK, 1996.
9. *Eco-Efficiency*; Organization for Economic Co-operation and Development (OECD): Paris, France, 1998; p. 88.
10. *Eco-Efficiency: Creating more Value with less Impact*; World Business Council for Sustainable Development (WBCSD): Conches-Geneva, Switzerland, 2000; p. 36.
11. Verfaillie, H.A.; Bidwell, R. *Measuring Eco-Efficiency: A Guide to Reporting Company Performance*; World Business Council for Sustainable Development: Conches-Geneva, Switzerland, 2000; p. 39.
12. Gómez-Limón, J.A.; Picazo-Tadeob, A.J.; Reig-Martínez, E. Eco-efficiency assessment of olive farms in Andalusia. *Land Use Policy* 2012, *29*, 395–406.
13. Picazo-Tadeo, A.J.; Gómez-Limón, J.A.; Reig-Martínez, E. Assessing farming eco-efficiency: A Data Envelopment Analysis approach. *J. Environ. Manag.* 2011, *92*, 1154–1164.
14. De Koeijer, T.J.; Wossink, G.A.A.; Struik, P.C.; Renkema, J.A.

Measuring agricultural sustainability in terms of efficiency: The case of Dutch sugar beet growers. *J. Environ. Manag.* 2002, *66*, 9–17.

15. Basset-Mens, C.; Ledgard, S.; Boyes, M. Eco-efficiency of intensification scenarios for milk production in New Zealand. *Ecol. Econ.* 2009, *68*, 1615–1625.
16. European Commission. European Competitiveness Report, Competitiveness and Benchmarking, 2002. Available online: ec.europa.eu/DocsRoom/documents/3007/attachments/1/translations/en/renditions/pdf(accessed on 26 July 2016).
17. Huppes, G.; Ishikawa, M. An introduction to quantified eco-efficiency analysis. In *Quantified Eco-efficiency. An Introduction with Applications*; Huppes, G., Ishikawa, M., Eds.; Springer Netherlands: Dordrecht, The Netherlands, 2007; pp. 1–39.
18. Wursthorn, S.; Poganietz, W.R.; Schebek, L. Economic-environmental monitoring indicators for European countries: A disaggregated sector-based approach for monitoring eco-efficiency. *Ecol. Econ.* 2011, *70*, 487–496.
19. Zentner, R.P.; Basnyat, P.; Brandt, S.A.; Thomas, A.G.; Ulrich, D.; Campbell, C.A.; Nagye, C.N.; Frick, B.; Lemke, R.; Malhi, S.S.; et al. Effects of input management and crop diversity on non-renewable energy use efficiency of cropping systems in the Canadian Prairie. *Eur. J. Agron.* 2011, *34*, 113–123.
20. Mousavi-Avval, S.H.; Mohammadi, A.; Rafiee, S.; Tabatabaeefar, A. Assessing the technical efficiency of energy use in different barberry production systems. *J. Clean. Prod.* 2012, *27*, 126–132.
21. Aigner, D.J.; Chu, S.F. On estimating the industry production function. *Am. Econ. Rev.* 1968, *58*, 826–839.
22. Aigner, D.J.; Lovell, C.A.; Schmidt, P. Formulation and estimation of stochastic frontier production functions. *J. Econ.* 1977, *6*, 21–37.
23. Meeusen, W.; Van Den Broeck, J. Efficiency Estimation from Cobb-Douglas production functions with composed errors. *Int. Econ. Rev.* 1977, *18*, 435–444.
24. Charnes, A.; Cooper, A.W.; Rhodes, E. Measuring the efficiency of decision-making units. *Eur. J. Oper. Res.*1978, *2*, 429–444.
25. Banker, R.D.; Charnes, A.; Cooper, W.W. Some models for estimating technical and scale inefficiencies in Data Envelopment Analysis. *Manag. Sci.* 1984, *30*, 1078–1092.

26. Farrell, M.J. The Measurement of productive efficiency. *J. R. Stat. Soc. Ser. A* 1957, *120*, 253–290.
27. Farrell, M.J.; Fieldhouse, M. Estimating efficient production functions under increasing returns to scale. *J. R. Stat. Soc. Ser. A* 1962, *125*, 252–267.
28. Hoang, V.N.; Prasada Rao, D.S. Measuring and decomposing sustainable efficiency in agricultural production: A cumulative exergy balance approach. *Ecol. Econ.* 2010, *69*, 1765–1776.
29. Latruffe, L.; Fogarasi, J.; Desjeux, Y. Efficiency, productivity and technology comparison for farms in Central and Western Europe: The case of field crop and dairy farming in Hungary and France. *Econ. Syst.* 2012, *36*, 264–278.
30. Picazo-Tadeo, A.J.; Castillo-Giménez, J.; Beltrán-Esteve, M. An intertemporal approach to measuring environmental performance with directional distance functions: Greenhouse gas emissions in the European Union. *Ecol. Econ.* 2014, *100*, 173–182.
31. Zhou, P.; Ang, B.W. Linear programming models for measuring economy-wide energy efficiency performance. *Energy Policy* 2008, *36*, 2911–2916.
32. Coelli, T.J.; Prasada Rao, D.S.; O'Donnell, C.J.; Battese, G.E. *An Introduction to Efficiency and Productivity Analysis*, 2nd ed.; Springer: Berlin, Germany, 2005.
33. Koopmans, T.C. Analysis of Production as an Efficient Combination of Activities. In *Activity Analysis of Production and Allocation*; Cowles Commission for Research in Economics; John Wiley and Sons, Inc.: New York, NY, USA, 1951; p. 13.
34. Latruffe, L. Competitiveness, Productivity and Efficiency in the Agricultural and Agricultural and Agri-Food Sectors. *OECD Food Agric. Fish. Work. Pap.* 2010.
35. European Comission, Natura 2000. Available online: http://ec.europa.eu/environment/nature/natura2000/index_en.htm (accessed on 26 July 2016).
36. BEGIP, Bretagne Environnement groupement d'intérêt public, France. 2013. Available online: http://www.bretagne-environnement.org (accessed on 26 July 2016).
37. Agreste. Ministère de l'agriculture, de l'agroalimentaire et de la forêt: Agreste La statistique, l'évaluation et la prospective agricole. Available

online: http://agreste.agriculture.gouv.fr/ (accessed on 26 July 2016).
38. Bravo-Ureta, B.E.; Solis, D.; Lopez, V.H.M.; Maripani, J.F.; Thiam, A.; Rivas, T. Technical efficiency in farming: A meta-regression analysis. *J. Product. Anal.* 2007, *27*, 57–72.
39. Blancard, S.; Boussemart, J.-P.; Briec, W.; Kerstens, K. Short- and long-run credit constraints in French agriculture: A directional distance function framework using expenditure-constrained profit functions. *Am. J. Agric. Econ.* 2006, *88*, 351–364.
40. Ciaian, P.; Swinnen, J. Credit market imperfections and the distribution of policy rents. *Am. J. Agric. Econ.*2009, *91*, 1124–1139.
41. Davidova, S.; Latruffe, L. Relationships between technical efficiency and financial management for Czech Republic farms. *J. Agric. Econ.* 2007, *58*, 269–288.
42. Zhu, X.; Oude Lansink, A. Impact of CAP subsidies on technical efficiency of crop farms in Germany, the Netherlands and Sweden. *J. Agric. Econ.* 2010, *61*, 545–564.
43. Minviel, J.J.; Latruffe, L. Effect of Public Subsidies on Farm Technical Efficiency: A Meta-Analysis of Empirical Results. *Appl. Econ.* 2016.
44. Serra, T.; Zilberman, D.; Gil, J.M. Farms' technical inefficiencies in the presence of government programs. *Aust. J. Agric. Resour. Econ.* 2008, *52*, 57–76.
45. Meul, M.; Nevens, F.; Reheul, D.; Hofman, G. Energy use efficiency of specialised dairy, arable and pig farms in Flanders. *Agric. Ecosyst. Environ.* 2007, *119*, 135–144.
46. Blancard, S.; Martin, E. Energy efficiency measurement in agriculture with imprecise energy content information. *Energy Policy* 2014, *66*, 198–208.

CHAPTER 2

HYBRID DECOMPOSITION-RECONFIGURATION MODELS FOR LONG-TERM SOLAR RADIATION PREDICTION ONLY USING HISTORICAL RADIATION RECORDS

Si-Ya Wang [1], Jun Qiu [2,3], and Fang-Fang Li [1,3]

[1]College of Water Resources & Civil Engineering, China Agricultural University; Beijing 100083, China

[2]State Key Laboratory of Hydroscience & Engineering, Tsinghua University, Beijing 100084, China

[3]State Key Laboratory of Plateau Ecology and Agriculture, Qinghai University, Xining 810016, China

ABSTRACT

Solar radiation prediction is significant for solar energy utilization. This paper presents hybrid methods following the decomposition-prediction-reconfiguration paradigm using only historical radiation records with dif-

Citation: Wang, S.-Y.; Qiu, J.; Li, F.-F. Hybrid Decomposition-Reconfiguration Models for Long-Term Solar Radiation Prediction Only Using Historical Radiation Records. Energies 2018, 11, 1376.

ferent combination of decomposition methods, Ensemble Empirical Mode Decomposition (EEMD) and Wavelet Analysis (WA), and the reconfiguration methods, regression model (RE) and Artificial Neural Network (ANN). The application in west China indicates that these hybrid decomposition-reconfiguration models perform well for monthly prediction, while the comparisons of the daily prediction show that the hybrid EEMD-RE model has a higher degree of fitting and a better prediction effect in long-term prediction of solar radiation intensity, which verifies (1) decomposition of original solar radiation data results in components with regular characteristics; (2) the relationship between the original solar radiation sequence and the derived intrinsic mode functions (IMFs) is linear; and (3) EEMD has strong adaptivity for non-linear and non-stationary series. The proposed hybrid decomposition-reconfiguration models have great application prospect for monthly long-term prediction of solar radiation intensity, especially in the areas where complex climate data is difficult to obtain, and the EEMD-RE model is recommended for the daily long-term prediction.

Keywords: long-term prediction, solar radiation, hybrid model, decomposition-reconfiguration

INTRODUCTION

Solar energy is one of the most favorable renewable energy sources, it has been continuously explored in recent years. Solar radiation data is the fundamental input for solar energy applications, and its reliability appears important to designing, developing and evaluating solar technologies [1]. Optimal design of solar power systems needs the expected long-term solar radiation on the horizontal plane. For example, sizing the projects is related to solar collector and PV systems [2]. Moreover, when solar energy is produced on large-scale and grid-connected, an accurate knowledge of long-term solar radiation makes a lot of sense for balancing the energy supply and demand [3].

Various researches on solar radiation forecasting methods have been reported, classified into physical models and statistical methods. Physical models are based on the physical state and dynamic motions of the atmosphere, also known as Numerical Weather Prediction (NWP) models [4], which was believed the most appropriate for day-ahead and ``multi-day forecast horizons [5]. However, the NWP models are greatly affected by weather factors, such as cloudiness, cloud evolution and optical properties in the forecast area [6]. Generally, such models result in good predictions in clear sky conditions, while with the effect of clouds, the prediction results become worse [7]. Besides, the application of such physical models [8] on long-term daily solar radiation prediction is also limited by their computational complexity. There are two types of statistical models: mathematical statistics and machine learning algorithms. Mathematical statistics mainly includes regression analysis [9], time series analysis [10], fuzzy theory [11], wavelet analysis [12] and Kalman filtering [13]. Regression analysis determines the best combination of the independent variables to predict the dependent variable, but the selection procedure is not always easy [14]. Nourani found that Auto Regressive Integrated Moving Average (ARIMA) model had a limited ability to capture non-stationarities and non-linearities [15]. In practice, the predictional accuracy of the statistical methods is not as high as the NWP models, as the parameters change over time due to various factors [16]. Typical machine learning algorithms include: Artificial Neural Networks (ANN) [17], Support Vector Machines (SVM) [18] and heuristic intelligent optimization algorithms [19]. Gala et al. believed that hybrid artificial intelligence systems are quite effective for solar energy prediction [20]. Lauret et al. found that the improvement of the machine learning techniques for hour ahead solar forecasting appears to be more pronounced in case of unstable sky conditions [18].

As for long-term solar radiation prediction, a limited number of related publications can be found, most of which focus on the characteristic analysis rather than prediction. The complexity of the relationship between solar radiation and meteorological, terrestrial, and extra-terrestrial variables makes it difficult to make long-term solar radiation prediction [21]. Coelho and Boaventura-Cunha [22] found even their proposed method combining support vector regression and Markov chains performed poorly when the prediction was sixty-step ahead after comparing linear autoregressive, nonlinear autoregressive, and support vector models on long-term solar prediction.

With the development of the big data-mining technology in recent decades, the machine learning algorithms have drawn much attention. As one of the most commonly used methods, ANN have been successfully applied to solar radiation prediction and solar systems design [3] since it has strong ability to solve non-linear function estimation, pattern detection and data sorting. Cao [16,23] predicted solar radiation in Shanghai and Baoshan by using a BP (back propagation) neural network after preprocessing the data with wavelet analysis, and found that the recursive BP network combined with wavelet analysis improves in both speed and accuracy. Paoli et al. [24] used mixed models to predict total daily solar radiation in three sites in France. They first used the seasonal index adjustment method to preprocess the original solar radiation sequence, and then applied daily multi-layer perceptive neural network (MLPNN) on daily solar radiation prediction. Amrouche and Le Pivert [25] used spatial models and ANNs to predict the daily solar radiation intensity at four US sites. Pedro and Coimbra [26] compared ARIMA, k-nearest neighbors (kNNs), ANN and Genetic Algorithm (GA) optimized neural networks (GA/ANN). It was found that the neural network optimized by GA is superior to other algorithms in hourly prediction. Khatib et al. [27] compared existing methods including linear, nonlinear and ANN models and pointed out that compared with linear and nonlinear models, ANNs are more accurate to predict solar energy. At the same time, it was found that the sunshine ratio, the ambient temperature and the relative humidity are the most relevant coefficients for predicting solar radiation. Yadav and Chandel [28] chose different ANN models based on different geographical locations for prediction, and found that the reasonable choice of model parameters had a great influence on the prediction results. Voyant et al. [29] found that the predictive effects of these methods were affected by the weather and seasonal factors by comparing the ARIMA model, an ANN using only endogenous inputs (univariate) of pretreatment and an ANN using both endogenous and exogenous inputs for pretreatment. Ozgoren et al. [30] used the ANN model based on Multiple Nonlinear Regression (MNLR) to predict the monthly average solar radiation in Turkey. The method requires the input of latitude, longitude, altitude, monthly temperature and monthly minimum temperature, maximum temperature, average temperature, soil temperature, relative humidity, wind speed, rainfall, barometric pressure, vapor pressure, cloud cover and sunshine duration and other variables, and the MNLR method is used to determine the most appropriate independent input variables. Koca et al. [31] applied ANN model to the prediction of the monthly mean solar radiation in the Mediterranean region of Anatolia in Turkey by inputting different parameters, and found that the number of the input parameters was the most effective parameter. Generally, the existing ANN model needs a lot of meteorological parameters when applied to radiation prediction to make the results more accurate [3]. The input parameters are basically a certain combination of meteorological and topographical data, which include day of the year, wind speed, rainfall, relative humidity, temperature, latitude, longitude, altitude and so on. [32,33,34]. Thus there exists great limitation when applying ANNs in some areas where meteorological data is hard to obtain. As to the long-term time sequence itself, the information in historical data needs to be explored fully. The Wavelet Analysis (WA) and the Ensemble Empirical Mode Decomposition (EEMD) are two typical decomposition methods to extract the regular components from a fluctuant time series.

WA was developed on the basis of the Fourier Transform (FT) in the early 1980s, overcoming the shortcomings of traditional spectral analysis methods and satisfying the local variation requirements in the time and frequency domains by a variable window [35]. Almasria et al. [36] applied WA to the empirical study of Swedish temperature data from 1850 to 1999. Kisi [37] predicted monthly runoff using wavelet regression instead of ANN. Nourani et al. [15] combined WA and ANN to predict the runoff in the Ligvanchai valley of Tabriz, Iran. Partal [38] conducted a reference evapotranspiration estimation using the wavelet transform and the feedforward neural network methods to evaluate climate data (temperature, solar radiation, wind speed, relative humidity) at two stations in the United States.

The EEMD is an improved version of the empirical mode decomposition (EMD) [39]. EEMD overcomes the essential defect of EMD modalities and is an adaptive data processing method adapted to nonlinear and nonstationary time series. EMD and EEMD have been widely used in some complex system models. Monjoly et al. [40] compared the data processing methods, EMD, EEMD and WA, using classical prediction model (Auto-Regression, AR) and nonlinear method to predict solar radiation intensity and found that the multi-step prediction hybrid approach led to additional improvements.

In this study, an attempt to rollingly predict long-term solar radiation by only using historical radiation data is carried out. WA and EEMD are firstly used to decompose the historical daily sequence of solar radiation into regular and predictable sub sequences, and then the relationships between these sub components and the original sequence are established by Regression Equation or ANN model. Different combination of the decomposition methods and the relational models are tested, including EEMD-RE, EEMD-ANN, WA-RE, WA-ANN. The Autoregressive Integrated Moving Average model (ARIMA) is also compared. The results show that the EEMD-RE model performs superior to the other ones, which is capable of capturing the main characteristics of solar radiation in the next year. With daily data of ten years, the monthly means prediction almost has the same accuracy as the published studies using diverse meteorological and topographic data. The method can be employed for the study and design of solar projects, particularly in underdeveloped areas where it is difficult to obtain complex data.

The rest of the paper is organized as follows: the method used in this study is explained in Section 2. Simulative experiments with different methods are presented in Section 3. Section 4 contains the comparison results. Section 5 and Section 6 present the discussion and conclusion, respectively.

METHODOLOGY

Empirical Mode Decomposition (EMD)

The EMD is efficient to analyze non-linear and non-stationary signal sequences with high signal-to-noise ratio, which decomposes a complex signal into a finite number of intrinsic mode functions (IMF) with local characteristics of different time scales.

Each IMF needs to meet the following two requirements: (1) Throughout the data sequence, the number of extremums and zero values across the entire sample dataset must be equal or differ by one; (2) the mean of the envelope formed by the local maximum and the local minimum is zero at any point of the sequence. Taking signal $s(t)$ as an example, the process of screening programs is summarized as follows:

Step 1: Find all the local maxima and local minima in $s(t)$, and connect all local maxima by a cubic spline line to configure the upper envelope; This process is repeated with a local minima to produce the lower envelope.

Step 2: Construct the mean envelope $m_1(t)$ with the average of the upper and lower envelopes.

Step 3: The average envelope is subtracted from the original signal $s(t)$ to derive the first component $h_1(t)$:

$$h_1(t) = s(t) - m_1(t) \tag{1}$$

Step 4: Check if $h_1(t)$ meets the IMF's conditions. If not, go back to Step 1 and use $h_1(t)$ as the original signal for the second screening:

$$h_2(t) = h_1(t) - m_2(t) \tag{2}$$

Repeat screening for k times, until $h_k(t)$ meets the IMF's conditions, when the first IMF component $c_1(t)$ is derived:

$$c_1(t) = h_k(t) \tag{3}$$

Step 5: Subtract $c_1(t)$ from the original signal $s(t)$ to get the residual $r_1(t)$:

$$r_1(t) = s(t) - c_1(t) \tag{4}$$

Step 6: Take $r_1(t)$ as the new original signal, and perform step 1 to step 5 to obtain a new residual $r_2(t)$. Repeat the steps above for n times. When the nth residual $r_n(t)$ becomes a monotonic function, the IMF cannot be decomposed anymore and the entire EMD is completed. The original signal $s(t)$ can be expressed as a combination of n IMF components and an average trend component $r_n(t)$, as shown in Equation (5):

$$s(t) = \sum_{k=1}^{n} c_k(t) + r_n(t) \tag{5}$$

With the Hilbert transform, the IMFs yield instantaneous frequencies as functions of time that give sharp identifications of imbedded structures. Each IMF can be either linear or non-linear with corresponding physical background.

Ensemble Empirical Mode Decomposition (EEMD)

Although the EMD shows great superiority in the analysis of non-linear and non-stationary signals, the mode mixing problem resulting from the intermittency of signals still exists. The EEMD adds white Gaussian noise to the EMD to solve such problem. The basic idea is to eliminate the intermittency of the original signal in the frequency domain by using the statistical characteristics of uniformly distributed Gauss white noise, so that the mode mixing can be avoided.

The specific decomposition steps of EEMD are as follows:

Step 1: A series of random Gauss white noise signals $w_i(t)$ are added to the original signal $s(t)$ to get a total signal $X(t)$:

$$X_i(t) = s(t) + kw_i(t) \tag{6}$$

where $w_i(t)$ indicates the total signal after the ith time adding noise. k is the amplitude coefficient of $w_i(t)$, usually $0.05 < k < 0.5$.

Step 2: Decompose $X_i(t)$ in accordance with step 1 to 6 in Section 2.1. However, it's necessary to replace spline interpolation with piecewise cubic Hermite interpolation in the first step of 2.1 section to obtain the maximum and minimum envelope of signal:

$$X_i(t) = \sum_{j=1}^{n} c_{ij}(t) + r_i(t) \quad (7)$$

where $c_{ij}(t)$ represents the jth IMF after the noise is added i times.
Step 3: To obtain the average values of all IMF and residuals obtained by the above steps, as in Equations (8) and (9):

$$c_j(t) = \sum_{i=1}^{M} c_{ij}(t)/M \quad (8)$$

$$(t) = \sum_{i-1}^{M} r_i(t)/i \quad (9)$$

where $c_j(t)$ and $r_j(t)$ stand for the jth IMF and jth residual component obtained by EMD technique. M denotes the number of the Gaussian white noise, usually $M = 100$.

Wavelet Analysis (WA)

The WA is a time-frequency localization analysis method with fixed time-frequency window area but changing time window and frequency window. Through the wavelet transforming of the original data sequence and mapping it to a different time-frequency domain, the inverse transforming of each frequency-domain component can be obtained. The separate analysis of these components helps understand their variation law in different frequency domains. Select the mother wavelet $Y(t)$, where t stands for time, and the wavelet sequence $Y_{j,k}(t)$ can be obtained by expanding and transforming $Y(t)$. In computation and practical application, a discrete wavelet sequence is usually used, which can be obtained by Equation (10):

$$Y_{j,k(t)} = A_0^{-j/2} Y\left[\frac{t - kA_0^j B_0}{A_0^j}\right] = A_0^{-j/2}\left(A_0^{-j} - kB_0\right) \; j,\, k = 0, \pm1, \pm2, \ldots \quad (10)$$

where A−j0 is a scale factor, kAj0B0 is a shift factor. When $A_0 = 2$, $B_0 = 1$, the above formula is a binary wavelet sequence. Let $\varphi(t)$ be the scaling function corresponding to the mother wavelet $Y(t)$; then the sequence of the binary functions φj,k(t) is:

$$\varphi_{j,k(t)} = 2^{-\frac{j}{2}} \varphi\left(2^{-j}t - k\right) \; j,\, k = 0, \pm1, \pm2, \ldots \quad (11)$$

After the decomposition of the original data $f(m)$, the corresponding low-frequency series a_N and the high-frequency series d_1, d_2, ..., d_N can be obtained.

The specific relationship is as follows:

$$f(m) = d_1 + d_2 + \ldots + d_N + a_N \quad (12)$$

The results of the wavelet decomposition vary according to the chosen mother wavelet; the resulting frequency domain alias also has different degrees. The more severe the alias in the frequency domain, the less obvious the variation of the components in the frequency domain. Therefore, the selection of the mother wavelet should be excluded from the frequency domain caused by the phenomenon of alias serious mother wavelet.

Back Propagation-Artificial Neural Network (BP-ANN)

Due to its strong ability of non-linear mapping, learning as well as fault tolerance, ANNs have been widely applied to nonlinear forecasting problems. The Kolmogorov continuity theorem guarantees the feasibility and validity of using neural networks for time series prediction mathematically. BP-ANN back propagation includes input layer, hidden layer and output layer. The existence theorem of Kolmogorov three-layer neural network has proved that any continuous function can be mapped to a three-layer BP network.

The output E_n of a neuron j on the BP-ANN hidden layer and output layer is given by Equation (13):

$$E_n = f_j(Net_j) = f_j\left(\sum w_{ij}e_i + \theta_j\right) \quad (13)$$

where f_j is the aviation function corresponding to the neuron j, and usually Sigmoid function $f(x) = 1/(1 + \exp(-x))$ is adopted; θ_j represents the threshold of the neuron j; e_i is the input of neuron j; w_{ij} indicates the connection weight of the corresponding input and the neuron.

Regression Model (RE)

Generalized linear models (GLM) are a unified class of regression methods for discrete and continuous response variables. There are some special cases, such as Logistic regression for binary responses, linear regression for continuous responses, log-linear models for counts, and some survival analysis methods. The systematic component and the random component compose a GLM. For the systematic component, one relates Y to x by assuming the average among individuals with a common value of x, $\eta=\lambda(Y)$, satisfing:

$$g(\eta) = x_1\alpha_1 + x_2\alpha_2 + \ldots + x_r\alpha_r \quad (14)$$

where g is a prespecified function known as the ‘link function’. α are regression coefficients. In this study, the liner regression model is selected, the coefficients of which are determined by the least square method.

Hybrid EEMD/WA-RE Model

The basic idea of the hybrid EEMD/WA-RE method proposed in this study follows the decomposition-prediction-reconfiguration paradigm. The main purpose of EEMD or WA decomposition is to better extract valid information from the data, simplify the original goal, and decompose it to more regular components for predictable sub-goals. First, the EEMD or the WA is used to decompose the long disordered sequence into several sub-sequences (IMFs for EEMD, and sequences with different frequencies for WA). Theoretically, EEMD can be applied to any time series without the requirements for stationarity, and does not require the default basic functionality For WA, the key technique to alleviate the aliasing phenomenon is the selection of the mother wavelet. In this study, db7 is set as the mother wavelet according to the previous study and testing results [41].

The daily average solar radiation sequence of 1~T year is decomposed into sub-sequences by EEMD or WA, which are used as the independent variables in RE, and the data of 2~(T + 1) year is taken as the dependent variable. The regression equation *g* is then established to predict the daily average solar radiation.

Hybrid EEMD/WA-ANN Model

The sub-sequences obtained by EEMD or WA using radiation data of 1~T years can also be used as the input to the ANN model, and the data of 2~(T + 1) year is the output. After training the ANN model, it can be used for prediction in the future.

The decomposition-prediction-reconfiguration idea derives four different combination of hybrid models in this study: EEMD-RE, WA-RE, EEMD-ANN, and WA-ANN, which are compared and evaluated. Figure 1 shows the flowchart of these four models. The step 1 and 2 with black circles aim at training the model, and establishing the relationship between $X_{1\sim T}$ and $X_{2\sim(T+1)}$; while step 3 to 5 with bule circle use such model to predict X_{T+2}. The part with blue background indicates predicting process in Figure 1.

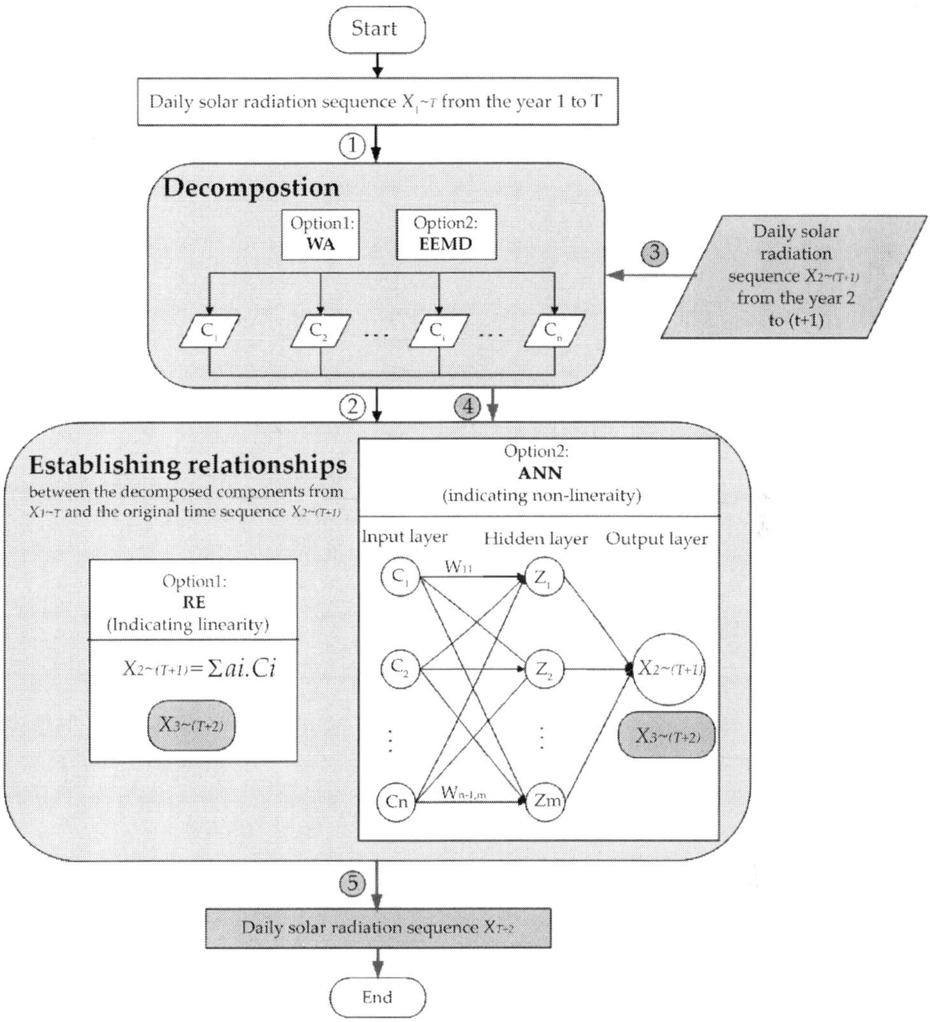

Figure 1: Flowchart of hybrid EEMD-RE, WA-RE, EEMD-ANN, and WA-ANN models.

CASE STUDY

Study Case

The Qinghai province is located in west China with an average elevation of above 3000 m. It has good atmospheric transparency, high sunlight transmittance, long sunshine duration and abundant solar energy resources.

The annual sunshine hours in eastern Qinghai Province are 3000 to 3200 h, and the annual solar radiation is 5860 to 6700 MJ/m^2 [42], ranking the second in the country. The whole Qinghai province has about 200,000 km^2 unused desert, which is suitable for the large-scale solar energy exploration [43].

In the past 10 years, the solar energy industry in Qinghai province has been developing vigorously, with a speed of 'one watt per watt'. By the end of 2017, the installation capacity of the photovoltaic (PV) power in the Qinghai Province had reached 7910 MW [44], and more projects are planned for construction.

The design of a PV power station needs accurate long-term radiation prediction. Gonghe County in Qinghai province, where a large scale PV power station is panned, is taken as the research area in this study. Solar radiation intensity data used in this experiment was obtained from NASA. The sample data is from 1st January 1984 to 31st December 1995.

Implementation

Using EEMD, daily average solar radiation intensity data of the area from 1 January of the year 1984 to 31 December of the year 1993 are decomposed to obtain 12 IMFs. The 12 IMFs are taken as independent variables, and the data from 1 January of the year 1985 to 31 December of the year 1994 is taken as dependent variables to establish the regression equation, as in Equation (15):

$$Y = \sum_{i=1}^{n} \zeta_i C_i \tag{15}$$

where ζ_i are the regression coefficients and Ci is the IMFs. Equation (15) is then used to predict the solar radiation of the year 1995.

The 12 IMFs derived from EEMD using the data from 1984 to 1993 can also be taken as the input to train an ANN model, and the data of the year 1985 to 1994 is the output. The number of hidden layer neurons of the ANN model in this study is 10 and the output layer neurons is 1.After training the BP-ANN model, it is used to predict the daily radiation sequence of the year 1995 with the data of the IMFs from 1985 to 1994.

Taking db7 as the mother wavelet, the three-scale Mallat pyramid wavelet decomposition of the solar radiation data series is carried out to obtain the low frequency sequence a_3 and the high frequency series d1, d2 and d3 of the solar radiation. Then the similar process is carried out as the EEMD-RE model and the EEMD-ANN model to establish the regression equation and the ANN model to predict the radiation for 1995.

Another typical prediction method for the ARIMA time series is also tested for comparison. The ARIMA model (3,0,4) × (0,0,1) is chosen after auto regression, partial regression and unit root test for daily data, while the ARIMA model (5,0,5) is chosen for monthly data.

To compare the predictive effect in different time scales, the daily, ten-day, and monthly results are calculated with the daily prediction. On the other hand, to verify the data mining effect by decomposition methods, the monthly data is also used for the four hybrid models to derive the monthly prediction, which is compared with the monthly statistics from daily predictions.

Model Evaluation Criteria

The standard root mean square error (RMSE), the mean absolute percentage error (MAPE), the correlation coefficient (r) and the coefficient of determination (R^2) are chosen as the evaluation criteria of the predictive value, as defined in Equations (16)–(19). RMSE reflects the extent to which the predicted data deviates from the true value. The smaller the RMSE value, the better the prediction. MAPE can be used to measure the quality of a model prediction; the smaller the MAPE value, the better the prediction. r and R^2 reflect the fitting degree of the model; the closer the r and R^2 to 1, the better the fitting degree of the model:

$$\text{RMSE} = \sqrt{\frac{1}{n}\sum_{i=1}^{n}\left(X_{hist,i} - X_{pred,i}\right)^2} \tag{16}$$

$$\text{MAPE} = \left(\frac{1}{n}\sum_{i=1}^{n}\left|\frac{X_{hist,i} - X_{pred,i}}{X_{hist,i}}\right|\right) \times 100\% \tag{17}$$

$$r = \frac{\sum_{i=1}^{n}\left(X_{hist,i} - \overline{X_{hist}}\right)\left(X_{pred,i} - \overline{X_{pred}}\right)}{\sqrt{\sum_{i=1}^{n}\left(X_{hist,i} - \overline{X_{hist}}\right)^2\left(X_{pred,i} - \overline{X_{pred}}\right)^2}} \tag{18}$$

$$R^2 = \frac{\sum_{i=1}^{n}\left(X_{pred,i} - X_{hist,i}\right)^2}{\sum_{i=1}^{n}\left(X_{hist,i} - \overline{X_{hist}}\right)^2} \tag{19}$$

where the subscript *hist* represents historical data, and the subscript *pred* represents the predictive results.

RESULTS

Figure 2 shows the 12 IMFs obtained from the EEMD decomposition and the subsequences with different frequencies obtained from the WA using daily solar radiation intensity from the year 1984 to 1993, and 1985 to 1994, respectively; Figure 3 shows the 7 IMFs and sequences with different frequencies derived from monthly data from 1984 to 1993, and from 1985 to 1994, respectively.

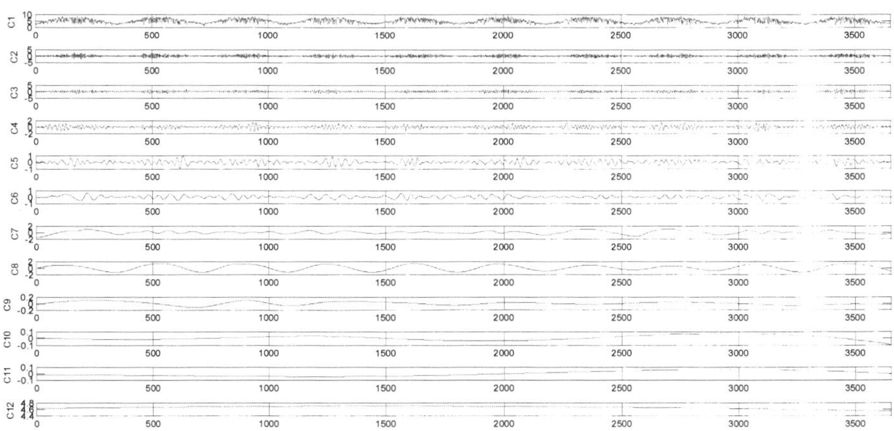

(a)

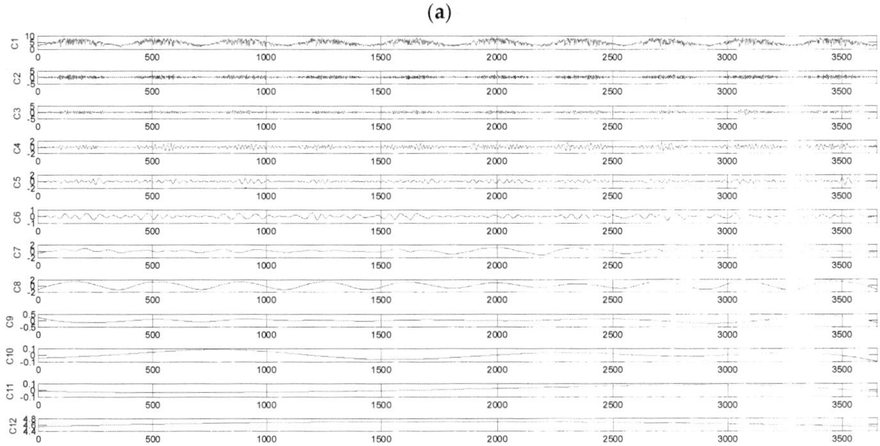

(b)

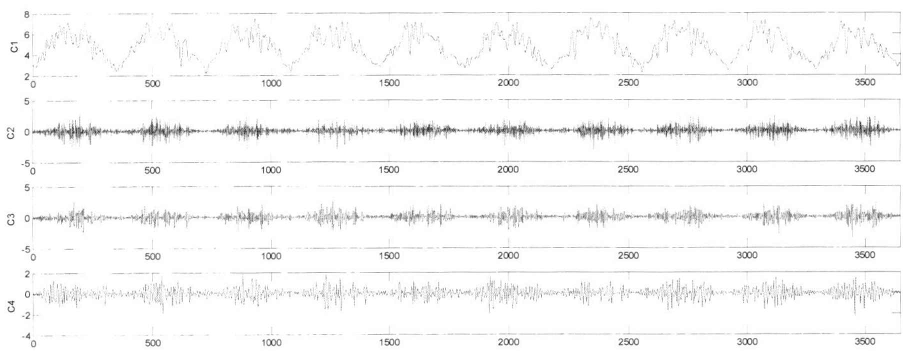

(c)

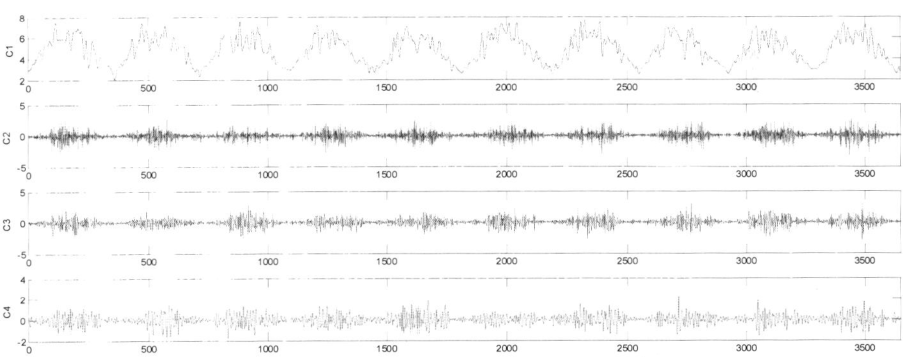

(d)

Figure 2: IMFs obtained from EEMD using daily solar radiation data of (a) 1984–1993, (b) 1985–1994; and sub sequences with different frequencies derived from WA using daily solar radiation data of (c) 1984–1993, (d) 1985–1994.

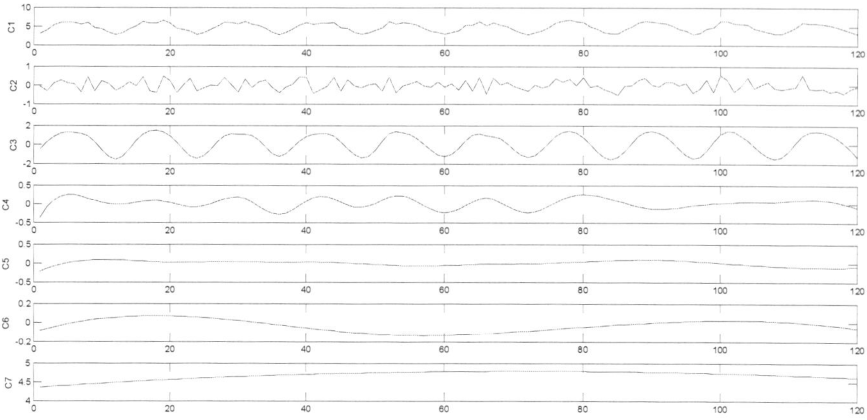

(**a**)

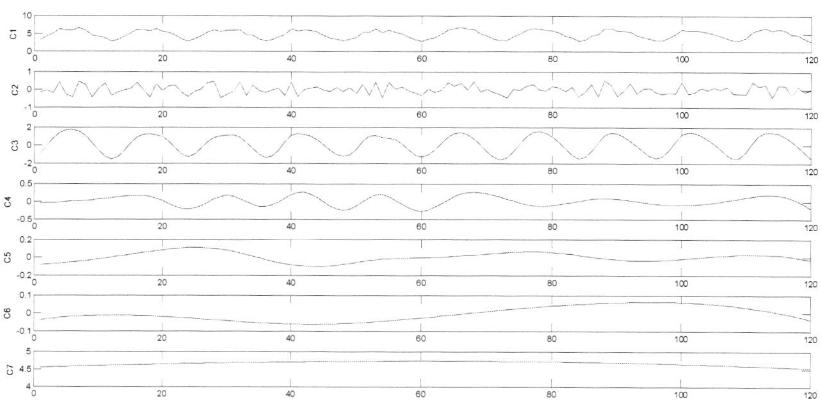

(**b**)

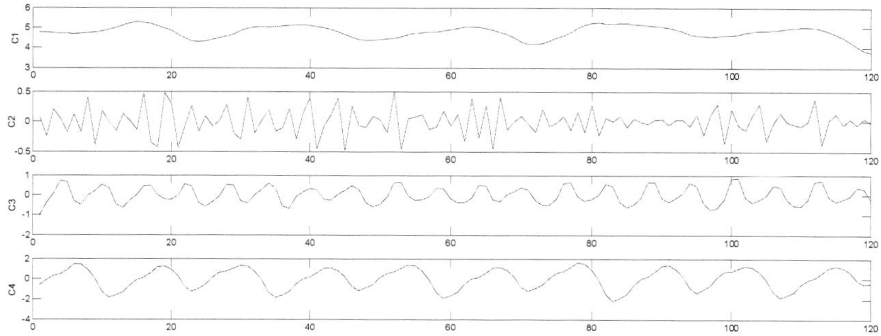

(**c**)

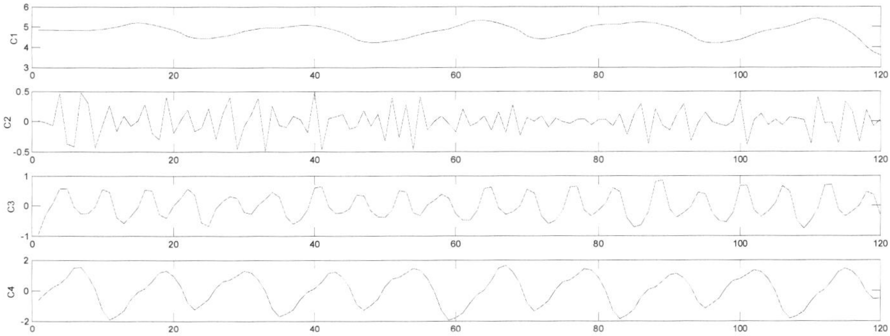

(d)

Figure 3: IMFs obtained from EEMD using monthly solar radiation data of (a) 1984–1993, (b) 1985–1994, sequences with different frequencies obtained from WA using monthly solar radiation data of (c) 1984–1993, (d) 1985–1994.

Figure 4 shows the daily prediction results by different prediction models, as well as the its statistical results with a time step of 10 days and 1 month.

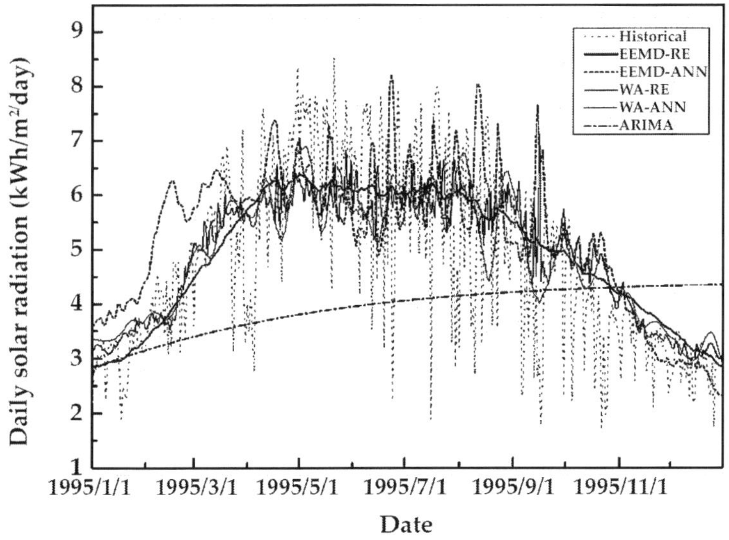

a

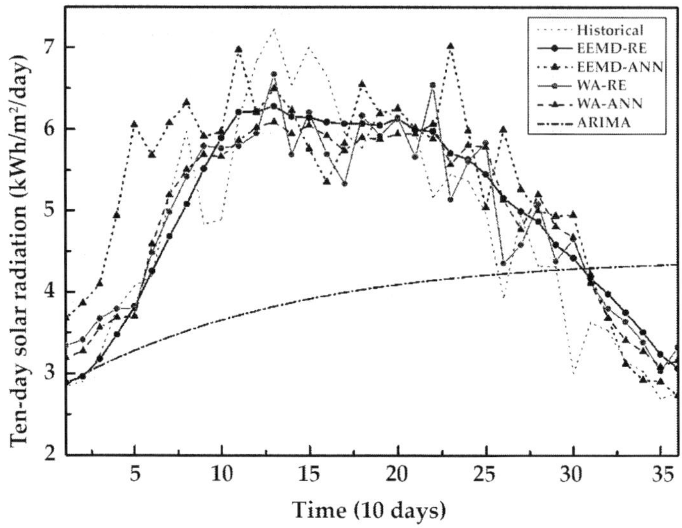

b

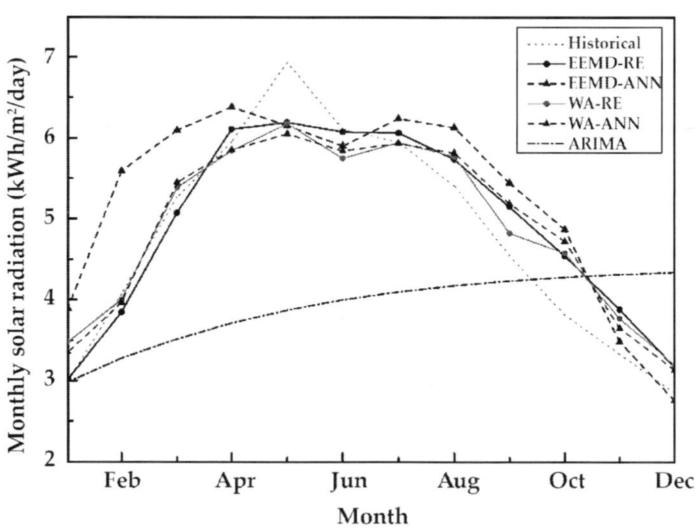

c

Figure 4: (a) Daily prediction results of the year 1995, and its statistcal results with a time step of (b) 10 days and (c) 1 month.

The predictive accuracy of the different models and time scales are shown in Table 1, Table 2, Table 3 and Table 4. Important cases where the computational definition of R^2 can yield negative values, depending on the definition used, arise where the predictions that are being compared to the corresponding outcomes have not been derived from a model-fitting procedure using those data, and where linear regression is conducted without including an intercept [43]. Additionally, negative values of R^2 may occur when fitting non-linear functions to data [44]. In cases where negative values arise, the mean of the data provides a better fit to the outcomes than do the fitted function values, according to this particular criterion [45,46].

Table 1: Predictive accuracy of different models with daily data

Methods	RMSE	MAPE (%)	*r*	R^2
EEMD-RE	**1.135**	**22.11**	**0.748**	**0.5484**
EEMD-ANN	1.474	29.47	0.590	0.2387
WA-RE	1.181	22.58	0.723	0.5116
WA-ANN	1.188	22.56	0.725	0.5052
ARIMA	1.948	33.21	0.035	-0.3097

Table 2: Predictive accuracy of statistical results using daily data with 10 days interval

Methods	RMSE	MAPE (%)	*r*	R^2
EEMD-RE	**0.571**	**10.23**	**0.918**	**0.8247**
EEMD-ANN	0.935	16.92	0.812	0.5297
WA-RE	0.637	12.03	0.897	0.7810
WA-ANN	0.619	11.34	0.904	0.7913
ARIMA	1.666	27.60	0.060	−0.4945

Table 3: Predictive accuracy of statistical results using daily data with 1 month interval

Methods	RMSE	MAPE (%)	*r*	R^2
EEMD-RE	**0.417**	**4.25**	**0.957**	**0.8973**
EEMD-ANN	0.775	14.87	0.883	0.6454
WA-RE	0.416	8.07	0.970	0.8979
WA-ANN	0.467	8.49	0.950	0.8712
ARIMA	1.607	27.01	0.040	0.5256

Table 4: Predictive accuracy of different models with monthly data

Methods	RMSE	MAPE (%)	*r*	R^2
EEMD-RE	0.339	6.00	0.979	0.9319
EEMD-ANN	0.362	5.66	0.966	0.9226
WA-RE	0.305	5.23	0.980	0.9450
WA-ANN	0.377	6.83	0.959	0.9161
ARIMA	0.368	6.79	0.967	0.9199

DISCUSSION

The decomposition results shown in Figure 2 show that there are 12 IMFs derived from EEMD, while four sub-sequences derived from WA as a comparison. It can be inferred that EEMD has stronger ability in mining more sufficient information with regularity. The comparison of the decomposed results by EEMD and WA with monthly data in Figure 3 also shows that the IMFs from EEMD for different time series seems more regular than the sub sequences from WA in Figure 3c,d, which vary distinctly for different years. It can be seen from Figure 4 that the predictive result by EEMD-RE model is the most stable one compared with other models, which indicates that EEMD-RE model can capture the stable information in the data instead of paying attention to the uncertainties. The statistical results in Table 1, Table 2 and Table 3 also confirm that the EEMD-RE model has the highest predictive accuracy with daily data compared to other models, with a smaller RMSE and higher model fitting degree. The comparison between the EEMD-RE and EEMD-ANN implies that the relationship between the original solar radiation sequence and the derived IMFs is linear. Thus the superiority of ANNs for complex non-linear problem does not work for solar radiation data. The comparison between EEMD and WA verifies the strong adaptivity of EEMD for non-linear and non-stationary series; while the WA relies greatly on the mother wavelet, while it may lead to virtual fluctuations. Especially in Figure 4c, which shows that the four kinds of hybrid models, including EEMD-RE, EEMD-ANN, WA-RE, and WA-ANN all perform well when predicting monthly solar radiation for the next year with historical daily data. Such results verify the validity and effectiveness of the idea that decomposing time series into sub sequences with more regularity is helpful for long-term prediction. Interestingly, the monthly predictive effects with monthly data in Figure 5 and Table 4 are better than those with daily data. Although we thought more information could be explored for data with smaller time intervals, it seems that more randomness and uncertainty were introduced for daily data compared to monthly data, and some errors in the prediction with shorter time interval might be smoothed in the statistical process for the longer time interval. The model fitting degree of the ARIMA model is low using daily data, indicating that it is not suitable for long-term prediction with large amount of data. The proposed EEMD-RE model is thus recommended for long-term solar radiation predictions.

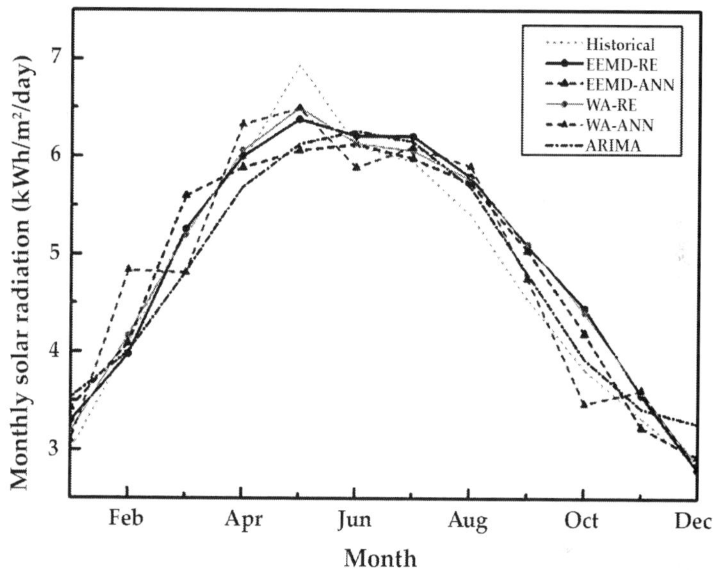

Figure 5: Predictive solar radiation results of the year 1995 using monthly data.

CONCLUSIONS

The solar radiation forecast is important for solar energy utilization. The causes of variations in solar radiation are various. There exists a complicated coupling relationship between the solar radiation intensity and the meteorological elements and terrain factors, but the data of complicated climate conditions is often difficult to obtain.

In this paper, hybrid methods following the decomposition-prediction-reconfiguration paradigm are proposed with different combination of EEMD, WA, RE, and ANN, which is only based on historical solar radiation data. The application on the west of China shows that basically these hybrid decomposition-reconfiguration models perform well for monthly prediction using monthly historical data; while for the daily prediction, the EEMD-

RE model outperforms other models, since (1) the decomposition results in components with regular characteristics; (2) the relationship between the original solar radiation sequence and the derived IMFs is linear; and (3) the EEMD has strong adaptivity for non-linear and non-stationary series. The proposed hybrid decomposition-reconfiguration models only relying on the historical radiation records have great practical value for long-term prediction of solar radiation intensity, especially in the areas where complex climate data is difficult to obtain.

AUTHOR CONTRIBUTIONS

F.-F.L. and J.Q. conceived and designed the experiments; S.-Y.W. performed the experiments; S.-Y.W. and F.-F.L. analyzed the data; J.Q. contributed reagents/materials/analysis tools; S.-Y.W. and F.-F.L. wrote the paper.

ACKNOWLEDGMENTS

This research was supported by National Key R&D Program of China (2017YFC0403600, 2017YFC0403602), the Open Project of State Key Laboratory of Plateau Ecology and Agriculture, Qinghai University.(Grant No. 2016-KF-03), and the Open Research Fund Program of State key Laboratory of Hydroscience and Engineering (Grant No. sklhse-2016-B-03).

REFERENCES

1. Bulut, H.; Buyukalaca, O. Simple model for the generation of daily global solar-radiation data in Turkey. *Appl. Energy* 2007, *84*, 477–491.
2. Kaplanis, S.; Kumar, J.; Kaplani, E. On a universal model for the prediction of the daily global solar radiation. *Renew. Energy* 2016, *91*, 178–188.
3. Qazi, A.; Fayaz, H.; Wadi, A.; Raj, R.G.; Rahim, N.A.; Khan, W.A. The artificial neural network for solar radiation prediction and designing solar systems: A systematic literature review. *J. Clean. Prod.* 2015, *104*, 1–12.
4. Diagne, M.; David, M.; Lauret, P.; Boland, J.; Schmutz, N. Review of solar irradiance forecasting methods and a proposition for small-scale insular grids. *Renew. Sustain. Energy Rev.* 2013, *27*, 65–76.
5. Perez, R.; Lorenz, E.; Pelland, S.; Beauharnois, M.; Van Knowe, G.; Hemker, K.; Heinemann, D.; Remund, J.; Muller, S.C.; Traunmuller, W.; et al. Comparison of numerical weather prediction solar irradiance forecasts in the US, Canada and Europe. *Sol. Energy* 2013, *94*, 305–326.
6. Mathiesen, P.; Collier, C.; Kleissl, J. A high-resolution, cloud-assimilating numerical weather prediction model for solar irradiance forecasting. *Sol. Energy* 2013, *92*, 47–61.
7. Zamora, R.J.; Dutton, E.G.; Trainer, M.; McKeen, S.A.; Wilczak, J.M.; Hou, Y.T. The accuracy of solar irradiance calculations used in mesoscale numerical weather prediction. *Mon. Weather Rev.* 2005, *133*, 783–792.
8. Wang, F.; Mi, Z.Q.; Su, S.; Zhao, H.S. Short-term solar irradiance forecasting model based on Artificial Neural Network using statistical feature parameters. *Energies* 2012, *5*, 1355–1370.
9. Trapero, J.R.; Kourentzes, N.; Martin, A. Short-term solar irradiation forecasting based on Dynamic Harmonic Regression. *Energy* 2015, *84*, 289–295.
10. Voyant, C.; Paoli, C.; Muselli, M.; Nivet, M.L. Multi-horizon solar radiation forecasting for Mediterranean locations using time series models. *Renew. Sustain. Energy Rev.* 2013, *28*, 44–52.
11. Chen, S.X.; Gooi, H.B.; Wang, M.Q. Solar radiation forecast based on fuzzy logic and neural networks. *Renew. Energy* 2013, *60*, 195–201.
12. Mellit, A.; Benghanem, M.; Kalogirou, S.A. An adaptive wavelet-

network model for forecasting daily total solar-radiation. *Appl. Energy* 2006, *83*, 705–722.

13. Akarslan, E.; Hocaoglu, F.O.; Edizkan, R. A novel M-D (multi-dimensional) linear prediction filter approach for hourly solar radiation forecasting. *Energy* 2014, *73*, 978–986.
14. Cevik, A. Unified formulation for web crippling strength of cold-formed steel sheeting using stepwise regression. *J. Constr. Steel Res.* 2007, *63*, 1305–1316.
15. Nourani, V.; Komasi, M.; Mano, A. A Multivariate ANN-Wavelet Approach for Rainfall-Runoff Modeling. *Water Resour. Manag.* 2009, *23*, 2877–2894.
16. Cao, S.H.; Cao, J.C. Forecast of solar irradiance using recurrent neural networks combined with wavelet analysis. *Appl. Ther. Eng.* 2005, *25*, 161–172.
17. Sharifi, S.S.; Rezaverdinejad, V.; Nourani, V. Estimation of daily global solar radiation using wavelet regression, ANN, GEP and empirical models: A comparative study of selected temperature-based approaches. *J. Atmos. Sol. Terr. Phys.* 2016, *149*, 131–145.
18. Lauret, P.; Voyant, C.; Soubdhan, T.; David, M.; Poggi, P. A benchmarking of machine learning techniques for solar radiation forecasting in an insular context. *Sol. Energy* 2015, *112*, 446–457.
19. Wang, J.Z.; Jiang, H.; Wu, Y.J.; Dong, Y. Forecasting solar radiation using an optimized hybrid model by Cuckoo Search algorithm. *Energy* 2015, *81*, 627–644.
20. Gala, Y.; Fernandez, A.; Diaz, J.; Dorronsoro, J.R. Support Vector Forecasting of Solar Radiation Values. *Hybrid Artif. Intell. Syst.* 2013, *8073*, 51–60.
21. Baser, F.; Demirhan, H. A fuzzy regression with support vector machine approach to the estimation of horizontal global solar radiation. *Energy* 2017, *123*, 229–240.
22. Coelho, J.P.; Boaventura-Cunha, J. Long term solar radiation forecast using computational intelligence methods. *Appl. Comput. Intell. Soft Comput.* 2015, *2014*, 21.
23. Cao, J.C.; Cao, S.H. Study of forecasting solar irradiance using neural networks with preprocessing sample data by wavelet analysis. *Energy* 2006, *31*, 3435–3445.
24. Paoli, C.; Voyant, C.; Muselli, M.; Nivet, M.L. Forecasting of

preprocessed daily solar radiation time series using neural networks. *Sol. Energy* 2010, *84*, 2146–2160.

25. Amrouche, B.; Le Pivert, X. Artificial neural network based daily local forecasting for global solar radiation. *Appl. Energy* 2014, *130*, 333–341.
26. Pedro, H.T.C.; Coimbra, C.F.M. Assessment of forecasting techniques for solar power production with no exogenous inputs. *Sol. Energy* 2012, *86*, 2017–2028.
27. Khatib, T.; Mohamed, A.; Sopian, K. A review of solar energy modeling techniques. *Renew. Sust. Energy Rev.* 2012, *16*, 2864–2869.
28. Yadav, A.K.; Chandel, S.S. Solar radiation prediction using Artificial Neural Network techniques: A review. *Renew. Sust. Energy Rev.* 2014, *33*, 772–781.
29. Voyant, C.; Muselli, M.; Paoli, C.; Nivet, M.L. Optimization of an artificial neural network dedicated to the multivariate forecasting of daily global radiation. *Energy* 2011, *36*, 348–359.
30. Ozgoren, M.; Bilgili, M.; Sahin, B. Estimation of global solar radiation using ANN over Turkey. *Expert Syst. Appl.* 2012, *39*, 5043–5051.
31. Koca, A.; Oztop, H.F.; Varol, Y.; Koca, G.O. Estimation of solar radiation using artificial neural networks with different input parameters for Mediterranean region of Anatolia in Turkey. *Expert Syst. Appl.* 2011, *38*, 8756–8762.
32. Celik, O.; Teke, A.; Yildirim, H.B. The optimized artificial neural network model with Levenberg-Marquardt algorithm for global solar radiation estimation in Eastern Mediterranean Region of Turkey. *J. Clean. Prod.*2016, *116*, 1–12.
33. Chiteka, K.; Enweremadu, C.C. Prediction of global horizontal solar irradiance in Zimbabwe using artificial neural networks. *J. Clean. Prod.* 2016, *135*, 701–711.
34. Renno, C.; Petito, E.; Gatto, A. ANN model for predicting the direct normal irradiance and the global radiation for a solar application to a residential building. *J. Clean. Prod.* 2016, *135*, 1298–1316.
35. Foufoula-Georgiou, E.; Kumar, P. (Eds.) *Wavelets in Geophysics*; Academic Press: San Diego, CA, USA, 1994.
36. Almasri, A.; Locking, H.; Shukur, G. Testing for climate warming in Sweden during 1850–1999, using wavelets analysis. *J. Appl. Stat.* 2008, *35*, 431–443.

37. Kisi, O. Wavelet regression model as an alternative to neural networks for monthly streamflow forecasting. *Hydrol. Process.* 2009, *23*, 3583–3597.
38. Partal, T. Modelling evapotranspiration using discrete wavelet transform and neural networks. *Hydrol. Process.* 2009, *23*, 3545–3555.
39. Huang, N.E.; Shen, Z.; Long, S.R.; Wu, M.C.; Shih, H.H.; Zheng, Q.; Yen, N.C.; Tung, C.C.; Liu, H.H. The empirical mode decomposition and the Hilbert spectrum for nonlinear and non-stationary time series analysis. *Proc. R. Soc. Lond. A Math. Phys. Eng. Sci.* 1998, *454*, 903–995.
40. Monjoly, S.; Andre, M.; Calif, R.; Soubdhan, T. Hourly forecasting of global solar radiation based on multiscale decomposition methods: A hybrid approach. *Energy* 2017, *119*, 288–298.
41. Shuanghua, C.; Jiacong, C.; Fengqiang, L. Study of application of wavelet analysis to neural networks for the forecast of solar irradiance. *J. Donghua Univ.* 2004, *6*, 18–22.
42. Meng, K.F.; Li, C.L.; Zhang, H.N.; Yang, L.B.; Zhang, S.C.; Wang, X.; Yang, J.; Li, Z.X. Analysis and discussion on related problems of solar energy resources development in Qinghai province. In Proceedings of the CSEE (Chineses Society for Electrical Engineering) Annual Meeting, Chengdu, China, 30 September 2015; pp. 1–3.
43. Yan, L.; Zhou, X.; Zhang, C.; Fei, W.Y.; Xia, X.C.; Zhou, F.Q.; Hu, X.H.; Xu, H.H.; Huang, C.G.; Liu, F.S.; et al. A proposal for planning and constructing a national integrated energy base combined with large-scale photo-voltaic power and hydropower in Qinghai province. *Adv. Technol. Electr. Eng. Energy* 2011, *29*, 1–9.
44. Photovoltaic leads the development of new energy and new materials industry in Qinghai Province Available online:. Available online: http://www.cpicorp.com.cn/zhxx/201803/t20180326_287170.htm (accessed on 26 March 2018).
45. Cameron, C.A.; Windmeijer, F.A.G. An R-squared measure of goodness of fit for some common nonlinear regression models. *J. Econ.* 1997, *77*, 1790–1792.
46. Muhammad, I. Coefficient of Determination. Available online: Itfeature.com (accessed on 29 April 2012).

CHAPTER 3

DESIGN AND CONCEPT OF AN ENERGY SYSTEM BASED ON RENEWABLE SOURCES FOR GREENHOUSE SUSTAINABLE AGRICULTURE

Ioan Aschilean [1,2], Gabriel Rasoi [1], Maria Simona Raboaca [1], Constantin Filote [3,4], and Mihai Culcer [1]

[1]National Research and Development Institute for Cryogenic and Isotopic Technologies—ICSI Rm. Valcea, Uzinei Street, No. 4, 240050 Ramnicu Valcea, Romania

[2]SC ACI Cluj SA, Avenue Dorobantilor, No. 70, Cluj-Napoca 400609, Romania

[3]Stefan cel Mare University of Suceava, Faculty of Electrical Engineering and Computer Science, 720229 Suceava, Romania

[4]MANSiD Integrated Center, PROTHILSYS Lab, Stefan cel Mare University, 720229 Suceava, Romania

ABSTRACT

Bio-organic greenhouses that are based on alternative resources for producing heat and electricity stand out as an efficient option for the sustainable

Citation: Aschilean, I.; Rasoi, G.; Raboaca, M.S.; Filote, C.; Culcer, M. Design and Concept of an Energy System Based on Renewable Sources for Greenhouse Sustainable Agriculture. Energies 2018, 11, 1201.

development of agriculture, thus ensuring good growth and development of plants in all seasons, especially during the cold season. Greenhouses can be used with maximum efficiency in various agricultural lands, providing ideal conditions of temperature and humidity for short-term plant growing, thereby increasing the local production of fruit and vegetables. This paper presents the development of a durable greenhouse concept that is based on complex energy system integrating fuel cells and solar panels. Approaching this innovative concept encountered a major problem in terms of local implementation of this type of greenhouses because of the difficulty in providing electrical and thermal energy from conventional sources to ensure an optimal climate for plant growing. The project result consists in the design and implementation of a sustainable greenhouse energy system that is based on fuel cells and solar panels.

Keywords: Hydrogen, energy, greenhouses, renewable

INTRODUCTION

Development of renewable energy as a primary global resource of clean energy is one of the main objectives of energy policies worldwide, which, in the general framework of sustainable development, aimed at reducing energy consumption, increasing security of supply, environmental protection, and friendly and sustainable energy technology development [1].

Renewable sources represent good alternatives to fossil resources, which are limited in quantity and are prone to exhaustion.

In this context, the use of the proposed hybrid system can be successfully used in areas where the connection to the grid is not possible or where the development of the electrical infrastructure is not technically feasible or in terms of investment costs. The area of greenhouses in Romania is 922 hectares, of which, about 450 hectares are heated greenhouses.

Agriculture is one of the most important sectors, which is characterized by the greatest potential for sustainable economic development [1,2].

In order to reduce production costs, it is necessary to implement a hybrid thermal power that is based on renewable energy designed and dimensioned according to local demand, so that the production costs to be reduced significantly, while considering that heating accounts for about 30% of the total energy used in the greenhouse [3,4]. Integration of renewable sources-based hybrid system in the greenhouses to provide heat and electricity is an important objective for sustainability and efficiency of commercial systems in order to increase the production and reduce costs that are associated to heat production in order to ensure an optimal climate for plant growing [5,6].

The concept of this project came as a result of the demand of a farming company that is specialized in greenhouse vegetable-growing, which has shown interest in developing a new greenhouse concept based on sustainable sources of energy-combustion piles and solar power. Major disadvantages for vegetable farmers occur in winter when it is necessary to provide specific environmental conditions: temperature (min 15 °C), humidity, sun exposure, water, and fertilizers. In this context, the design of a sustainable energy system is the first Romanian initiative to implement renewable energy sources in the agricultural field [7].

A combination of renewable sources by creating a mixed system is a sustainable and economic solution that could address these issues [8,9,10].

Reducing fossil fuel consumption by using solar energy can contribute to global climate change as a result of reducing greenhouse gas emissions and the impact of energy use on the environment [11].

Photovoltaic energy is a valuable energy source that comes from renewable sources that are inexhaustible and non-polluting. To be used in a wide range of applications and to meet cost constraints, the implemented energy system must feature a good optimization of photovoltaic cells with a practical validation [12].

The total amount of solar energy received at ground level for one week exceeds the energy that is produced by oil, coal, natural gas, and uranium in the world.

In most cases, it is necessary to convert solar energy into electricity [13].

Solar energy is a source of green and inexhaustible energy and its production cost is zero, thus successfully replacing the conventional energy that we buy, and, consequently, reducing the production costs considerably [14]. These initiatives were adopted in some Smart Island north European [15].

Photovoltaic energy is the product of direct conversion of solar light into electricity using solar cells that are connected to produce the desired electrical energy [16].

Using solar energy that is provided through solar panels and solar collectors is an efficient and environmentally friendly way that can help to reduce production costs in stand-alone greenhouses [17,18,19].

Free energy of the sun can be used to heat greenhouses by collecting and storing heat during the hot summer season and using it during the cold season.

Moreover, solar energy can be used to generate electricity by integrating a system of photovoltaic panels that are mowing on the roof of the greenhouse [19].

As an alternative to fossil fuels that are expensive, farmers use renewable energy sources, such as solar heat pumps [20], geothermal heating systems, thermo-solar and photovoltaic panels, and biomass-derived fuels for greenhouse heating.

Greenhouses are covered with transparent materials since they mainly use solar energy, being designed to provide optimum growth conditions for plants [21].

There are two types of greenhouses that use solar energy for heating.

First, passive greenhouses are designed to maximize solar heat gains by using special coating and structural materials that are used as solar collectors [22].

Secondly, there are active greenhouses that are equipped with solar systems using an independent heat collection and storage system, supplying the greenhouse with additional thermal energy when compared to the heat that is generated by direct heating [23,24].

The main objective of the research partnership made up of Stefan cel Mare University of Suceava and National Research and Development Institute for Cryogenic and Isotopic Technologies (ICSI) Ramnicu Valcea is to develop a functional and durable energy system that is aimed at greenhouse bio-organic farming.

HYBRID ENERGY SYSTEM: A CASE STUDY

This paper identifies specific elements of a case study on the concept of sustainable development of organic greenhouses by integrating a hybrid energy system that is based on renewable sources [25].

In the warmer months, the excess energy that is produced by photovoltaic panels is stored in a hydrogen tank using an electrolyzer, and in the cold season, hydrogen is used by the fuel cell to generate energy when the photovoltaic panels are unable to cover the demand energy.

This concept of hybrid energy system based greenhouse was designed, built, and implemented in a research project having ICSI as partner. The project beneficiary provided a greenhouse having the parameters that are specified in Table 1.

Table 1: Calculation of greenhouse heat demand

Climatic Zone	**II**	**Wind Area**	**II**
Outside temperature	−15 (°C)	Wind speed	7.00 (m/s)
Sealed greenhouse		Outside the village	
Insulation material: double-layer polyethylene film			
Indoor temperature	18 (°C)		
1	d-the greenhouse wall thickness (mm)		
1000	ë-Thermal conductivity greenhouse walls		
3	Embodiment of the greenhouse		
90	Surface land that is located greenhouse (m^2)		
190	A-Total greenhouse area (m^2)		
284	V-Greenhouse volume (m^3)		
79.23	Q-heat requirement for calculating (kW)		
9.68	Kconv-total coefficient of heat transfer by convection through the surface (W/m^2 K)		
1.70	n-tightness coefficient greenhouse		
0.10	πn-penetration coefficient (kJ/kg K)		
0.32	ξ-coefficient that takes account of indoor and outdoor air enthalpy		
11.60	αi-heat transfer coefficient of surface to the inside (W/m^2 K)		
32.58	αe-heat transfer coefficient on the outside surface (W/m^2 K)		
8.48	KET-total coefficient of heat transfer by convection through the surface of the greenhouse, considered sealed (W/m^2 K)		
0.61	ΨA-coefficient that depends on the area of land that is located greenhouse		
10.00	L-greenhouse length (m)		
9.00	l-width greenhouse (m)		
4.00	H-maximum height (m)		

4.50	r_1-circle's radius = l/2 (m)
4.00	r_2-circle's radius = H (m)
90.00	S-greenhouse area (m^2)
205	A1-total area when r_1 = l/2 (m^2)
176	A2-total area when r_1 = H (m^2)
190.4	A~total area (m^2): average between A1 and A2
318	V1-greenhouse volume when r_1 = l/2 (m^3)
251.3	V2-greenhouse volume when r_1 r_2 = H (m^3)
284.6	V~greenhouse volume (m^3): average between V1 and V2

Table 1 shows the calculation of heat demand of a modular greenhouse.

The hybrid energy system is able to produce cost efficient heat and electricity at any time, having good efficiency and a low level of environmental pollution.

An important requirement is to investigate the feasibility of the equipment installed in experimental greenhouse, and to evaluate the mutual benefits that are arising from this integration. The case study refers to a modular greenhouse with an area of 90 m^2, airfoil shaped tunnel with steel structure and round arches.

Gauge dimensions of the greenhouse are: L = 10 m, l = 9 m, H = 4 m.

Figure 1 shows the concept and design of the greenhouse and Figure 2 shows a functional greenhouse [26].

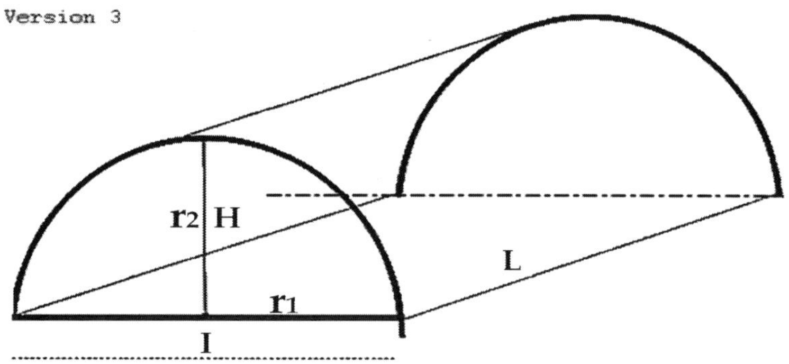

Figure 1: Elliptic design of the greenhouse.

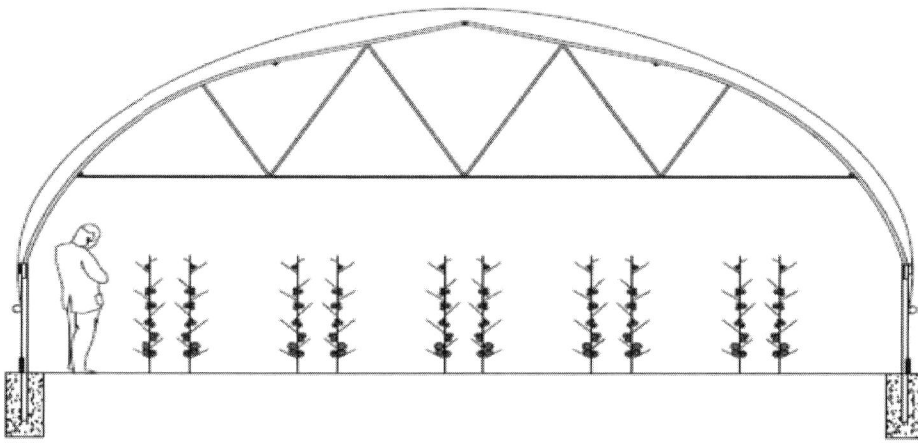

Figure 2: Functional greenhouse.

This greenhouse model was selected because it displays good strength and durability, being able to resist winds of 90 km/h and snow layer (80 kg/ m^2 + 25 kg/m^2 internal load) [27].

According to Romanian standard SR 1907-3, the energy for heating the greenhouses was calculated, while considering the type of material that I sused for insulation and coatings [28].

Efforts to decrease energy consumption have led the researchers to use alternative energy sources for greenhouse heating.

Several types of passive solar systems and techniques have been proposed and used for the substitution of conventional fuels with solar energy as available low-cost technology [29,30].

Because the sunlight may be insufficient in winter, then a combination of renewable energy sources is very useful to be used in this situation.

Table 2 shows the calculation of greenhouse heat loss.

Table 2: Calculation of greenhouse heat loss

Calculation of Greenhouse Heat Loss	**Data**	**U/M**
Double-layer sheet losses	15.28	(kW)
Loss through ground	1.02	(kW)
Other losses	16.30	(kW)

Effective thermal calculation	41.1	(kW)
Thermal calculation for heat generator choosing	46	(kW)
Heat generator efficiency = 0.9	41.078	(kcal/h)
Overall thermal power	37	(kW)

Collecting solar radiation is more efficient when the greenhouse is oriented East-West, which may be performed both in summer and winter [31,32,33,34,35,36].

HYBRID ENERGY SYSTEM COMPONENTS

The project research team identified the following components of the hybrid energy system, in accordance with the specific technical requirements of the beneficiary.

The use of the fuel cell in the proposed hybrid system will have a high economic profitability, as it will be implemented by as many users as possible, thus reducing the cost of fuel cell production.

In the next level of this research, a multicriterial analyse taking into account economic criteria for all equipment from hybrid energy will be realized.

Thermal Energy Production System

-Thermal heating generator that is based on fuel wood and biomass is used to produce heat for the greenhouse needs. This equipment has a nominal heat output of 38 kW, it works very efficiently, gasification has low fuel consumption, and it shows superior performance, which is up to 93%.

From thermal calculation performed, it results that this model of power with thermal power of 38 kW is sufficient to provide the energy requirements of the greenhouse at a rate of up to 70%, when considering that its use is done mainly in winter.

-Thermal solar collector panels with vacuum tubes are a great alternative to produce hot water using solar energy in summer.

The total area is 3.5 m^2, P_{max} = 1260 kWh, 666.34 kWh/m^2 (63 kWh/tube), 67% optical efficiency, maximum temperature 239 °C.

Using thermal solar panels for a period of 4–6 months per year in the greenhouse can bring in significant savings on heat production.

Mixed hybrid heating system based on solar-hydrogen energy and biomass allows for a saving of up to 30% of annual fuel that is used for heating and domestic hot water [37].

Electricity Generation System

-The assembly of photovoltaic panels, Off Grid.

-Polymer electrolyte membrane fuel cells (PEMFC), P_{max} = 9 kW, T = 14.4 V, I = 35 A, hydrogen consumption 6.5 L/min.

-Proton excenge membrane (PEM) electrolyzer, with capacity of 1.05 Nm3H2/h at a pressure up to 30 bar, U = 230 V, P_{max} = 2 kW.

The hydrogen produced by the electrolyzer is very efficient when it is converted into electricity using fuel cells with proton exchange membrane, which are actually electrochemical energy converters [38].

This equipment has the advantage that it can be used to produce electricity at any time using stored hydrogen, but only when it is necessary.

PEM fuel cells are the most promising type of power generation, due to its advantages, such as simplicity, low operating temperature, and easy maintenance [39,40].

PEM fuel cells are the future of generators that provide electricity and portable station types, using renewable energy sources for this purpose.

The implementation of a combined electrolysis fuel system for the production and storage of hydrogen in a demonstration greenhouse is a good alternative to traditional power solutions, given that it provides reliable equipment and it generates electricity at all times [41].

Electrical and Thermal Energy Storage System

-Mixed boiler for hot water heating and storage, with a capacity of 500 L and thermal energy storage power up to 42 kWh.

-Pressure hydrogen storage cylinders, with a capacity of 50 L and volume of 10 m^3.

-Solar batteries with gel solution, U = 12 V, I = 200 Ah, are designed for photovoltaic systems and kits, and they are used to store electricity.

This type of battery uses innovative technology "Absorbent Glass Mat", which gives them the property to provide significant energy reserves that can feed many electrical consumers throughout its service life [42]. A schematic diagram of the constructed experimental system is illustrated in Figure 3.

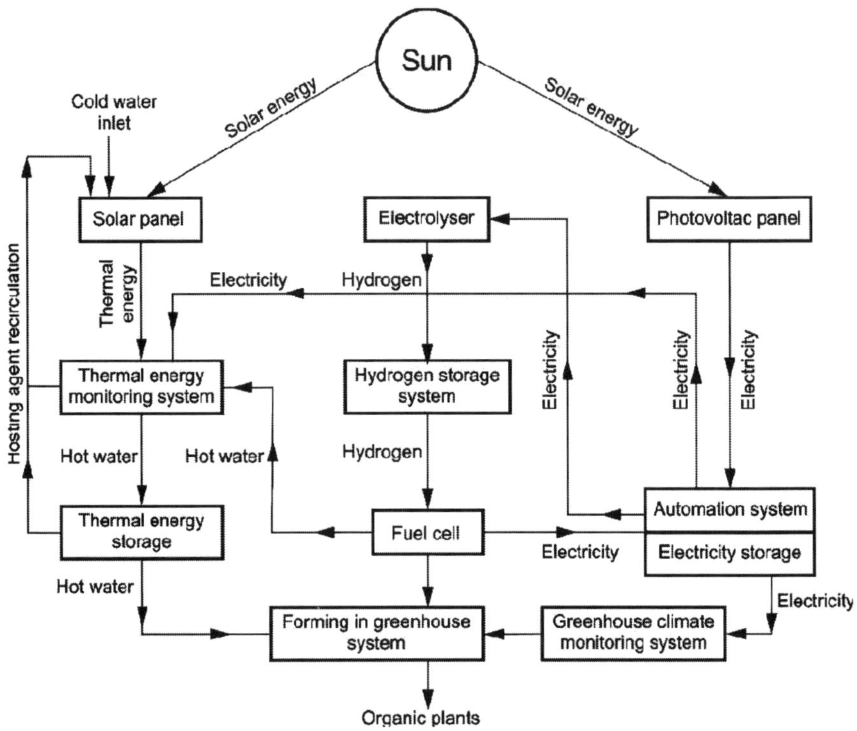

Figure 3: Schematic diagram of the hybrid greenhouse system.

The output power of a solar PV panel changes in accordance with change in solar radiation and temperature level.

This makes it impossible to use the direct-coupled method to automatically track the maximum power point.

These changes in weather conditions are shown by the P-V curves that are displayed in Figure 4 and Figure 5, respectively.

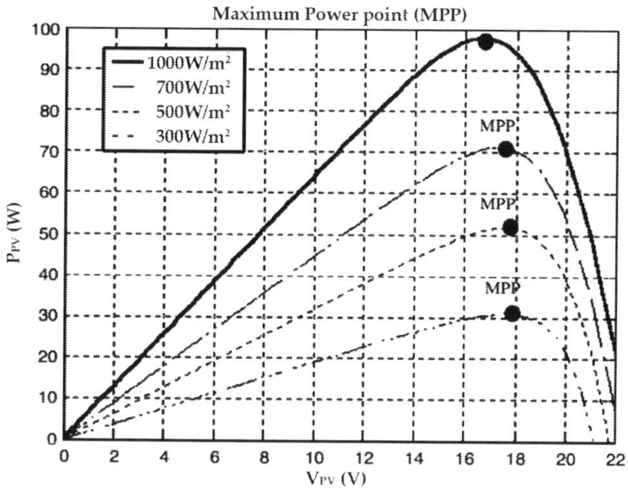

Figure 4: Characteristic curves at four different irradiances.

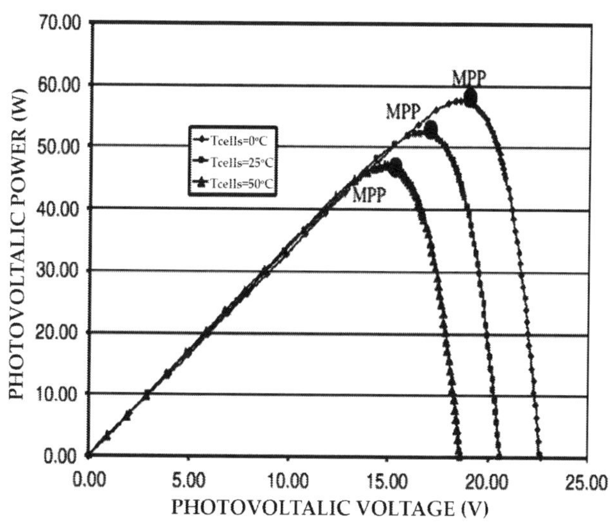

Figure 5: Characteristic curves at different temperatures.

Figure 4 shows the characteristic curves at four different irradiances. A Maximum Power Point Tracking (MPPT) system needs to be implemented in order to extract maximum power during the operation of solar panel and to be able to track the changes in power due to changes in the atmospheric conditions [43].

Figure 5 presents the characteristic curves at different temperatures.

Modularity is one major advantage of this sustainable type of renewable sources based greenhouse. Once the hybrid system is sized and implemented by resizing individual components, an unlimited number of various constructive structural elements can be added: photovoltaic panels, thermal solar modules, fuel cells heating systems, ventilation systems, etc. [44].

In the event that there are conditions for biogas production to complement the energy requirement of the greenhouse, a small system to produce the fuel gas by decomposing organic matter can also be integrated.

Renewable energy resources, such as wind, sunlight, geothermal, and biomass are mostly used.

They are working together and their integration into the energy market can improve the sustainability and the reliability of the power systems [45].

A microgrid is an autonomous electric distribution system that combines one or more energy resources with the loads, having its own management and control system, and working as an independent controllable entity [46].

The communication and control responsible device collects the data from the microgrid and manages the system.

In Figure 6 is represented the functional scheme of biogas system.

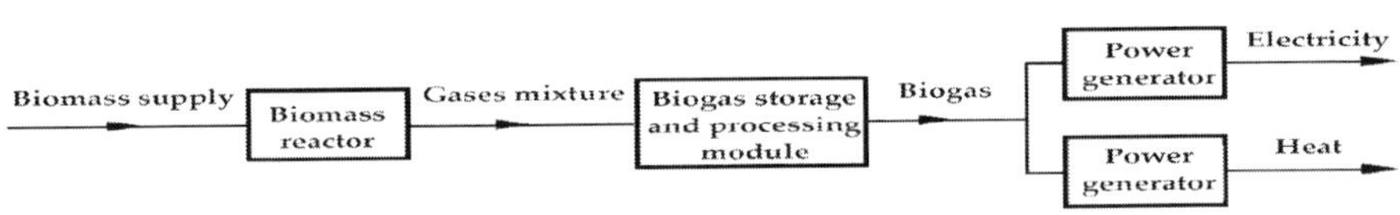

Figure 6: Schematic diagram of the biogas system.

The algorithm that is used to test the system is based on the state of charge of the batteries. The read values are the following: BatSoc, PacSI, ExtPwrAt, GnManStr, PacSB, and Pbio (power of the biomass generator).

The microgrid setup has emulators for the geothermal and biomass generators and a photovoltaic system with storage capability and two inverters, a grid forming capable, and a grid follower.

That proper energy production of the microgrids is also a substantial issue.

The balance between the energy flow and the load demands is their basic rule considering the availability of the resources. A properly functioning energy management system can ensure the best solution and meet the load requirements continuously and in short time.

The gradient-based systems are too slow to be used in real-time energy management systems, so the articles from this area focus on the off-line application [47].

The use of biogas to produce heat and electricity in the case of greenhouses is one of the most effective solutions to ensure their sustainability.

Heating and cooling systems are major costs involved in plant production in greenhouses.

Normally, heat-generating generators that imply the high consumption of energy are normally used to heat greenhouses, which are usually supplied by combustion of fossil fuels (diesel fuel, oil, oil, gas).

In view of the above, an effective solution for the sustainable development of greenhouse farming is the replacement of fossil fuels with alternative energy sources [48].

The main alternative energy sources to be implemented in the greenhouse for the supply of heat and power are the following:

- thermo-solar energy;
- energy from biomass and solid wood; and,
- energy from hydrogen energy.

The innovative combination of these renewable sources and the use of local air conditioning systems [49,50] will create an energy system that can meet the energy needs of an agricultural greenhouse, thus achieving an optimal climate for plant growing [51].

DEVELOPMENT AND PERSPECTIVES

The greenhouse systems that are provided with heating systems which are usually used during cold nights and during the winter season have a significant advantage over the quality of the products that are obtained, as well as a significant reduction in the planting and harvesting time.

In order to maintain the ambient temperature at optimal parameters for plant growing, it is necessary to consume large amounts of heat, which is usually supplied by fossil fuel energy systems [52].

As a result, the average temperature difference between the inlet and outlet of the earth-air heat exchanger (EAHE) was 8.29 °C.

The total electricity consumption of this system was 8.10 kWh, operating approximately 11 h/day, when 34.55% of this energy demand was provided by photovoltaic cells [53].

An Indian study [54] analyzed and modeled an integrated energy system for a greenhouse consisting of solar PV, a PEM polymeric membrane electrolyzer, and fuel cell assemblies. This study demonstrated that 51 PV modules, each modulus with a power of about 75 W together with a 3.3 kW electrolyzer and two PEM fuel cell assemblies, each 480 W power unit can cover the energy requirement of a flower greenhouse of 90 m^2.

Solar radiation in the greenhouse depends on its orientation and positioning, but East-West orientation is more effective in collecting solar radiation in winter than in summer collection [31,40].

A key factor in the proper functioning of the greenhouses is the implementation of an efficient irrigation system to ensure the effective hydration of the plants.

In this respect, ensuring the supply of electricity for pumping, transport, and water storage equipment is one of the most important objectives.

A solar water pumping system has many important advantages, for example, besides any fuel and maintenance costs; there is no environmental pollution hazard.

There are very frequent cases where rural settlements, which are made up of villages and communes, are not connected to the conventional power distribution lines because they are not located near them, being located at considerable and relatively isolated distances, so that it is almost impossible to connect them to classic energy distribution systems due to huge costs.

In this respect, the best solution to solve these problems is the use of small-scale energy applications consisting of combinations of photovoltaic panels with thermo-solar panels, which ensures the supply of greenhouses with heat, electricity, and water, representing an efficient and cost-effective solution for these isolated areas [55].

CONCLUSIONS

This paper presents the case study of a research project that had as the main objective the development and implementation of an experimental model of functional greenhouse, an integrated energy system for the production of thermal and electric energy using low-pollution renewable sources.

The case study that was analyzed in this paper shows that the development of a sustainable greenhouse concept that implements an integrated hydraulic energy system, based exclusively on renewable sources, such as solar energy, hydrogen energy, biomass with possible applicability in the future, along with the development of production technologies and the development of the production capacities of fuel cells.

The amount of additional energy that was produced by photovoltaic panels during the period 20 March–10 November, exceeding of technological consumption and lighting is 6775.84 kW, of this amount of energy between 20 March and 10 November, from using the electrolyzer we produce 203.28 kg of hydrogen. During 11 November–19 March, from 203.28 kg of hydrogen using fuel cell, we produce 3374.44 kW, which we provide 3 kW of energy for lighting and technology per day, with a total consumption of 387 kW and the difference of 2997.44 kW is used for heating the greenhouse as an additional energy, to the thermal energy produced by the wood, covering the peak load in the days with temperatures lower than −3 °C, with an average of 31 days per year.

New energy generation systems use all of the systems at renewable sources contribute to reducing overall energy consumption, increasing energy supply security, and protecting the environment, thus reducing the polluting emission [56,57].

In recent years, the use of food, water, and energy resources has become an essential issue, especially in rural areas, with some being unable to connect to electricity, water or gas networks but having very high potential for solar, wind and biomass renewable sources.

Due to different socio-economic obstacles, these renewable sources are under exploited and are poorly used by mankind.

Globally, due to population growth, and, implicitly, food and water needs, significant increases in energy consumption in agriculture are estimated.

The implementation of renewable energy sources, especially solar energy and biomass, will solve these problems by ensuring the provision of cheap and environmentally-friendly energy, especially for greenhouses, which use an appreciable amount of energy for the proper functioning of the cooling systems, heating, lighting, and irrigation.

Moreover, the use of green energy will lead to the sustainability of greenhouses, increased energy efficiency, increased food production, and the provision of cheap and clean energy.

AUTHOR CONTRIBUTIONS

Conceptualization, I.A. and M.S.R.; Methodology, C.F.; Software, I.A.; Validation, I.A., C.F. and M.C.; Formal Analysis, G.R.; Investigation, M.S.R.; Resources, G.R.; Data Curation, I.A.; Writing-Original Draft Preparation, M.S.R.; Writing-Review & Editing, I.A. and C.F.; Visualization, G.R.; Supervision, I.A.; Project Administration, C.F.; Funding Acquisition, G.R.

ACKNOWLEDGMENTS

This work was supported by a grant of the Romanian Ministery of Research and Innovation, CCCDI-UEFISCDI, project number PN-III-P1-1.2-PCCDI-2017-0776/No. 36 PCCDI/15.03.2018, within PNCDI III.

REFERENCES

1. The Institute for Diversification and Saving of Energy (IDAE). *Renewable Energy Plan 2011–2020*; IDAE; Spanish Ministry of Industry, Energy and Tourism: Madrid, Spain, 2010; Volume 1, p. 542.
2. Odeim, F.; Roes, J.; Heinzel, A. Power Management Optimization of an Experimental Fuel Cell/Battery/Supercapacitor Hybrid System. *Energies* 2015, *8*, 6302–6327.
3. Boulard, T.; Baille, A. Analysis of thermal performance of a greenhouse as a solar collector. *Energy Agric.* 1987, *6*, 17–26.
4. Kim, M.; Kim, J.; Bae, S. Design of a Sustainable and Efficient Transportation Station (SETS) Based on Renewable Sources and Efficient Electric Drives. *Symmetry* 2016, *8*, 146.
5. Nachidi, M.; Rodriguez, F.; Tadeo, F.; Guzman, J.L. Takagi-Sugeno control of nocturnal temperature in greenhouses using air heating. *ISA Trans.* 2011, *50*, 315–320.
6. Lopez, J.C.; Baille, A.; Bonachela, S.; Perez-Parra, J. Analysis and prediction of greenhouse green bean (*Phaseolus vulgaris* L.) production in a Mediterranean climate. *Biosyst. Eng.* 2008, *100*, 86–95.
7. Palander, T.; Kärhä, K. Adaptive Procurement Guidelines for Automatic Selection of Renewable Forest Energy Sources within a Sustainable Energy Production System. *Energies* 2016, *9*, 155.
8. Erdinc, O.; Uzunoglu, M. Optimum design of hybrid renewable energy systems: Overview of different approaches. *Renew. Sustain. Energy Rev.* 2012, *16*, 1412–1425.
9. Deshmukh, M.K.; Deshmukh, S.S. Modeling of hybrid renewable energy systems. *Renew. Sustain. Energy Rev.* 2008, *12*, 235–249.
10. Nema, P.; Nema, R.K.; Rangnekar, S. A current and future state of art development of hybrid energy system using wind and PV-solar: A review. *Renew. Sustain. Energy Rev.* 2009, *13*, 2096–2103.
11. Ozgener, O.; Hepbasli, A. Experimental Performance Analysis of a Solar Assisted Ground-Source Heat Pump Greenhouse Heating System. *Energy Build.* 2005, *37*, 101–110.
12. Zerhouni, F.Z.; Zegrar, M.; Benmessaoud, M.T.; Stambouli, A.B.; Midoun, A. A novel method to optimize photovoltaic generator operation. *Int. J. Energy Res.* 2008, *32*, 1444–1453.
13. Torres-Morenzo, J.L.; Gimenez-Fernandez, A.; Perez-Garcia, M.;

Rodriquez, F. Energy Management Strategy for Micro-Grids with PV-Battery Systems and Electric Vehicles. *Energies* 2018, *11*, 522.

14. Ibrahim, H.; Ilinca, A.; Perronv, J. Energy storage systems—Characteristics and comparisons. *Renew. Sustain. Energy Rev.* 2008, *12*, 1221–1250.
15. Cannistraro, G.; Cannistraro, M.; Trovato, G. Islands "Smart Energy" for eco-sustainable energy a case study "Favignana Island". *Int. J. Heat Technol.* 2017, *35*, 87–95.
16. Rosa, C.B.; Rediske, G.; Rigo, P.D.; Wendt, J.F.M.; Michels, L.; Siluk, J.C.M. Development of a Computational Tool for Measuring Organizational Competitiveness in the Photovoltaic Power Plants. *Energies* 2018, *11*, 867.
17. Von Zabeltitz, C. *Integrated Greenhouse Systems for Mild Climates*; Springer: Berlin/Heidelberg, Germany, 2011.
18. Sethi, V.P.; Sharma, S.K. Survey and evaluation of heating technologies for worldwide agricultural greenhouse applications. *Sol. Energy* 2008, *82*, 832–859.
19. Carreno-Ortega, A.; Galdeano-Gomez, E.; Perez-Mesa, J.C.; Galera-Quiles, M.C. Policy and Environmental Implications of Photovoltaic Systems in Farming in Southeast Spain: Can Greenhouses Reduce the Greenhouse Effect? *Energies* 2017, *10*, 761.
20. Esen, M.; Yuksel, T. Experimental evaluation of using various renewable energy sources for heating a greenhouse. *Energy Build.* 2013, *65*, 340–351.
21. Panwar, N.L.; Kaushik, S.C.; Kothari, S. Solar greenhouse an option for renewable and sustainable farming. *Renew. Sustain. Energy Rev.* 2011, *15*, 3934–3945.
22. Bot, G.P.A.; van de Braak, N.J.; Challa, H.; Hemming, S.; Rieswijk, T.; van Straten, G.; Verlodt, I. The solar greenhouse: State of the art in energy saving and sustainable energy supply. *Acta Horticult.* 2005, *691*, 501–508.
23. Sethi, V.P.; Sumathy, K.; Lee, C.; Pal, D.S. Thermal modeling aspects of solar greenhouse microclimate control: A review on heating technologies. *Sol. Energy* 2013, *96*, 56–82.
24. Santamouris, M.; Balaras, C.A.; Dascalaki, E.; Vallindras, M. Passive solar agricultural greenhouses: A worldwide classification and evaluation of technologies and systems used for heating purposes. *Sol.*

*Energy*1994, *53*, 411–426.

25. Nižetić, S.; Papadopulos, A.M.; Tina, G.M.; Rosa-Clot, M. Hybrid energy scenarios for residential applications based on the heat pump split air-conditioning units for operation in the Mediterranean climate conditions. *Energy Build.* 2017, *140*, 110–120.
26. Available online: www.seretransilvania.com (accessed on 17 November 2017).
27. Short, G.D. *Proiectarea si Constructia Serelor-Manual Tehnologic*; United States Agency for International Development (USAID); Agricultural Competitiveness and Enterprise Development (ACED); Millennium Challenge Corporation: Chisinau, Moldova, 2013.
28. Shen, Y.; Wei, R.; Xu, L. Energy Consumption Prediction of a Greenhouse and Optimization of Daily Average Temperature. *Energies* 2018, *11*, 65.
29. Fabrizio, E. Energy reduction measures in agricultural greenhouses heating: Envelope, systems and solar energy collection. *Energy Build.* 2012, *53*, 57–63.
30. Bargach, M.N.; Tadili, R.; Dahman, A.S.; Boukallouch, M. Survey of thermal performances of a solar system used for the heating of agricultural greenhouses in Morocco. *Renew. Energy* 2000, *20*, 415–433.
31. Abdel-Ghany, A.M. Solar energy conversions in the greenhouses. *Sustain. Cities Soc.* 2011, *1*, 219–226.
32. Castilla, N.; Hernandez, J. Greenhouse technological packages for high-quality production. *Acta Horticul.*2007, *761*, 285–297.
33. Barbir, F. *PEM Fuel Cells: Theory and Practice*; Academic Press: Cambridge, MA, USA, 2005; ISBN 978-0-12-078142-3.
34. Larminie, J.; Dicks, A. *Fuel Cell Systems Explained*, 2nd ed.; John Wiley & Sons Ltd.: Hoboken, NJ, USA, 2003; ISBN 978-0-47-084857-9.
35. Nižetić, S.; Grubišić-Čabo, F.; Marinić-Kragić, I.; Papadopoulos, A.M. Experimental and numerical investigation of a backside convective cooling mechanism on photovoltaic panels. *Energy* 2016, *111*, 211–225.
36. Ay, M.; Midilli, A.; Dincer, I. Exergetic performance analysis of a PEM fuel Cell. *Int. J. Energy Res.* 2006, *30*, 307–321.
37. Available online: www.esolar.ro (accessed on 29 October 2017).

38. Pascuzzi, S.; Anifantis, A.S.; Blanco, I.; Scarascia Mugnozza, G. Electrolyzer Performance Analysis of an Integrated Hydrogen Power System for Greenhouse Heating. A Case Study. *Sustainability* 2016, *8*, 629.

39. González, I.; Calderón, A.J.; Andújar, J.M. Novel remote monitoring platform for RES-hydrogen based smart microgrid. *Energy Convers. Manag.* 2017, *148*, 489–505.

40. Rosa, R.; Silva, A.M.; Miguel, A. Solar irradiation inside a single span greenhouse. *J. Agric. Eng. Res.* 1989, *43*, 221–229.

41. Mengelkamp, E.; Gärttner, J.; Rock, K.; Kessler, S.; Orsini, L.; Weinhardt, C. Designing microgrid energy markets: A case study: The Brooklyn Microgrid. *Appl. Energy* 2017, *210*, 870–880.

42. Smaoui, M.; Krichen, L. Control, energy management and performance evaluation of desalination unit based renewable energies using a graphical user interface. *Energy* 2016, *114*, 1187–1206.

43. Abbasi, M.A.; Zia, M.F. Novel TPPO Based Maximum Power Point Method for Photovoltaic System. *Adv. Electr. Comput. Eng.* 2017, *17*, 95–100.

44. Luque, A.; Hegedus, S. *Handbook of Photovoltaic Science and Engineering*, 2nd ed.; John Wiley & Sons: Hoboken, NJ, USA, 2011.

45. Whiteman, Z.S.; Bubna, P.; Prasad, A.K.; Ogunnaike, B.A. Design, Operation, Control, and Economics of a Photovoltaic/Fuel Cell/Battery Hybrid Renewable Energy System for Automotive Applications. *Processes*2015, *3*, 452–470.

46. Ahmadi, H.; Marti, J.R.; Dommel, H.W. A Framework for Volt-VAR Optimization in Distribution Systems. *IEEE Trans. Smart Grid* 2014, *6*, 1473–1483.

47. Lazar, E.; Petreus, D.; Etz, R.; Patarau, T. Software Solution for a Renewable Energy Microgrid Emulator. *Adv. Electr. Comput. Eng.* 2018, *18*, 89–94.

48. Jaramillo-Nieves, L.; Del Río, P. Contribution of Renewable Energy Sources to the Sustainable Development of Islands: An Overview of the Literature and a Research Agenda. *Sustainability* 2010, *2*, 783–811.

49. Cannistraro, G.; Cannistraro, M.; Restivo, R. The local media radiant temperature for the calculation of comfort in areas characterized by radiant surfaces. *Int. J. Heat Technol.* 2015, *38*, 111–118.

50. Cannistraro, G.; Cannistraro, M.; Restivo, R. Smart Control of Air

Climatization System in Function on the Values of Mean Local Radiant Temperature. *Smart Sci.* 2015, *3*, 157–163.

51. Chai, L.; Ma, C.; Ni, J.Q. Performance evaluation of ground source heat pump system for greenhouse heating in northern China. *Biosyst. Eng.* 2012, *111*, 107–117.
52. Chinese, D.; Meneghetti, A.; Nardin, G. Waste-to-energy based greenhouse heating: Exploring viability conditions through optimisation models. *Renew. Energy* 2005, *30*, 1573–1586.
53. Yıldız, A.; Ozgener, O.; Ozgener, L. Energetic performance analysis of a solar photovoltaic cell (PV) assisted closed loop earth-to-air heat exchanger for solar greenhouse cooling: An experimental study for low energy architecture in Aegean Region. *Renew. Energy* 2012, *44*, 281–287.
54. Ganguly, A.; Misra, D.; Ghosh, S. Modeling and analysis of solar photovoltaic-electrolyzer-fuel cell hybrid power system integrated with a floriculture greenhouse. *Energy Build.* 2010, *42*, 2036–2043.
55. Meah, K.; Fletcher, S.; Ula, S. Solar photovoltaic water pumping for remote locations. *Renew. Sustain. Energy Rev.* 2008, *12*, 472–487.
56. Cannistraro, G.; Cannistraro, M.; Cannistraro, A.; Galvagno, A. Analysis of Air Pollution in the Urban Center of Four Cities Sicilian. *Int. J. Heat Technol.* 2016, *2*, 219–225.
57. Cannistraro, G.; Cannistraro, M.; Galvagno, A.; Trovato, G. Technical and Economic Evaluations about the Integration of CoTrigeneration Systems in the Dairy Industry. *Int. J. Heat Technol.* 2016, *34*, 332–336.

CHAPTER 4

TOTAL-FACTOR ENERGY EFFICIENCY IN CHINA'S AGRICULTURAL SECTOR: TRENDS, DISPARITIES AND POTENTIALS

Zhihai Yang [1], Dong Wang [2,4,5], Tianyi Du [3,4], Anlu Zhang [6], and Yixiao Zhou [7]

[1]College of Economics and Management, Huazhong Agricultural University, No. 1 Shizishan Street, Wuhan 430070, China

[2]UWA School of Agriculture and Environment, The University of Western Australia, M089, 35 Stirling Highway, Crawley, WA 6009, Australia

[3]School of Business and Management, Murdoch University, 90 South Street, Murdoch, WA 6150, Australia

[4]Australia-China Sustainable Research and Development Center, Perth, WA 6009, Australia

[5]Inner Mongolia Honghe Energy and Environment Consultancy, Huhhot 010020, China

[6]College of Land Management, Huazhong Agricultural University, No. 1 Shizishan Street, Wuhan 430070, China

[7]School of Economics and Finance, Faculty of Business and Law, Curtin University, Rm 3016, Bldg 408, Bentley, Perth, WA 6004, Australia

Citation: Yang, Z.; Wang, D.; Du, T.; Zhang, A.; Zhou, Y. Total-Factor Energy Efficiency in China's Agricultural Sector: Trends, Disparities and Potentials. Energies 2018, 11, 853.

ABSTRACT

This paper investigates total-factor energy efficiency and analyses the trends of the efficiency changes in China's agricultural production across 30 provinces and three regions from 2001 to 2011, based on data envelopment analysis (DEA) approach. The potential amount of energy savings and five potential factors for energy efficiency improvement are also empirically studied by Tobit regression model. The findings show that (1) total-factor energy efficiency in China's agricultural sector is increasing over years but performs heterogeneously across regions; (2) agriculture intensive regions and energy abundant provinces tend to be relatively energy inefficient in agricultural production; and (3) economic structure, agricultural production structure, technological progress and income effect are major potentials for improving energy efficiency, whereas energy price is not a significant factor. This phenomenon results from the divergence of economic development, endowment effects as well as the scale of agricultural production. Policy implications drawn from this research are to upgrade industrial structure and promote agricultural transformation to enhance farmers' income as well as to establish a land market with entitling land property rights to farmers. This conclusion can assist to form more scientific rural energy policy decision-making in China and also can be extended to other developing economies for sustainable agriculture.

Keywords: energy efficiency, data envelopment analysis, agriculture transformation

INTRODUCTION

China is an ancient and grand agricultural economy. Energy consumption of agricultural sector has been increasing in recent decades so that the efficiency issue has emerged and attracted attention of economists and policymakers. China produces food for 23% of the world's population with only 7% of the world's farms, along with agricultural modernization and rapid urbanization [1]. This progress is closely connected with energy consumption in the agricultural sector given that energy is to some degree a foundation and key input of industrialization. As the largest energy consumer and the largest food producer in the world, China recently has launched a national strategy to decrease energy intensity and increase energy efficiency in order to tackle environmental degradation as well as climate change [2], from 42 million tons of standard coal (Mtce) to 75 Mtce. Concerning the accelerative trend, the central government announced that the agricultural sector will share equal responsibilities in improving energy efficiency with other sectors in the Twelfth National Five-Year Plan (2011–2015) [2]. Agriculture plays an important role in achieving energy efficiency targets mainly because the quantity of energy consumption would not decrease in China's predictable future given the substantial economic growth in rural areas. In fact, the absolute quantity of energy consumption is enormous and should not be ignored by policymakers. In many years, energy consumption within China's agricultural sector exceeds the total energy consumption of some industrialized or Organization for Economic Co-operation and Development (OECD) countries. For instance, according to the China Energy Statistical Yearbook and the China Energy Databook (Version 8.0), China's agricultural sector used 75 Mtce energy in 2011, which is 3.75 times the total energy consumption of Israel (20 Mtce), 2.7 times that of Switzerland (27 Mtce), 1.6 times that of Sweden (46 Mtce), and 1.2 times that of Belgium (61 Mtce) as well as nearly equal to the Netherlands (85 Mtce). Moreover, China's energy mix relies a lot on traditional energy compared to those developed countries in terms of rural underdevelopment. This implies that an effective regulation on agriculture energy efficiency would have a more significant effect on global greenhouse gases (GHG) mitigation, whilst the premise is to understand the state of energy efficiency in the agricultural sector delicately. Although some researchers have done a lot related to China's industry energy efficiency [3,4] and found that China's energy efficiency is much lower than some developed countries [5,6,7,8], and seldom has investigated energy efficiency in the agricultural sector. In this paper, we will investigate the trends of China's agriculture energy efficiency over the years, exploring the disparities across regions and analysing potentials for improving energy efficiency in the future. To our knowledge, this research will contribute a first glance to energy efficiency in China's agriculture by means of data envelop analysis (DEA) and the Tobit regression model.

Generally, we find that energy efficiency in China's agricultural sector increased during the past decades and some agriculture intensive provinces have lower efficiency in energy use. That is, more food in total output usually means less energy efficiency in the agricultural sector across provinces. This disparity between developed regions and underdeveloped regions further suggests that an interaction with divergent industrialization and economies of scale in agriculture. Thus, promotion of energy efficiency is connected with agriculture transformation as well. This is the primary contribution of this paper.

The article is structured as follows. Section 2 will review the main literature on energy efficiency. Section 3 will describe the methodologies we use. Section 4 will provide data and variable description. In addition, we will present and discuss results in Section 5 followed by policy implications in Section 6 and conclusions in Section 7.

LITERATURE REVIEW

An earlier relevant research was conducted by Catania [9] and he firstly stated that some characteristics of China's rural energy system include dispersed resource distribution, lower population density, lower energy technologies and investment; in addition, he pointed out that the energy consumption in the agricultural sector is closely related to agricultural production and ecological environment. However, this paper did not study energy efficiency in agriculture production. Some Chinese scholars reviewed the overall pattern and policy in rural China by provinces and fuels [10], but they did not calculate energy efficiency as well.

Basically, there are three dimensions on energy efficiency research. The traditional energy technical efficiency refers to total output over energy consumption, with the same meaning as energy productivity [11]. However, this partial-factor energy efficiency treats energy as a single input of production so that it ignores potential relations of substitution effect with other inputs, say, labor and capital [6,8]. To overcome this inaccuracy, Hu and Wang [6] have proposed an index named total-factor energy efficiency (TFEE) based on data envelopment analysis (DEA), which defines as a ratio of optimal to actual energy input under a multi-factor framework). This efficiency index presents an exact version of energy technical efficiency. Following this direction, an increasing studies emerged in total-factor energy efficiency analysis by using DEA [3,12].

At an international level, some researchers examined TFEE for the European agricultural sector [13,14,15,16]. Some other studies compare the energy efficiency across countries within China. Hu and Kao [17] investigate energy efficiency across 17 Asia-Pacific Economic Cooperation (APEC) countries. Wei and Ni [18] compare energy efficiency among 156 countries. These researchers found that China's energy efficiency is considerably lower than other countries but has a great potential to improve. Furthermore, Zhang, Cheng, Yuan and Gao [8] found that China experienced the most rapid increase between the periods of 1980–2005 among 23 developing countries and, more importantly, they suggest that the effective energy policy plays a crucial role in energy efficiency improvement. However, China's government usually targets different energy-saving goals for different provinces in terms of disparities of natural resources' endowments and priorities of development strategies [7]. That is to say, a detailed investigation on drivers of efficiency improvement may assist policy decision-making at the provincial level [6,12]. These researchers find that most provinces in eastern China have higher TFEE than provinces in central or in western areas since the eastern region experienced the most rapid economic growth and also attracted the advanced technology and managerial knowledge in China.

Owing to the fact that the industry sector takes the highest share of energy consumption, several studies also examine TFEE by the industry sector. For instance, Wei and Liao [19] explore energy efficiency for China's iron and steel sector; Shi and Bi [20] conduct it in the provincial industry sector; and Zhao and Yang [5] extend this research to assess energy efficiency both by provinces and sectors. These researchers have drawn two conclusions: (1) the TFEE gap among industrial sectors is found to be narrowed in the eastern region but expanded into the central and western regions; (2) and technological change, energy price and economic development are factors of energy efficiency performance.

On the side of the literature examining energy use in the agricultural sector, there are essentially two research strands. The first one focuses on investigating the relationship between energy consumption and agricultural growth using econometric methods [21,22,23,24]. Their results suggest that energy input is sensitive to agricultural productivity and support the energy-led growth hypothesis in the agricultural sector. These studies do not discuss the energy efficiency; however, they highlight the importance of energy input in agriculture. The second strand utilizes the energy content of farm inputs and converts it into joules to measure exergy efficiency at the farm level [25,26,27,28,29]. Although their investigations are meaningful, they do not provide information on TFEE and potentials of energy efficiency promotion by provinces.

In general, research on energy efficiency in agriculture production at the provincial level is rare and needed. This paper seeks to fill the gaps by evaluating the TFEE of China's provincial agricultural sector.

METHODOLOGY

Two mainstream methods have been commonly used for assessing efficiency: data envelopment analysis (DEA) and stochastic frontier analysis (SFA).

The major difference between them is that DEA is a non-parametric method, whereas SFA is a parametric method. In contrast with SFA, no assumptions on functional form of production function or distribution of the error term are in the DEA model [30,31]. Thus, DEA is a sufficiently powerful analytical tool for efficiency analysis, and has been widely used by previous studies to estimate energy efficiency [6]. We will apply the DEA approach to analyse TFEE in the provincial agricultural sector and then do a regression model to examine what factors would contribute to improve energy efficiency.

Total-Factor Energy Efficiency Based on DEA

Using linear programming, DEA creates a piecewise linear frontier that envelops the input and output data of each decision-making unit (DMU). Efficiency measurements are obtained relative to this frontier [30]. There are two different assumptions of returns to scale that are widely used in the DEA model including the constant returns to scale (CRS) and the variable returns to scale (VRS) [30]. Compared with the CRS DEA model imposing hypotheses on all DMUs operate at their optimal scale, the VRS model is more appropriate in the agricultural sector since, in agriculture, increasing inputs do not proportionally increase outputs [32]. Therefore, the VRS DEA model is employed in this study. In addition, an input-oriented model under variable returns to scale (VRS) is applied because we focus on using a minimum level of inputs to produce a given level of outputs.

Let us consider JDMU that produce N outputs using M inputs. The technical efficiency of DMU0 can be computed by solving the following linear programming problem:

$$\min \theta, \text{ s.t.,}$$

$$\sum_{j=1}^{J} \lambda_j y_{nj} - s_n^{+} = y_{no},\ n = 1, ..., N,$$

$$\sum_{j=1}^{J} \lambda_j y_{mj} + s_m^{-} = \theta x_{mo},\ m = 1, ..., M,$$

$$\sum_{j=1}^{J} \lambda_j = 1,$$

$$s_n^{+} \geq 0, s_m^{-} \geq 0, \lambda_j \geq 0,\ j = 1, ..., J, \tag{1}$$

where θ is a scalar, ranging from 0 to 1; s_n^{+} and s_m^{-} represent the output and input non-radial slacks, respectively; y and x are the output and input variables, respectively; J is the number of DMU; N and M are the types of output and inputs, respectively; and λ_j are nonnegative constants. The results obtained from solving the above linear programming can be interpreted as follows:

1. If $\theta=1$ and all slacks $s_n^+ = 0$ and $s_m^-=0$, the jth DMU is considered to be strongly efficient.
2. If $\theta=1$ and $s_n^+ \neq 0$ and/or $s_m^- \neq 0$, the jth DMU is considered to be weakly efficient.
3. If $\theta<1$, the jth DMU is considered to be inefficient.

The value of $(1-\theta)_{xmj}$ is called radial adjustments for the m-th input of jth DMU. The sum of the radial adjustments and non-radial adjustments s_m^- is called the total adjustments that can be reduced for the observed DMU to reach the production frontier, without compromising the outputs. Therefore, the target energy input can be defined as dividing the actual energy input by the total adjustments, which represents the minimum feasible energy input for the observed DMU to produce the given output. Thus, the TFEE indicator of jth DMU at time t can be specified as [6]

$$TFEE_{j,t} = \frac{Target\ Energy\ Input(j,t)}{Actual\ Energy\ Input(j,t).} \tag{2}$$

Since the target energy input is the optimal quantities of energy input in a region, the actual energy input is thus always larger than or equal to this target energy input. Hence, the TFEE score always ranges from 0 to 1. When the actual energy input is equal to the target energy input for a region, a TFEE score of unity is achieved. The TFEE score is lower than unity if the actual energy input is larger than the target energy input, implying that there are redundant energy inputs, which are needed to be reduced. A lower score of TFEE implies a higher potential for the region to reduce the application rate of energy.

Factors Influencing the Possibility of Improving TFEE

To examine the factors that may influence improvement of TFEE, we conduct a regression model in the second stage. However, given that the efficiency scores are censored values and bounded between 0 and 1, the binary Probit model is inappropriate in this study. In addition, the ordinary least squares (OLS) method does not provide unbiased and consistent estimates. Instead, a panel Tobit regression is used in this study. The Tobit model is proposed by Tobin [33], and it is widely employed by economists to do regression analysis when the dependent variable is incontinuous and/or subjected to a known upper or lower bound [5,8]. This model can be specified as:

$$TFEE_{it} = \alpha + \beta x_{it} + \varepsilon_{it}, \; i = 1, ..., N, \quad (3)$$

where $TFEE_{it}$ is total-factor energy efficiency score of the ith region in the agricultural sector, which is obtained from stage one; α is an intercept; β is a parameter vector that needs to be estimated; and x_{it} is a vector of independent variables that may affect efficiency.

VARIABLE AND DATA

This study involves 30 provinces categorized by three regions according to their economic and geographic characteristics. The eastern region includes 11 provinces (Beijing, Tianjin, Shanghai, Hebei, Liaoning, Jiangsu, Zhejiang, Fujian, Shandong, Guangdong, and Hainan). The central region includes eight inland provinces (Shanxi, Jilin, Heilongjiang, Anhui, Jiangxi, Henan, Hubei, and Hunan). The western region includes 11 provinces (Inner Mongolia, Guangxi, Chongqing, Sichuan, Guizhou, Yunnan, Shaanxi, Gansu, Qinghai, Ningxia, and Xinjiang). By and large, the eastern region is most developed and the west is underdeveloped in terms of per capita GDP. In addition, central provinces produce more food than the others.

Input and Output Variables of the DEA Model

We use agricultural added value (AAV) as the output variable to measure the agricultural sector's TFEE, which is consistent with available studies. Labour, capital stock, and energy consumption are selected as input variables [3,6]. The years range from 2001 to 2011. Labour input is measured by total number of employees in the agricultural sector in each province. Capital stock data in the agricultural sector are obtained from Wu [34] and our estimation through the perpetual inventory method, since the data of capital stock are not available in any statistical yearbooks of China. AAV and labour input are collected from the China Statistical Yearbook, and all the monetary variables are converted into the constant price in 2001 according to the index of AAV of each region. Energy consumption includes coal, oil, and electricity. All energy data are collected from *China Energy Databook (Version 8.0)* and convert to the standard coal equivalent. Table 1 illustrates the descriptive statistics of the input and output data. To compare the relative importance of agriculture in each province, we also present the agricultural share of total output in Table 1.

Table 1: Summary statistics of input and output variables by region

Region	Input Variables						Output Variable		Agri. Share (%)
	Labour Input		Capital Stock		Energy Usage		AAV		
	(10,000 Persons)		(100 Million RMB)		(Mtce)		(100 Million RMB)		
	Mean	SD	Mean	SD	Mean	SD	Mean	SD	
Beijing (E)	61.45	3.23	550.21	33.36	1.09	0.08	91.88	8.17	1.02
Tianjin (E)	77.89	2.88	677.58	69.19	0.86	0.11	88.84	9.07	1.98
Hebei (E)	1542.24	87.90	6798.39	1031.91	5.89	1.52	1255.50	291.78	13.05
Liaoning (E)	668.05	11.74	1912.33	460.65	2.71	0.60	725.49	168.07	9.67
Shanghai (E)	56.27	17.37	1228.77	206.19	0.90	0.21	112.14	21.76	0.78
Jiangsu (E)	1054.15	213.32	4607.41	882.04	3.85	0.43	1373.50	333.92	6.93
Zhejiang (E)	762.39	127.04	3431.27	126.38	3.54	0.23	843.80	150.06	5.43
Fujian (E)	679.73	55.64	3056.10	865.74	2.17	0.87	794.86	146.19	10.52
Shandong (E)	2109.87	173.58	8260.35	1958.76	6.22	1.71	1876.15	449.05	9.64
Guangdong (E)	1519.87	46.15	3769.90	212.64	4.38	0.67	1254.23	247.16	5.44
Hainan (E)	195.64	9.97	409.78	38.49	0.73	0.33	237.80	35.24	28.41
Shanxi (C)	641.37	9.40	1904.72	355.42	3.41	0.35	233.26	72.71	5.92
Jilin (C)	501.42	7.44	804.18	229.47	1.94	0.89	479.81	89.18	14.15
Heilongjiang (C)	700.73	27.62	3444.98	531.67	3.31	0.33	547.70	132.21	13.57
Anhui (C)	1716.69	165.84	3900.36	593.08	1.82	0.19	977.34	210.61	15.39
Jiangxi (C)	921.26	51.99	1516.46	214.52	1.69	0.38	643.50	121.10	14.65
Henan (C)	3040.34	289.93	8572.38	2021.19	5.01	0.76	1585.65	316.22	14.89
Hubei (C)	1043.06	92.24	2402.78	640.26	3.85	0.79	1055.22	314.60	14.57
Hunan (C)	1935.17	69.80	2721.92	494.75	5.54	2.19	1163.20	304.00	15.48
Inner M. (W)	530.15	9.03	2497.07	894.84	3.60	1.31	467.46	105.19	11.43
Guangxi (W)	1533.30	22.04	1455.01	342.54	1.10	0.25	787.74	196.65	19.72
Chongqing (W)	738.56	94.50	731.00	287.71	2.35	0.26	370.16	67.38	10.03
Sichuan (W)	2290.46	161.46	2640.35	448.00	2.24	0.56	1306.47	303.02	16.82
Guizhou (W)	1255.85	69.79	1437.97	34.78	2.18	0.74	344.75	67.51	15.15
Yunnan (W)	1673.97	18.96	3506.37	259.74	2.16	0.62	558.54	99.81	17.32
Shaanxi (W)	927.04	58.48	2203.29	427.05	1.88	0.52	392.22	118.02	10.23
Gansu (W)	737.58	20.97	899.10	160.98	2.72	0.35	272.49	61.69	14.71
Qinghai (W)	127.01	8.09	565.62	102.89	0.14	0.04	63.18	17.33	10.47
Ningxia (W)	137.87	9.74	382.15	109.54	0.51	0.12	63.69	14.68	10.02
Xinjiang (W)	351.63	23.82	2320.66	480.59	3.28	0.86	429.55	114.18	18.22

East region	793.41	68.07	3154.74	535.03	2.94	0.61	786.74	169.13	8.44
Central region	1312.51	89.28	3158.47	635.05	3.32	0.74	835.71	195.08	13.58

Note: SD is for standard deviation; AAV is for agricultural added value; agricultural share is defined as total agricultural output value divided by gross regional domestic product; E indicates eastern region; C indicates central region; W indicates western region.

Variables in the Regression Model

To explore the possibilities of TFEE improvement, five factors are included in the regression model, including energy price, economic structure, agricultural production structure, technological change, and income level. Previous studies have suggested that energy price [5], economic structure [35,36], and technological change [37,38] are important in influencing energy efficiency. In this study, we assume that agricultural production structure (the ratio of farming added value to total agricultural added value for each province) also affects energy efficiency in agriculture since different agricultural sectors (farming, livestock husbandry and fishery) usually have different energy intensity and energy utilization patterns. Moreover, farmers' income is also included in the model. Income usually reflects farmers' ability to utilize farm resources including energy in agriculture and upgrading their farm machinery. In the context that new farm equipment tends to be highly efficient and gets energy savings due to the government's energy policy, upgrading farm machinery may help to improve energy efficiency. To a large extent, income level also represents the managerial skills of utilizing resources, which could contribute to using resources including energy at their optimal level.

Energy price is measured by the purchasing price of fuel and power at the provincial level with 2001 as the base year. Economic structure is by the ratio of agricultural added value to GDP (gross domestic product). Technological change is measured by the per capita total power of agricultural machinery as farm machinery usually determines the direct energy consumption in agricultural production, and also reflects the level of modernization in agriculture. Income is represented by per capita annual net income of rural households in each province. All these data are collected from statistical yearbooks of each province. Table 2 describes the independent variables used in the empirical analysis.

Table 2: Summary statistics of variables in regression model

Variable	Whole Country		East Region		Central Region		West Region	
	Mean	**SD**	**Mean**	**SD**	**Mean**	**SD**	**Mean**	**SD**
Energy price	156.23	53.10	165.96	55.72	152.48	43.75	149.22	55.51
Agricultural production structure (%)	56.45	8.81	52.23	7.03	58.22	7.14	59.37	9.88
Economic structure (%)	13.12	6.62	9.41	8.11	14.98	4.35	15.47	4.35
Technological change (KW per capita)	1.01	0.57	1.07	0.64	1.10	0.46	0.90	0.55
Income level (RMB per capita)	3772.33	2062.26	5465.00	2391.11	3201.45	883.83	2494.83	776.39

Note: SD for standard deviation; energy price is using price indices.

RESULTS AND DISCUSSION

Using the DEA model (1) and definition (2), the total-factor energy efficiency in the agricultural sector for each province of China is evaluated. Then, the Tobit model (3) is applied to examine the factors influencing the possibilities to improve energy efficiency.

Estimated Results of Total-Factor Energy Efficiency

The TFEE scores in the agricultural sector for each province from 2001 to 2011 are presented in Table 2. On average, the TFEE score for China's agricultural sector is 0.657 during the period, implying that about 34% of energy can be saved given the level of output. Thus, China's agricultural sector could contribute effectively to energy savings and GHG mitigation.

Overall, the eastern provinces are more efficient than the west while the central provinces perform worst. Only eastern provinces (0.750) exceed the average efficiency of China (0.657). This tells us that an agriculture-intensive economy generally performs the worst in agricultural energy efficiency.

In the east, four provinces (Beijing, Tianjin, Shanghai and Hainan) exhibit a high TFEE with average efficiency scores above 0.8. Other seven provinces (Hebei, Liaoning, Jiangsu, Fujian, Zhejiang, Shandong, and Guangdong) have the average efficiency scores less than 0.8. Hainan shows the best TFEE with 0.939 and Hebei is the lowest (0.468). Hainan is a special case here as Hainan is a typical island agriculture economy and served as the national agriculture base. The agricultural share of Hainan province is always the highest (average 28.4% all years) within the country and exceeds the others much more (average 11.8% all years). However, for the continental provinces: the lowest score in the east (Hebei) is the most intensive agriculture province compared to other continental eastern provinces.

In the west, Qinghai is the most efficient with an average score of 0.989, followed by Guangxi (0.935), but the economic structures of these two provinces are distinguished. An agricultural share of Guangxi is highest in the western group and the number (19.7%) is also the highest across continental provinces; instead, Qinghai is one of the smallest agricultural shares in the west. Furthermore, the four lowest efficiency score provinces (Gansu 0.319, Shaanxi 0.374, Yunnan 0.397, and Guizhou 0.387) are all agriculture intensive provinces (14.7%, 10.2%, 17.3%, and 15.2%). Generally, the more agriculture in economic structure, the less energy efficiency in agriculture production.

In the central region, only one province (Jilin) has mean efficiency scores above 0.8 and three provinces (Anhui, Jiangxi and Hubei) around 0.7 to 0.8. The other four provinces (Shanxi, Heilongjiang, Henan and Hunan) show low TFEE with average scores less than 0.6, in which Shanxi is particularly noteworthy since it is the lowest average TFEE score in the whole country (0.231). In terms of output structure, we also find that, except for Jilin, the more agricultural output one province produces, the lower TFEE it has.

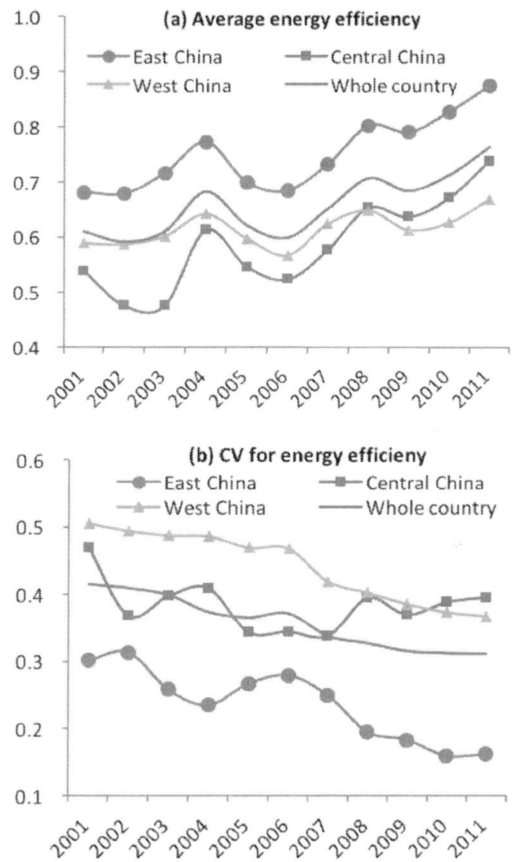

Figure 1: Energy efficiency and differences in the agricultural sector by regions and year (2001–2011). (a) Average energy efficiency; (b) coefficient of variation(CV) for energy efficiency.

Moving to the disparity of TFEE across three regions, we find that the average TFEE of each region differed and fluctuated dramatically during the period from 2001 to 2011, but the TFEE varying among regions is narrowing in terms of the trends of the coefficient of variation (CV) of

TFEE in Figure 1. The efficiency score of the west was exceeded thoroughly that of the central region after 2008 and nearly reached an average level of the country by 2011. The overall patterns of all lines are rising, but two turning points appear in 2004 and 2008 followed by a slight decline. The first turning point is associated with the subsidy policy of purchasing farm machinery that was provided by the Chinese government since 2004. The subsidy policy dramatically increased the appliance of farm machinery and thereby stimulated energy consumption in agriculture. It is reasonable to conjecture that the sharp rises of energy input are likely to cause a great deal of energy to be wasted. For the turning point in 2008, one explanation is that China's government invests massive resources into agriculture production to guarantee the domestic food security by avoiding the world food crisis in 2008. This may reduce energy efficiency performance in the agricultural sector in the next year, 2009. The Chinese government started to subsidize agricultural machinery in 2005. The huge capital inputs on machines may significantly decrease TFEE in the following year. This is why, in 2006, neither province is on the frontier.

By and large, the TFEE differences among the 30 provinces in China have decreased significantly between 2001 and 2011. From Figure 1b, the trends of the CV of TFEE indicate that the efficiency differences between provinces in the east are the smallest, followed by those in the central region. However, the disparity in the central region widened after the year 2007. Reminding readers that the central region is the main agricultural production region of this country, the increase in CV can draw an attentive policy implication. Although the gap in the western region is larger than other regions, it experienced a declining trend during the period.

Finally, the TFEE of all 30 provinces in the agricultural sector displays different trends in 2001, 2006 and 2011. The following can be seen from Figure 2: (1) 16 out of 30 provinces' TFEE experienced an increase from 2001 through 2006 and 2011, in which Shanghai, Jiangsu, Shandong, Anhui, Jiangxi and Hubei present a more significant improvement; (2) nine provinces' TFEE decreased from 2001 to 2006, and then increased in 2011. However, the efficiencies of two out of nine provinces in 2011 is lower than those in 2001 (Fujian, and Yunnan), and the other six provinces exceed their initial level (Hebei, Liaoning, Jilin, Guangxi, Sichuan and Shaanxi), whereas Hainan recovers to its initial level; and (3) the remaining five provinces' TFEE fluctuated slightly across the three time points.

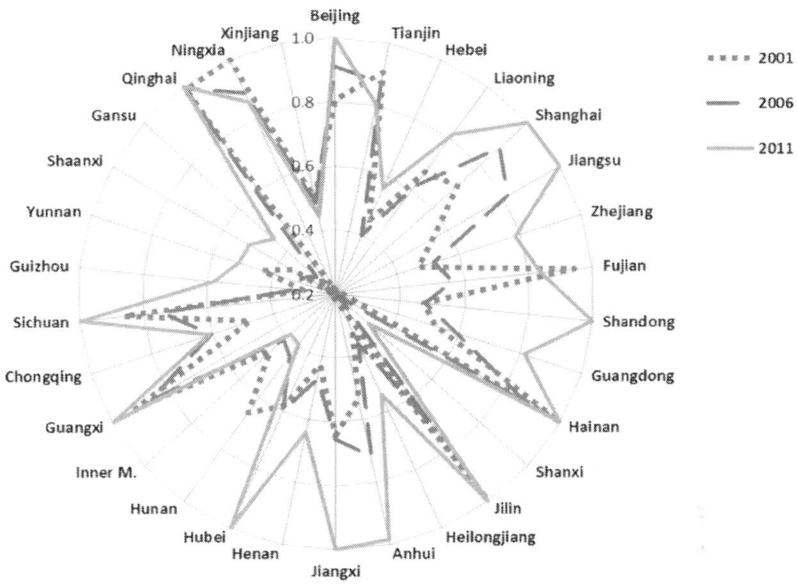

Figure 2: Provincial level energy efficiency in 2001, 2006 and 2011.

Potential Energy Savings of Provinces in China

The DEA theory suggests that inefficient decision making units (DMUs) can become efficient and reach the frontier by adjusting their inefficient application of inputs [30]. We find that some provinces exhibit full efficiency and thus are located at the frontier in the period. For instance, two provinces (Hainan and Ningxia) constituted the energy efficiency frontier in 2001, while Hainan and Qinghai provinces constituted the frontier in 2002. Overall, Qinghai is located at the TFEE frontier in five out of eleven years, followed by Beijing, Hainan, Jilin and Guangxi, which are constituting the frontier for three years. There are also some other provinces located at the frontier for two years or one year. Those provinces that are not yet at the frontier can project themselves onto the energy use frontier to optimize their inputs' portfolio through adjusting their technology and the production method. Thus, the DEA method can be used to measure the energy savings potential for each province during the surveyed period (see Table 3).

Table 3: Total-factor energy efficiency for each province from 2001 to 2011

Region	2001	2002	2003	2004	2005	2006	2007	2008	2009	2010	2011	Mean
Beijing (E)	0.806	0.897	0.861	1.000	0.987	0.911	1.000	0.837	0.841	0.897	1.000	0.912
Tianjin (E)	0.907	0.872	0.947	0.934	0.888	0.873	0.892	0.908	0.879	0.855	0.806	0.887
Hebei (E)	0.452	0.406	0.444	0.513	0.449	0.400	0.456	0.473	0.477	0.517	0.559	0.468
Liaoning (E)	0.670	0.678	0.676	0.754	0.657	0.617	0.698	0.722	0.736	0.775	0.816	0.709
Shanghai (E)	0.704	0.725	0.748	0.772	0.805	0.880	0.857	0.932	0.939	1.000	1.000	0.851
Jiangsu (E)	0.546	0.546	0.598	0.695	0.759	0.813	0.884	0.989	0.957	0.952	1.000	0.794
Zhejiang (E)	0.470	0.466	0.468	0.525	0.543	0.518	0.579	0.625	0.653	0.744	0.784	0.580
Fujian (E)	0.944	0.897	0.941	0.998	0.587	0.561	0.628	0.800	0.736	0.789	0.837	0.793
Shandong (E)	0.472	0.455	0.675	0.737	0.487	0.478	0.516	0.881	0.873	0.923	1.000	0.682
Guangdong (E)	0.524	0.518	0.587	0.606	0.590	0.586	0.635	0.708	0.697	0.766	0.813	0.639
Hainan (E)	1.000	1.000	0.930	0.961	0.931	0.888	0.903	0.941	0.893	0.881	1.000	0.939
Shanxi (C)	0.143	0.127	0.132	0.239	0.242	0.212	0.240	0.240	0.309	0.324	0.337	0.231
Jilin (C)	0.993	0.543	0.375	1.000	0.775	0.715	0.706	1.000	0.964	0.930	1.000	0.818
Heilongjiang (C)	0.330	0.333	0.363	0.350	0.365	0.372	0.429	0.464	0.446	0.466	0.543	0.406
Anhui (C)	0.533	0.552	0.569	0.680	0.742	0.718	0.817	0.854	0.838	0.895	0.985	0.744
Jiangxi (C)	0.643	0.663	0.631	0.735	0.635	0.654	0.753	0.843	0.813	0.883	1.000	0.750
Henan (C)	0.426	0.424	0.411	0.491	0.516	0.472	0.535	0.601	0.591	0.677	0.642	0.526
Hubei (C)	0.584	0.560	0.688	0.826	0.652	0.600	0.673	0.777	0.723	0.862	1.000	0.722
Hunan (C)	0.659	0.611	0.641	0.582	0.445	0.443	0.460	0.446	0.404	0.333	0.393	0.492

Inner M. (W)	0.470	0.437	0.454	0.454	0.441	0.417	0.459	0.473	0.403	0.356	0.388	0.432
Guangxi (W)	0.919	0.892	0.980	1.000	0.894	0.864	0.912	1.000	0.896	0.924	1.000	0.935
Chongqing (W)	0.476	0.511	0.545	1.000	0.785	0.605	0.671	0.638	0.625	0.562	0.611	0.639
Sichuan (W)	0.862	0.830	0.783	0.859	0.771	0.743	0.941	1.000	0.872	0.889	1.000	0.868
Guizhou (W)	0.297	0.299	0.302	0.333	0.327	0.326	0.392	0.473	0.450	0.485	0.577	0.387
Yunnan (W)	0.437	0.464	0.426	0.300	0.316	0.325	0.372	0.397	0.404	0.416	0.513	0.397
Shaanxi (W)	0.351	0.385	0.331	0.340	0.325	0.303	0.346	0.392	0.386	0.451	0.506	0.374
Gansu (W)	0.173	0.158	0.204	0.297	0.259	0.279	0.364	0.434	0.428	0.461	0.455	0.319
Qinghai (W)	0.988	1.000	1.000	0.975	1.000	0.992	1.000	0.995	0.962	0.970	1.000	0.989
Ningxia (W)	1.000	0.982	0.983	0.944	0.912	0.884	0.860	0.854	0.856	0.859	0.855	0.908
Xinjiang (W)	0.503	0.494	0.602	0.555	0.538	0.491	0.540	0.471	0.449	0.508	0.447	0.509
East region	0.681	0.678	0.716	0.772	0.698	0.684	0.732	0.802	0.789	0.827	0.874	0.750
Central region	0.539	0.476	0.476	0.613	0.546	0.523	0.576	0.653	0.636	0.671	0.737	0.586
West region	0.589	0.586	0.601	0.642	0.597	0.566	0.623	0.648	0.612	0.626	0.668	0.614
Whole country	0.609	0.591	0.610	0.682	0.621	0.598	0.651	0.706	0.683	0.712	0.762	0.657

Note: E indicates eastern region provinces; C indicates central region provinces; W indicates western region provinces; the last four rows are calculated by simple average.

Table 3 presents the potential energy savings for each province as well as the total potential energy savings for the east, central, west and the whole country from 2001 to 2011. We find that the average energy savings potential for six provinces is more than 2.0 Mtce during the period. Hebei has the largest potential to reduce energy consumption, on average, the energy consumption of 3.1 Mtce could be saved in Hebei to improve energy efficiency, followed by Hunan (3.016 Mtce), Shanxi (2.619 Mtce), Henan (2.319 Mtce), Shandong (2.125 Mtce) and Inner Mongolia (2.083 Mtce). Nine provinces could save 1.0–2.0 Mtce of their energy consumption in the agricultural sector , of which Heilongjiang's energy savings potential ranks first (1.950 Mtce), followed by Gansu (1.823 Mtce), Xinjiang (1.636 Mtce), Guangdong (1.549 Mtce), Zhejiang (1.470 Mtce), Guizhou (1.393 Mtce), Yunnan (1.328 Mtce), Shaanxi (1.161 Mtce) and Hubei (1.034 Mtce). Amongst the other 15 provinces that have lower energy savings potential (less than 1.0 Mtce), Qinghai can save the lowest amount of energy (0.002 Mtce).

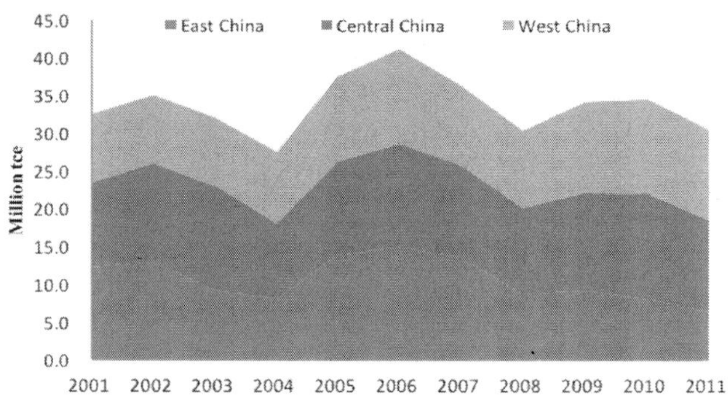

Figure 3: Total energy savings potential in the east, central and west.

The energy savings potential for the whole country is reported in the last row of Table 3. We find that the energy savings potential in China's agricultural sector ranging from 27.581 Mtce in 2004 to 41.182 Mtce in 2006, with an average of 33.803 Mtce during 2001 and 2011. The results imply that a substantial energy can be reduced through improving energy efficiency in agriculture. This gives rise to a meaningful implication for the country to mitigate GHG emission and manage energy conservation. At the regional level, we find that the central region has the largest energy savings potential of 12.349 Mtce, compared to the east (10.758 Mtce) and west regions (10.696 Mtce). The slight differences across the three regions in terms of potential energy savings imply that the energy savings effort and energy efficiency-improved polices should be emphasized equally in all regions of China. As it can be seen from Figure 3, the total energy savings potentials for the three regions during 2001 and 2011 exhibit similar trends, decreased from 2002 to 2004, then increased dramatically until 2006, and at last fluctuated and declined from 2007 to 2011. This finding further confirms that the energy policies in China could have similar effects in all regions of China.

Regression Results of Tobit Model and Discussions

Table 4 reports which factors can improve energy efficiency. We find that energy price does not show a significant relationship with TFEE in the agricultural sector. Although it is a general belief that higher energy price could reduce the energy intensity and improve energy efficiency, our finding is reasonable and can be explained when we consider that China's energy market does not function well [36] and the abundant subsidy for energy consumption exists in the agricultural sector.

Significantly, a negative relationship between agricultural production structure and TFEE suggests that higher farming share is associated with lower TFEE compared with livestock husbandry and fishery. In other words, more farming means less energy efficiency in the agricultural sector. This could be attributed to the small and scattered scale of operation of farming in China. The scale of operation for farming is limited to adjusting for smallholder producers to a large extent due to the land tenure system, the informal and less developed land rental market in rural China. In contrast, it is easy for the livestock husbandry and fishery to enlarge scale of operations and thereby could gain scale economy. Moreover, the coefficient of agricultural production structure is found to be the largest in the central region (−0.700), followed by the east region (−0.613), implying that if agricultural production structure increases by 1%, the possibility of improving TFEE decreases 0.7% for the central region and 0.6% for the east region.

Table 4: Energy savings potential for each province from 2001 to 2011 (Mtce)

Region	2001	2002	2003	2004	2005	2006	2007	2008	2009	2010	2011	Mean
Beijing (E)	0.228	0.119	0.156	0.000	0.012	0.092	0.000	0.180	0.182	0.120	0.000	0.099
Tianjin (E)	0.072	0.114	0.037	0.047	0.090	0.107	0.095	0.079	0.108	0.142	0.210	0.100
Hebei (E)	2.033	2.606	2.374	2.162	3.124	3.578	3.497	3.596	3.852	3.799	3.477	3.100
Liaoning (E)	0.678	0.619	0.611	0.496	0.965	1.191	0.977	0.873	0.814	0.714	0.620	0.778
Shanghai (E)	0.344	0.305	0.290	0.257	0.195	0.098	0.115	0.045	0.041	0.000	0.000	0.154
Jiangsu (E)	1.849	1.761	1.564	1.216	0.810	0.641	0.403	0.038	0.162	0.202	0.000	0.786
Zhejiang (E)	1.643	1.966	1.868	1.611	1.549	1.655	1.512	1.321	1.220	0.972	0.855	1.470
Fujian (E)	0.060	0.115	0.065	0.002	1.162	1.288	1.096	0.485	0.690	0.581	0.492	0.549
Shandong (E)	3.525	3.626	1.046	0.934	3.917	4.321	4.149	0.643	0.738	0.482	0.000	2.125
Guangdong (E)	1.891	1.826	1.279	1.382	2.054	2.009	1.757	1.426	1.398	1.068	0.943	1.549
Hainan (E)	0.000	0.000	0.030	0.020	0.038	0.080	0.063	0.050	0.104	0.145	0.000	0.048
Shanxi (C)	3.064	3.265	3.168	2.406	2.386	2.561	1.977	2.531	2.445	2.469	2.538	2.619
Jilin (C)	0.009	1.306	2.582	0.000	0.441	0.599	0.651	0.000	0.046	0.093	0.000	0.521
Heilongjiang (C)	2.125	1.894	1.967	2.136	1.932	2.015	1.942	1.679	1.963	1.944	1.849	1.950
Anhui (C)	0.874	0.784	0.775	0.559	0.396	0.447	0.304	0.273	0.314	0.218	0.032	0.452
Jiangxi (C)	0.474	0.457	0.589	0.381	0.879	0.793	0.480	0.275	0.311	0.173	0.000	0.437
Henan (C)	2.395	2.452	2.589	2.325	2.234	2.631	2.330	2.062	2.213	1.854	2.429	2.319
Hubei (C)	1.293	1.429	0.856	0.459	1.308	1.657	1.415	1.001	1.319	0.637	0.000	1.034
Hunan (C)	1.016	1.356	1.129	1.600	2.844	2.859	3.023	3.632	4.202	6.446	5.073	3.016
Inner M. (W)	1.089	1.218	1.361	1.669	1.705	1.911	1.837	2.043	2.868	3.713	3.500	2.083
Guangxi (W)	0.064	0.095	0.015	0.000	0.105	0.148	0.116	0.000	0.128	0.105	0.000	0.070

Chongqing (W)	1.197	1.172	1.055	0.000	0.461	0.930	0.766	0.877	0.871	1.167	1.090	0.871
Sichuan (W)	0.183	0.236	0.329	0.313	0.524	0.676	0.145	0.000	0.343	0.294	0.000	0.277
Guizhou (W)	1.721	1.807	1.965	1.894	1.937	1.994	1.426	0.694	0.768	0.645	0.477	1.393
Yunnan (W)	0.702	0.627	0.734	1.969	1.846	1.747	1.513	1.475	1.495	1.387	1.108	1.328
Shaanxi (W)	0.814	0.663	0.897	1.011	1.165	1.423	1.365	1.376	1.435	1.326	1.295	1.161
Gansu (W)	1.936	1.984	1.828	1.602	2.026	2.061	1.795	1.681	1.744	1.663	1.733	1.823
Qinghai (W)	0.002	0.000	0.000	0.003	0.000	0.001	0.000	0.001	0.007	0.006	0.000	0.002
Ningxia (W)	0.000	0.007	0.007	0.025	0.041	0.059	0.074	0.087	0.088	0.092	0.099	0.053
Xinjiang (W)	1.228	1.234	0.918	1.102	1.345	1.609	1.511	2.038	2.269	2.072	2.665	1.636
East region (total)	12.322	13.058	9.319	8.127	13.914	15.061	13.664	8.736	9.309	8.226	6.598	10.758
Central region (total)	11.250	12.944	13.656	9.866	12.420	13.562	12.123	11.453	12.813	13.834	11.920	12.349
West region (total)	8.935	9.044	9.110	9.588	11.154	12.559	10.549	10.272	12.015	12.468	11.966	10.696
Whole country (total)	32.507	35.045	32.085	27.581	37.488	41.182	36.335	30.461	34.137	34.528	30.484	33.803

Note: E for eastern region; C for central region; W for western region; the last four rows are calculated by summing up the corresponding numbers.

In contrast, the coefficient for the west region is lowest and not significant (−0.257). It is worth noting that most provinces in the central region are major grain producing regions of China, and its scale of operation for farming are less than that in the west region, which is not helpful to reach economies of scale including energy consumption. Therefore, the higher agricultural production structure in the central region can be seen as a major reason that it has the lowest TFEE score compared with the other two regions. Moreover, the reason that Hainan in the east region and Guangxi in the west region exhibit a higher TFEE score compared to other agriculture provinces can also be seen, since the two provinces have the lowest and the second lowest share of farming in agriculture within their groups. Particularly, Qinghai province in the west region shows the highest TFEE score throughout the whole country and also has the lowest share of farming in agriculture.

The coefficient of economic structure for the whole sample (0.185) indicates that a province with a higher share of primary industry is more likely to exhibit a higher TFEE score in the agricultural sector. If economic structure increases by 1%, and the possibility of enhancing TFEE increases 0.185%. However, this story is not true for the east and central regions. In contrast, a 1% increase of economic structure in the west region can increase the possibility of TFEE improvement by 0.861% correspondingly. This result is contradictory with the findings that most of the agriculture heavier provinces are associated with lower TFEE scores but is consistent with the findings that Hainan, Jilin and Guangxi provinces have the relatively higher efficiency scores. To some extent, this result confirms the findings above referring to the heterogeneous effects of economic structure across regions, implying that the contributions of economic structure on increasing the possibility of enhancing TFEE among the east and central regions are weakened by the economic structure.

Although the effect of technological change on TFEE is not significant on a country level, it is pronounced for the central and west regions. However, the coefficient of technological change is negative in the central region but positive in the west region. The technological change in this study is defined by the application of farm machinery. Given the fact that the dramatically increased use of farm machinery significantly stimulates the energy consumption in agriculture in a relatively short period, the consumed energy is more likely to be wasted and thus leads to energy inefficiency. The significant and negative influence of the application of farm machinery in the central region (−0.467) can be attributed to the higher agricultural production structure and smaller scale of operation to a large extent, which could weaken the efficiency of mechanization. However, farm size in the west region is usually larger than that in the east and central regions due to its geographic and population factors. Thus, the application of farm machinery in the west region could be more efficient at energy consumption. Moreover, as mentioned above, the west region is the most underdeveloped in the economy among the three regions, implying that its technological level in agriculture also ranks the worst. In other words, the marginal benefits of wide application of farm machinery in the west region can be larger than the other two regions.

The significant influences of income level suggest that increasing a farmer's income can improve the possibility of enhancing TFEE in the agricultural sector at the country level (0.592). However, the story is not true for the west region. The coefficient of the income level in the west region (0.079) is insignificant and is the smallest compared to that in the other two regions. In contrast, the coefficient for central region is pronounced to be larger than that for the other two regions (0.943), followed by the east region (0.593). Improving income level in the central and east regions by 1%, the possibilities to increase TFEE can increase by 0.943% and 0.593% in these two regions, respectively. Thus, the different influences of income level in the central and west regions can be seen as the reason that the TFEE in the west region has been exceeded by the central region since 2009. Additionally, the heterogeneous influences of technological change and income level on TFEE among the three regions also indicate that the possibility of energy efficiency improvement can be stimulated by technology investment at an early stage of agricultural development, whereas the income effect will be greater than the technology effect at a late stage of development.

POLICY IMPLICATION

The improvement in energy efficiency is one of the most effective methods to save energy and mitigate GHG emission. Although the industry sector consumes a large amount of primary energy, energy consumption in the agriculture production should not be ignored, especially for big agriculture economies like China. Under the total-factor framework, this study uses the DEA method to examine the TFEE performance of China's provincial agricultural sector during the period of 2001–2011. We further calculate how much energy can be saved potentially. In addition, Tobit regression is applied to explore the determinants of TFEE (see Table 5).

Table 5: Results of Tobit regression (Dependent variable: TFEE)

Independent Variable	**All Sample**	**East Region**	**Central Region**	**West Region**
	Coefficient	**Coefficient**	**Coefficient**	**Coefficient**
Energy price	−0.02	−0.086	0.338	0.001
	−0.087	−0.125	−0.266	−0.136
Agricultural production structure	−0.573 ***	−0.613 ***	−0.700 **	−0.257
	−0.178	−0.234	−0.349	−0.321
Economic structure	0.185 **	0.089	0.241	0.861 ***
	−0.073	−40.083	−0.216	−0.209
Technological change	−0.076	−0.056	−0.467 ***	0.513 ***
	−0.055	−0.066	−0.16	−0.139
Income level	0.592 ***	0.593 ***	0.943 ***	0.079
	−0.122	−0.193	−0.32	−0.228
Constant	1.27	1.931	−3.084	2.283
	−1.224	−1.831	−2.619	−1.908
Log likelihood	26.66	34.79	5.32	12.15

Note: *, **, *** indicate significant at 10%, 5% and 1% respectively. Standard errors in parentheses.

Firstly, our results indicate that China's agricultural sector has large energy-saving potential and 34% of current energy consumption could be reduced without reduction of agriculture output if energy efficiency increases. Secondly, the overall trend of TFEE is considerably upward moving during the period from 2001 to 2011, though two turning points appear in 2004 and 2008, respectively. Thirdly, agricultural energy efficiency is heterogeneous across regions, but the disparities are narrowing during the observed period. By and large, the developed region (the east) has the highest efficiency scores followed by the west, and the TFEE score of the central region is the lowest, whereas TFEE of the central region exceeds that of the western region after 2008. Fourthly, most agriculture intensive provinces usually have lower TFEE except for Hainan, Jilin and Guangxi provinces. In addition, provinces with abundant energy resource are found to be more inefficient in energy use in agriculture production, especially for Shanxi and Inner Mongolia provinces.

This research also shows how much energy can be saved in each region yearly. We find that agricultural structure, economic structure, technological change and income level are the main factors that can influence TFEE performance across provincial agricultural sectors, but energy price does not affect TFEE significantly. Moreover, these influences are found to differ across regions. From the findings above, several policy implications can be derived to assist the improvment of China's agriculture energy efficiency.

This research suggests that provinces with higher primary industry share generally consume energy inefficiently in the agricultural sector; furthermore, considering the agricultural production structure, if the amount of farming grows, then energy efficiency tends to be worse. That is to say, the performance of energy efficiency in the agricultural sector is highly connected with China's industrial upgrading as well as structural transformation. This delivers hints for improving energy efficiency: the promotion of energy efficiency in the agricultural sector not only depends on the structural change between sectors but also that within the agricultural sector. China should insist on upgrading industrial structure towards a service-oriented economy, and meanwhile adjusting agricultural production structure with a stronger emphasis on diversified agriculture instead of traditional farming.

Secondly, there is a positive relationship between energy efficiency promotion and farmers' income increase. Put differently, we can improve them simultaneously without any contradiction. Without a doubt, this is a crucial conclusion for policymakers while making rural development policy. The roots would result from the fact that farmers' awareness of energy savings will enhance energy utilization as living standard continuously improves. They tend to apply more advanced technologies in production and save more energy for avoiding environmental degradation. Consequently, the government should accelerate the application of energy-saving technologies opportunely while farmers become rich.

Lastly, from the perspective of institutional economics, changing scattered farmland into a larger scale area would be beneficial to energy efficiency improvement as large farms usually enjoy increasing return to energy consumption by using energy more efficiently and wisely. This requires a deepening of land transition policy. The central government should release the regulation of rural land property and entitle the same rights for rural land and city land. Furthermore, a nationwide land market is needed for farmers to conduct land transactions freely. This will result in an agglomeration of farms, which can assist with improving energy efficiency and food productivity.

CONCLUSIONS

In conclusion, China's energy consumption in the agricultural sector is huge in absolute quantity; thus, energy efficiency issues stay at the center of rural energy policy decision-making and should be paid more attention to by economists and policymakers in order to tackle global GHG emission. To some degree, China's agriculture energy efficiency is a development issue in addition to being a technical one. From the perspective of economics and policy, energy efficiency can be improved by further development, which has seen China grow from an ancient agriculture economy to a modern industry economy. This is the major message conveyed by this paper. More importantly, by increasing total-factor energy efficiency, the energy intensity of the agricultural sector will be expected to decline over time, which would be much more important for China's energy transition to a sustainable development pathway. China has a huge potential to enhance the performance of energy efficiency in the agricultural sector if they can make proper policies for structural transformation of sectors, income growth of farmers and institutional changes of property rights. All of these socioeconomics elements provide a deeper policy thinking beyond technology. These also provide an alternative pathway for energy transition and sustainable growth in other developing worlds.

ACKNOWLEDGMENTS

We gratefully acknowledge the financial support from the National Natural Science Foundation of China (No. NSFC 71573101), (No. NSFC71373095) and the National Social Science Foundation of China (No. 16CGL038) and are thankful for the financial support from the China Scholarship Council (CSC). Dong Wang thanks the research funding from UWA School of Agriculture and Environment, the University of Western Australia. Tianyi Du thanks the PhD scholarship from Murdoch University. We thank Rui Hou for his generous help in LaTeX techniques. The authors are also grateful for valuable comments from the editor and two reviewers.

AUTHOR CONTRIBUTIONS

Zhihai Yang and Dong Wang developed the original idea of the paper, convinced and designed the research project. They completed the most of modeling work and drafted the original version of the paper. Tianyi Du contributed to the results presentation, some writing parts and coded the whole paper into Latex. Anlu Zhang navigated the research direction from beginning to the end and also made a significant contribution on the policy discussion part. Yixiao Zhou contributed to the data collecting and also conducted the task of writing, editing, revising and proofreading.

REFERENCES

1. Tong, S.; Berry, H.L.; Ebi, K.; Bambrick, H.; Hu, W.; Green, D.; Hanna, E.; Wang, Z.; Butler, C.D. Climate change, food, water and population health in China. *Bull. World Health Organ.* 2016, *94*, 759–765.
2. Government, C.C. *The 13th Five-Year Plan for Economic and Social Development of the People's Republic of China*; Central Compilation and Translation Press: Beijing, China, 2016.
3. Wang, Z.H.; Zeng, H.L.; Wei, Y.M.; Zhang, Y.X. Regional total factor energy efficiency: An empirical analysis of industrial sector in China. *Appl. Energy* 2012, *97*, 115–123.
4. Wang, K.; Wei, Y.M. China's regional industrial energy efficiency and carbon emissions abatement costs. *Appl. Energy* 2014, *130*, 617–631.
5. Xiaoli, Z.; Rui, Y.; Qian, M. China's total factor energy efficiency of provincial industrial sectors. *Energy* 2014, *65*, 52–61.
6. Hu, J.L.; Wang, S.C. Total-factor energy efficiency of regions in China. *Energy Policy* 2006, *34*, 3206–3217.
7. Wang, K.; Yu, S.; Zhang, W. China's regional energy and environmental efficiency: A DEA window analysis based dynamic evaluation. *Math. Comput. Model.* 2013, *58*, 1117–1127.
8. Zhang, X.P.; Cheng, X.M.; Yuan, J.H.; Gao, X.J. Total-factor energy efficiency in developing countries. *Energy Policy* 2011, *39*, 644–650.
9. Catania, P. China's rural energy system and management. *Appl. Energy* 1999, *64*, 229–240.
10. Zhang, L.; Yang, Z.; Chen, B.; Chen, G. Rural energy in China: Pattern and policy. *Renew. Energy* 2009, *34*, 2813–2823.
11. Wang, D. A dynamic optimization on economic energy efficiency in development: A numerical case of China. *Energy* 2014, *66*, 181–188.
12. Wei, C.; Ni, J.; Shen, M. Empirical analysis of provincial energy efficiency in China. *China World Econ.* 2009, *17*, 88–103.
13. Błażejczyk-Majka, L. Application of DEA for evaluating the efficiency of economic policy as exemplified by EU agriculture. *Stud. Hist. Oecon.* 2017, *35*, 163–176.
14. Błażejczyk-Majka, L.; Radosław, K. Concentration and productivity of livestock and mixed farms in new and old EU member states. A regional level approach. *J. Cent. Eur. Agric.* 2015, *16*, 159–176.
15. Yan, Q.; Yin, J.; Baležentis, T.; Makutėnienė, D.; Štreimikienė,

D. Energy-related GHG emission in agriculture of the European countries: An application of the Generalized Divisia Index. *J. Clean. Prod.* 2017, *164*, 686–694.

16. Vlontzos, G.; Niavis, S.; Pardalos, P. Testing for Environmental Kuznets Curve in the EU Agricultural Sector through an Eco-(in) Efficiency Index. *Energies* 2017, *10*, 1992.
17. Hu, J.L.; Kao, C.H. Efficient energy-saving targets for APEC economies. *Energy Policy* 2007, *35*, 373–382.
18. Wei, C.; Ni, J.; Sheng, M. China's energy inefficiency: A cross-country comparison. *Soc. Sci. J.* 2011, *48*, 478–488.
19. Wei, Y.M.; Liao, H.; Fan, Y. An empirical analysis of energy efficiency in China's iron and steel sector. *Energy*2007, *32*, 2262–2270.
20. Shi, G.M.; Bi, J.; Wang, J.N. Chinese regional industrial energy efficiency evaluation based on a DEA model of fixing non-energy inputs. *Energy Policy* 2010, *38*, 6172–6179.
21. Karkacier, O.; Gokalp Goktolga, Z.; Cicek, A. A regression analysis of the effect of energy use in agriculture. *Energy Policy* 2006, *34*, 3796–3800.
22. Zaman, K.; Khan, M.M.; Ahmad, M.; Rustam, R. The relationship between agricultural technology and energy demand in Pakistan. *Energy Policy* 2012, *44*, 268–279.
23. Sebri, M.; Abid, M. Energy use for economic growth: A trivariate analysis from Tunisian agriculture sector. *Energy Policy* 2012, *48*, 711–716.
24. Zhao, Q.; Chen, Q.; Xiao, Y.; Tian, G.; Chu, X.; Liu, Q. Saving Forests through Development? Fuelwood Consumption and the Energy-Ladder Hypothesis in Rural Southern China. *Transform. Bus. Econ.* 2017, *16*, 199–219.
25. Pahlavan, R.; Omid, M.; Akram, A. Energy use efficiency in greenhouse tomato production in Iran. *Energy*2011, *36*, 6714–6719.
26. Khoshroo, A.; Mulwa, R.; Emrouznejad, A.; Arabi, B. A non-parametric Data Envelopment Analysis approach for improving energy efficiency of grape production. *Energy* 2013, *63*, 189–194.
27. Unakitan, G.; Hurma, H.; Yilmaz, F. An analysis of energy use efficiency of canola production in Turkey. *Energy* 2010, *35*, 3623–3627.
28. Rahman, S.; Rahman, M.S. Energy productivity and efficiency of maize accounting for the choice of growing season and environmental

factors: An empirical analysis from Bangladesh. *Energy* 2013, *49*, 329–336.

29. Luo, Y.; Long, X.; Wu, C.; Zhang, J. Decoupling CO2 emissions from economic growth in the agricultural sector across 30 Chinese provinces from 1997 to 2014. *J. Clean. Prod.* 2017, *159*, 220–228.
30. Coelli, T.J.; Rao, D.S.P.; O'Donnell, C.J.; Battese, G.E. *An Introduction to Efficiency and Productivity Analysis*; Springer: New York, NY, USA, 2005; pp. 47–61.
31. Błażejczyk-Majka, L.; Kala, R. On the combined estimation of technical efficiency and its application to agriculture. *Agric. Econ. Zeměd. Ekon.* 2015, *61*, 441–449.
32. Speelman, S.; D'Haese, M.; Buysse, J.; D'Haese, L. A measure for the efficiency of water use and its determinants, a case study of small-scale irrigation schemes in North-West Province, South Africa. *Agric. Syst.* 2008, *98*, 31–39.
33. Tobin, J. Estimation of Relationships for Limited Dependent Variables. *Econometrica* 1958, *26*, 24–36.
34. Wu, Y. China's Capital Stock Series by Region and Sector. *Front. Econ. China* 2016, *11*, 156–172.
35. Fan, Y.; Xia, Y. Exploring energy consumption and demand in China. *Energy* 2012, *40*, 23–30.
36. Song, F.; Zheng, X. What drives the change in China's energy intensity: Combining decomposition analysis and econometric analysis at the provincial level. *Energy Policy* 2012, *51*, 445–453.
37. Ma, C.; Stern, D.I. China's changing energy intensity trend: A decomposition analysis. *Energy Econ.* 2008, *30*, 1037–1053.
38. Eom, J.; Clarke, L.; Kim, S.H.; Kyle, P.; Patel, P. China's building energy demand: Long-term implications from a detailed assessment. *Energy* 2012, *46*, 405–419.

CHAPTER 5

TESTING FOR ENVIRONMENTAL KUZNETS CURVE IN THE EU AGRICULTURAL SECTOR THROUGH AN ECO-(IN)EFFICIENCY INDEX

George Vlontzos [1], Spyros Niavis [2], and Panos Pardalos [3]

[1]Department of Agriculture Crop Production and Rural Environment, University of Thessaly, Fytoko, 38446 Volos, Greece

[2]Department of Planning and Regional Development, University of Thessaly, Pedion Areos, 38334 Volos, Greece

[3]Department of Industrial and Systems Engineering, University of Florida, 401 Weil Hall, Gainesville, FL 32611-6595, USA

ABSTRACT

Studies on agricultural production practices advance within international literature and new methods are proposed in order to assess the agricultural sustainability, either at farm or macro level. The present paper builds on these advancements and develops a synthetic Eco-(in)efficiency index by employing a directional distance function—data envelopment analysis

Citation: Vlontzos, G.; Niavis, S.; Pardalos, P. Testing for Environmental Kuznets Curve in the EU Agricultural Sector through an Eco-(in)Efficiency Index. Energies 2017, 10, 1992.

(DEA) model. This index is used in order to assess the sustainability of the EU agricultural sector for the period 1999–2012 on a country level. Furthermore, Eco-(in)efficiency, together with the energy use and greenhouse gas (GHG) emissions are regressed on the gross domestic product (GDP) of EU countries, in order to check for any environmental Kuznets curve relationship existence. Results signify that efficiency improvements are possible, both towards output development and GHG emissions reduction. In addition, the potential of each country in adopting more sustainable production practices is not totally connected with its economic development, as Eco-(in)efficiency and GDP levels of EU countries seem to be linked with an N-Shaped curve.

Keywords: Kuznets curve, sustainability, agriculture, data envelopment analysis, energy, emissions

INTRODUCTION

Uncontrolled economic growth imposes severe risks for the environment as natural resources used for economic activities are not infinite [1]. Ice melting and the increase of temperature are considered as tangible examples of environmental degradation due to human activities [2]. After a long period of uncontrolled economic growth, policy makers, mostly representing matured economies, started to realize the risk imposed because of environmental degradation and initiated global actions in order to set the basis for sustainable development. Undoubtedly, the benchmark point towards sustainability has been the United Nations (UN) World Conference on the Environment which was held in Stockholm in 1972. Although the debate did not succeed in a strong commitment towards more environmental-friendly economic growth, its 26 common-agreed principles regarding environment and development could be regarded as the first common achievement towards global sustainability [3,4].

In the following years, a heated debate around the limits between environmental protection and economic growth evolved across academics, policy makers and practitioners. A common belief is that environmental degradation increases during the early stages of economic development. After a certain point of growth the trend is reversed, in a sense that development becomes more sustainable [5]. The rationale behind this trend is that in early stages of development, pollution-generating economic sectors are developed without concrete control, since economic targets surpass the respective environmental ones, and environmental degradation is an accepted byproduct of the desired economic outcomes. In addition, as income rises most people start to put a larger weight on their quality of life and thus environmental issues are addressed by legislation in a more systematic context [1].

This kind of functional relationship between environmental degradation and economic growth is graphically expressed by an inverted U-shape slope called the environmental Kuznets curve (EKC) [6]. Many scholars have tried to empirically verify the existence of EKC for different environmental dimensions (i.e., CO_2 emissions, particulate matter, energy use) and in various geographical and time contexts. The papers of [7,8] provide two comprehensive and detailed reviews of the relevant past empirical studies.

As far as the agricultural sector is concerned, production comes up with several environmental damages that mainly have to do with land–take, habitat loss, soil erosion and degradation, water contamination and climate change [9]. Nevertheless, research on the relationship between negative environmental impacts and economic growth is still limited. More precisely, the majority of studies focus their interest on developing countries seeking empirical evidence on the relationship between economic development and agricultural land expansion, particularly against forests [10,11]. Besides these studies, to the best of our knowledge, there is only one study examining environmental risks of agriculture in advanced economies. More precisely, ref. [12] analyzed the longitudinal fluctuations of environmental risks arising from pesticides used in agriculture within US states against real gross state product, abatement efforts and an environmental productivity index. Managi [12] argued that although environmental risks and economic growth seem to validate the EKC relationship, the risk of observing an N relationship still exists due to scale effects. Thus, even for advanced economies, such as USA, EKC in agriculture could not be taken for granted.

The existing literature testify that, for agriculture, there is a lack of studies regarding advanced economies' dynamics, such as those of EU countries, and other environmental impacts such as GHG emissions remain rather unexplored. The importance of examining the atmospheric impacts of agriculture is extracted from the fact that global agriculture accounts for about 24% of the total GHG emissions. In addition, the Organisation for Economic Co-operation and Development (OECD) countries account for more than one third of global emissions [13]. In a European context, the share of agriculture in total GHG emissions is exceeding 10%, with a total production of 470.6 million tons of CO_2 in 2012. Nevertheless, the contributions vary significantly across countries. To illustrate, Irish, Latvian and Lithuanian agricultural sectors account for more than 20% of the total national emissions, whereas the share of agricultural GHG emissions in countries such as Malta, Luxemburg and the Czech Republic do not exceed 7% [14]. In addition, focusing on agricultural nutrients, EU countries have managed to reduce their nitrogen and phosphorus balances during the period 2000–2012. Nitrogen Balance for EU-28 has been reduced from 63 to 51 and phosphorus from six to two. Nevertheless, efficiency in nutrient utilization, as this is expressed by the ratio of total nutrient outputs to total nutrient inputs, also varies profoundly across countries. For the period 2009–2014, the most efficient countries in nitrogen efficiency was Romania, Ireland and Bulgaria and the least efficient ones were Cyprus, Luxemburg, Greece, Malta and Croatia [15].

Undoubtedly, the variations among EU countries reflect, in a straightforward context, the composite nature of agricultural production sustainability issue. Countries such as Malta and Luxemburg present positive figures regarding GHG emissions whilst being inefficient in nutrient utilization. The opposite stands for countries such as Ireland. Such variations have been also identified by the recent study of [16] who employed the generalized Divisia Index to evaluate the effect of energy use and agricultural growth in energy-related GHG emissions of 17 EU countries from 1995 to 2012. These variations portray, that in order to extract a comprehensive picture regarding the sustainability levels of national agricultural practices, the analysis should focus on all aspects of agricultural production. That is to say, not only the amount of inputs or the produced outputs could define sustainability, but the level of efficiency of how inputs are utilized in order to produce certain outputs that lead to a comprehensive evaluation index of agricultural sustainability [16].

The link between sustainability and efficiency is reflected in the recent reforms of the common agricultural policy (CAP) which precisely focus on increasing the efficiency of use of agricultural inputs, and on reducing the production of undesirable outputs, such as GHG emissions [17]. Nevertheless, this goal requires the development of various technologies on both the production and application of agricultural inputs, like fertilizers, agrochemicals, water and fuels [18]. Additionally, efficiency of inputs use is an integral part of sustainability where economic and environmental parameters merge, in order for production cost reduction and mitigation of contamination to occur [19]. It is obvious that the economic and the environmental performance of agricultural production are controversial parameters, being at the core of a continuous debate between farmers and various stakeholder groups, with the latter requiring from the former to adjust the production process towards an environmentally friendly approach [20]. The experience of other economic sectors has shown that the costly environmental-friendly amendments to production are more easily applied in advanced economies [21,22]. By extension, and regarding the agriculture sector, mature economies are expected to present more sustainable agricultural production practices, such as those having resolved basic dietary issues, are usually characterized by increased environmental awareness and their demand focus on safe and clean, in environmental terms, agricultural products.

The justification of this debate necessitates the need for empirical evidence. As the relationship between agricultural sustainability and economic development is still rather unexplored by international literature, the present paper seeks to fill this gap by exploring the potential existence of an EKC for the European agricultural sector. In order to account for all dimensions of agricultural sustainability, a synthetic Eco-(in)efficiency indicator is developed and applied to the 28 EU countries in order to evaluate sustainability variations for the period 1999–2012. At a later stage, the countries' figures are regressed against GDP p.c. in order to test for an EKC relationship. The remainder of the paper is as follows. Section 2 describes the composition of the eco-efficiency indicator and the regression details for testing for the EKC function. In Section 3, the variables employed by the present study are explained and the results of the analysis are presented and discussed. The paper ends up with the main conclusions and some policy implications towards more sustainable agricultural practices on the European level.

MATERIALS AND METHODS

Defining the Eco-(In)Efficiency Index

Considerations Regarding DEA Eco-Efficiency Indicators Development

The challenge of conceptualizing and measuring sustainability has captured the interest of scholars across various disciplines in developing synthetic quantitative indices which could be used as the basis of sustainability evaluation [23]. Towards this direction, eco-efficiency could be regarded as an effective approximation of sustainability since it measures the efficiency of a production unit in maximizing its desirable outputs whilst keeping the production of its undesirable outputs and the use of natural resources at the lowest possible levels [24,25]. That is, eco-efficiency could produce integrated assessments of various dimensions of production efficiency such as technical, economic, ecological, energy and environmental [26].

A large part of literature is based on production frontier theory, firstly introduced by [27], in order to develop eco-efficiency indices. The main methodologies based on the frontier theory are data envelopment analysis (DEA) and stochastic frontier analysis (SFA) [25,28]. The two methods are based on different assumptions concerning their application. DEA is a non-parametric method which involves the solution of a series of linear programming problems in order to extract the relative efficiency of each Decision Making Unit (DMU). In addition, SFA is based on the specification of a production function and efficiency estimates are extracted with the use of statistical regression models [29]. Between the two methods, DEA could be regarded as more easily applicable and thus has been rendered rather popular towards eco-efficiency estimations [30].

Among the various specifications of eco-efficiency DEA models there are some basic issues that should be taken into account before its empirical application. The first refers to the reference set and the type of desirable and undesirable output relationship that would be chosen for the formulation of the model. On the one hand, there are studies that assume weak disposability among outputs meaning that reductions of undesirable outputs result in desirable production losses [31,32]. On the other, when strong disposability is assumed, then reductions of undesirable outputs could be achieved without any reduction in the desirable outputs [33]. In addition, another issue that should be defined is whether the models are based on simple distance functions or directional distance functions. In the first case of studies, the adjustments of desirable and undesirable outputs are headed to the same direction after all undesirable outputs have been transformed mainly by using their reciprocals [34,35,36], whilst in the second case these could be realized to opposite directions [37,38,39]. In addition, before the formulation of a DEA model, one should also consider if the adjustments of inputs or outputs would be proportional or non-proportional. In the first case, radial DEA models are used [40,41] and in the latter, non-radial measures are employed [42,43]. The papers of [44,45] deal extensively with the aforementioned issues providing comprehensive reviews regarding the application of the DEA method in sustainability assessments taking into account the energy and environmental dimensions of production.

DEA Applications on Agricultural Sector

As far as agricultural studies are concerned, the latest relevant literature has shown a remarkable progress in developing synthetic methods towards sustainability assessment, lying far from a simple measurement of output production or input use [12,13,16]. More precisely, until recently, most of the attention was given to the technical efficiency of agricultural sectors. In a cross-country context, ref. [46] assessed the technical efficiency of 93 countries for the 1980–2000 period based on DEA, ref. [47] analyzed the impact of CAP subsidies on the technical efficiency of crop farms in Germany, the Netherlands and Sweden and [29] assessed the longitudinal technical efficiency of EU countries by comparing the results of DEA and SFA methods. Martinho [48] employed DEA to assess the efficiency and total factor productivity in the European agricultural sector at both farm and regional level. Finally, ref. [49] extracted technical efficiency scores for EU countries and used a bootstrap method in order to extract their statistical properties.

In recent years, the notion of efficiency estimation through frontier deviations has been extended to account for environmental and energy dimensions of agricultural production. The flexibility of DEA provided a fertile ground for such modifications. In this second category of agricultural studies, it was [50] who firstly decomposed agricultural efficiency in its technical, economic and environmental dimensions. In addition, ref. [26] assessed the sustainability of Dutch sugar beet growers by taking into account indices of technical efficiency, environmental efficiency and a composite index of sustainable efficiency. Managi [12] modified the total factor productivity Malmquist–Luenberger index, in order to account for environmental factors. Later, ref. [51] developed a composite sustainability indicator consisting of the dimensions of technical efficiency and cumulative exergy allocative efficiency and evaluated the sustainability of the agricultural sector of OECD countries. Picazo-Tadeo et al. [24] extended the basic DEA model and used a directional distance function (DDF) in order to extract different estimates of the efficiency of olive-growing farms in Spain. Since both undesirable inputs and outputs were considered, authors regarded these estimations as representing the eco-efficiency of the DMUs' sample. Vlontzos et al. [30] employed a non-radial DEA model in order to extract different estimations of the environmental and energy efficiency of

EU countries. Moreover, Tian et al. [52] used a slacks-based DEA model in order to incorporate undesirable agricultural emissions in the efficiency evaluation of rice farmers of China. The slacks of energy inputs have been also assessed by [53] within the technical efficiency measurement of French farms. Finally, Vlontzos et al. [35] used Window DEA to assess the longitudinal technical efficiency of EU countries taking into consideration their GHG emissions. GHG emissions factor was entered into estimations with two ways. The first involved its consideration as input and the second as output after being rescaled.

The preceding studies provide useful insights into the efficiency of the agricultural sector. Since the aim of this paper is to test for an EKC relationship between agricultural sector and economic growth, estimations of pure technical efficiency could not be used as the basis of the comparisons, as these do not encompass the environmental dimension of production. Thus, the present paper will mostly focus on the models of the second category of studies which can provide a more holistic assessment of agricultural sustainability. These kinds of models have been extensively used for testifying an EKC curve in other disciplines [33,54,55] but have never been adjusted for the agricultural sector. Keeping in mind, that for an EU cross-country evaluation context both non-radial [23] and radial [35] models have been used to quantify environmental efficiency, the present paper follows the rationale of [12,32] in order to enrich the methodological portfolio regarding environmental efficiency in EU countries. It should be noted that the difference between the present study and the study of [12] lies in the scope of efficiency estimation. On this, efficiency in the present paper is estimated in order to be used as a dependent variable in EKC testing, whereas within the context of [12], this was used as an independent variable capturing the environmental technology differences across US states. Taking into account the preceding remarks, the contribution of the present paper is twofold. Initially, it extends the previous sustainability assessments in the agricultural sector with a novel composite DEA indicator based on the DDF rationale. Secondly, it uses this indicator to test, for the first time, for an EKC relationship of agriculture sector in a group of advanced economies such as that of EU members. By doing so, the paper enriches the agricultural literature but also contributes to the global debate around the relationship between environmental degradation and economic growth.

Developing the DEA Eco-(In)Efficiency Indicator for EU Agricultural Sector Sustainability Assessment

In contrast to the basic DEA formulations which seek to maximize outputs or minimize inputs, DDF allows for simultaneous adjustments on both inputs and outputs. This rationale has been also incorporated in order to provide estimations of productive (in)efficiency when both desirable and undesirable outputs are taken into consideration [32].

Given the presence of x inputs, g desirable outputs and b undesirable outputs, the technology set is defined as:

[(x,g,b;ecoEff):x can produce (g,b)]

In addition, all the feasible output vectors sets for a given input vector are the following:

P(x)=[(g,b):(x,g,b)∈T]

Then, the directional distance function has the following form:

$$\vec{D}_T(x,g,b;ecoEff) = \sup\delta\colon \left(g+\delta ecoEff^{g}, b-\delta eff^{b}\right) \in P(x+\delta eff^{x})$$

Moreover, we suppose that we have *n* national agriculture sectors expressed as NAS_j (j=1,2,…,n), and $x_{ij}>0$ is the amount of input i used in total by National Agricultural Sector (NAS) NAS_j, $g_{rj}>0$ is the amount of desirable output r of NAS_j and $b_{cj}>0$ the amount of undesirable output c of NAS_j. Then, the inefficiency score for the NAS_i is extracted by solving the following model [56]:

$$\begin{aligned}
&\max_{Eff,\lambda} ecoEff \\
\text{s.t.}\quad &\sum_{j=1}^{n} x_{ij}\lambda_j + ecoEff^{x}_{io} \le x_{io} && i=1,2,\dots,m \\
&\sum_{j=1}^{n} b_{cj}\lambda_j + ecoEffd^{b}_{co} = b_{co} && c=1,2,\dots,k \\
&\sum_{j=1}^{n} g_{rj}\lambda_j - ecoEffd^{g}_{ro} \ge y_{ro} && r=1,2,\dots,d \\
&\sum_{j=1}^{n} \lambda_j = 1 \\
&\lambda_j \ge 0j = 1,2,\dots,n \\
&\mathrm{d} = (0,\mathrm{g},-\mathrm{b})
\end{aligned} \tag{1}$$

The objective of the model is to maximize the desirable outputs and minimize the undesirable outputs whilst keeping the inputs stable. The value of ecoEff portrays the Eco-(in)efficiency, meaning that for a fully efficient

NAS, the value of ecoEff will be 0. The constraints of the model portray that a weak disposable production process has been selected for the present dataset, which assumes that reductions in CO_2 emissions could not be realized without desirable output losses. This kind of model formulation seems more realistic than a strong disposable approach which allows for CO_2 emissions reductions without any loss in desirable outputs [51]. In addition, the model assumes variable returns of scale in the production process, thus eliminating the scale effects and resulting in more reliable targets for each NAS.

Defining the Model for EKC Testing

After extracting the ecoEff scores for each country, a regression analysis will be conducted in order to check for an EKC relationship. For comparison reasons, three different models will be tested. The first will test for EKC between Eco-(in)efficiency and GDP. The second will focus on energy use and the third on total GHG emissions. The review of previous studies has shown that different modifications of regression models, such as time-series, cross-section and panel-data have been employed in order to check for EKC in other sectors [7]. Among them, panel-data is the most widely used method for acquiring cross-country empirical evidence [57]. The fundamental consideration regarding the adaptation of a panel-data regression on a study is the functional form of the relationship among dependent and independent variables. Three basic forms could be found in the literature. The first is the linear specification, the second is the log-log specification in which both dependent and independent variables are logged before being entered into the model and the third is the semi-log function in which only the dependent or the independent variables are logged. In the present paper, the linear form is preferred, as it leads to a direct approximation of the turning point of the curve.

The second consideration is the choice between fixed and random effects specification of the model. In fixed effects, the intercept is gathered as fixed whereas in random effects specification, it is assumed that it is extracted through the function $a_i = \overline{a} + \mu_i$, where $\overline{a}$ is an unknown parameter and μ_i are unknown *iid* variables. For the selection of the appropriate specification it is advisable to run a Hausman test before running the statistical model [55]. Finally, it should be noted that since DEA estimations portray the relative efficiency of each NAS, it is expected that a cross panel correlation will emerge in the respective panel data model which could result in biased estimations. To overcome this difficulty we employ a regression with panel-corrected standard errors (PCSE) which controls for heteroscedasticity and contemporary correlation of disturbances across panels [58]. The models specifications are given below:

$$Y_{it} = a_i + \beta_{GDP} \cdot GDP_{it} + \beta_{GDP^2} \cdot GDP_{it}^2 + \beta_{GDP^3} \cdot GDP_{it}^3 + \varepsilon_{it}$$
$$i = 1, 2, \ldots, N \text{ countries}, \; t = 1, 2, \ldots, T \text{ years}$$
$$Y = ecoEff \text{ for model 2.1,}$$
$$Y = ENERpc \text{ for model 2.2,}$$
$$Y = GhGpc \text{ for model 2.3} \qquad (2)$$

RESULTS AND DISCUSSION

Eco-(In)Effiiency Assessment Results

In order to depict the production process of EU agricultural sectors, respective inputs and outputs have been selected. The inputs used in this study are expressed as follows: land variable is expressed in total hectares, capital as the fixed capital consumption in monetary terms and labor as the employees' annual working units (AWU). In addition, total energy cost and the total cost of chemicals and fertilizers used in the production process have been also incorporated into the model as significant contributing factors to the agricultural output of each country. Moreover, output has been segmented to desirable output (total crop and animal output) and undesirable output (total GHG Emissions). The basic descriptive statistics of the input and output variables of model 1 are presented in Table 1.

The data covers the period between 1999 and 2012. In order to extract the longitudinal Eco-(in)efficiency scores of each country, Model 1 is executed 14 times. The main results for each NAS are presented in Table 1, whereas the detailed annual Eco-(in)efficiency scores for each NAS are given in Table A1 of the Appendix A. Regarding Table 1, the mean Eco-(in)efficiency score for each NAS is given in the first column, the years for which each country presented zero (in)efficiency are presented in the second column, whereas the highest and lowest Eco-(in)efficiency score of each country through the whole 1999–2012 period are given in the third and fourth columns, respectively. As can be seen in Table 2, the average Eco-(in) efficiency score for the EU agricultural sector is 0.11. This finding implies that with the given inputs, EU agricultural could produce 11% more output and less GHG emissions. Nevertheless, variations exist among countries.

Table 1: Descriptive statistics of DEA inputs and outputs

Statistic	**Land (1000 Ha)**	**Energy (mil. $)**	**Chemicals and Fertilizers (mil. $)**	**Fixed Capital Consumption (mil. $)**	**Labour (1000 AWU)**	**Output (mil. $)**	**GHG Emissions (1000 tonnesCO_2)**
Mean	6642.64	786.57	880.12	1815.24	1210.88	11,720.45	17.69
St Dv	7901.59	933.03	1314.67	2730.81	1729.06	15,644.37	22.99
Max	35,177.80	4502.70	7599.10	12,377.39	7307.35	70,394.90	100.46
Min	9.70	5.44	1.73	3.76	3.59	115.18	0.08

Source: [59,60,61]; Own elaboration. More precisely, only Malta and the Netherlands kept a zero Eco-(in)efficiency for all the considered period, followed by Italy which was found to be rather inefficient only in one year (2005), and France, which was found to be inefficient in three years (1999, 2005, 2006). On the contrary, the least efficient group of countries consists of Lithuania, Slovakia and Finland, whose average Eco-(in)efficiency scores for the period 1999–2012 exceeded 0.3. A second group of inefficient countries is shaped by Latvia, Hungary, the United Kingdom, the Czech Republic and Slovenia with Eco-(in)efficiency scores ranging between 0.2 and 0.3.

Table 2: Eco-(in)efficiency scores and longitudinal performance in EU agricultural sector (1999–2012)

Country	Average Eco-(In) Efficiency	Years Fully Efficient	Highest Eco-(In) Efficiency	Lowest Eco-(In)Efficiency
Netherlands	0.000	14	0.000	0.000
Malta	0.000	14	0.000	0.000
Italy	0.003	13	0.043	0.000
France	0.004	11	0.030	0.000
Spain	0.005	12	0.050	0.000
Austria	0.009	9	0.055	0.000
Belgium	0.010	10	0.068	0.000
Portugal	0.010	10	0.058	0.000
Poland	0.013	13	0.188	0.000
Germany	0.017	10	0.098	0.000
Greece	0.018	7	0.077	0.000
Bulgaria	0.021	10	0.133	0.000
Denmark	0.022	8	0.088	0.000
Luxembourg	0.025	11	0.136	0.000
Ireland	0.026	10	0.153	0.000
Cyprus	0.046	11	0.255	0.000
Romania	0.050	8	0.216	0.000
Croatia	0.075	8	0.241	0.000
Sweden	0.159	0	0.286	0.029
Estonia	0.175	6	0.479	0.000
Slovenia	0.199	1	0.312	0.000

Czech Republic	0.219	5	0.451	0.000
United Kingdom	0.220	6	0.492	0.000
Hungary	0.281	0	0.422	0.113
Latvia	0.282	3	0.634	0.000
Finland	0.311	0	0.375	0.239
Slovakia	0.351	0	0.513	0.167
Lithuania	0.402	0	0.551	0.213
Mean	**0.105**	-	-	-
St.Dv	**0.150**	-	-	-

Source: Own elaboration.

Apart from the top-ranked countries, a number of other national sectors achieved remarkable figures in terms of times being evaluated as efficient during the recursive application of the DEA models. Poland, Spain, Luxemburg and Cyprus were found fully efficient for more than ten years and Belgium, Portugal, Germany, Bulgaria and Ireland for exactly ten years. In addition, Lithuania, Slovakia, Finland Hungary and Sweden have not acquired a zero Eco-(in)efficiency score in any of the annual estimations. Moreover, examining the last two columns, it is interesting to highlight the large variability of Latvia figures, as within the period under consideration its highest Eco-(in)efficiency score was 0.634 whilst the lowest 0. Rather high variability is found also for the United Kingdom, Estonia and the Czech Republic. As it is evident, the least efficient countries present a larger range of scores during the 1999–2012 period.

Having examined the individual records of the countries, Figure 1 presents the annual Eco-(in)efficiency estimations for three categories of countries against the respective scores of the whole sample. The first category of countries is composed of countries that joined the EU before 2004, the second of the countries that entered into the EU in 2004 and the third of the three countries that joined the EU after 2007. This division is critical in order to check the performance of countries with reference to the CAP policy guidelines. In general, average EU Eco-(in)efficiency presents variable trends through the considered period. The least efficient scores are observed over the years 1999 and 2005 whereas the most efficient scores are found in years 2007, 2011 and 2012. Among the group of countries, these that have accessed EU in 2004 are characterized with the highest average Eco-(in)efficiency scores during the whole period. It should be noted that average inefficiency of these countries exceeds the average EU inefficiency at all annual estimations. The other two groups of countries present lower inefficiency records, which is reflected in the fact that in any year their mean inefficiency exceeded the EU average. Finally, comparisons between the inefficiency trends of the two groups of countries, could lead to any safe conclusions, as these present high variability. The only remarkable observation is the significant better performance of late-joining countries during the period 2005–2007.

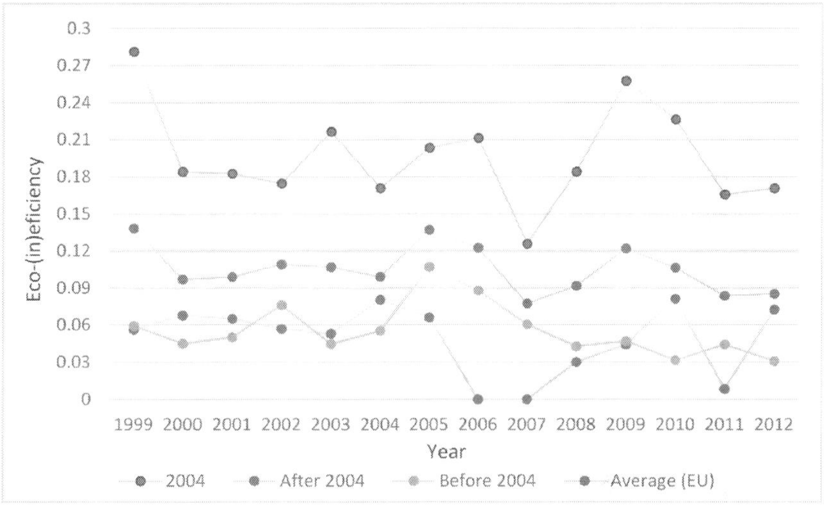

Figure 1: Average Eco-(in)efficiency of EU agricultural sector and average Eco-(in)efficiency by groups of countries according to their date of accession (1999–2012) (Source: Own elaboration).

EKC Testing Results

As it was stated in Section 2, the scores of Eco-(in)efficiency will be used as the basis for fitting an EKC curve on the relationship of agriculture environmental efficiency and economic growth in EU countries. In addition, the existence of an EKC will be also examined for p.c. energy use of agriculture and p.c. GHG emissions. The annual GDP p.c. of EU countries for the years 1999–2012 will be used as the variable to represent the economic growth of each country. In order to achieve an unbiased representation of the economic level of each country, the GDP will be expressed in purchasing power parity (PPP), as this measure allows for more accurate cross-country comparisons and has been employed in several empirical studies regarding EKC [5]. A cubic function has been selected in order to check for any

N-shape relationships. The empirical recognition of an EKC or N-shaped pattern is based on the estimations of the regression parameters and, more precisely, on their signs and their statistical significance. In this sense, an EKC curve would be confirmed only under the following signs of estimated parameters $\beta GDP>0$, $\beta GDP2<0$, $\beta GDP3=0$. Finally, it should be mentioned that the final sample includes 27 EU countries. Luxemburg data was omitted from estimations because it presents GDP figures that are exceptionally higher than the respective figures of other countries, and thus could lead to biased estimations.

Table 3 presents the results for the three models. As far as the goodness-of-fit tests are concerned, the statistical significance of the Wald test result denotes that all three models are performing better than the respective models with zero parameters. In addition, for models 2.2 and 2.3, the results of the Hausman test indicate that for both models, a random effect specification could be used with the available data. For model 2.1, estimations present statistical significance for all parameters at the 0.01 level, except for the intercept. The estimations for the βGDP and βGDP3 are positive, whereas the estimation of βGDP2 is negative. This finding signifies that eco(in) efficiency and economic growth are linked through an N-Shaped curve. More specifically, in early stages of development, agricultural production is rather inefficient with desirable output shortages and undesirable output excesses. As the national income increases, production becomes more efficient in input utilization but after a certain point of income, inefficiency presents increasing trends again. This fact could be attributed to the ability of wealthier countries to afford more undesirable inputs such as fertilizers and energy. Hints for this explanation are provided by the examination of the estimated coefficients of model 2.2. As for model 2.1, all estimations acquire statistical significance except for the constant term. The pattern of parameter sign change is exactly the same as that of model 2.1. This finding provides hints that Eco-(in)efficiency is strongly connected to energy use. Finally, model 2.3 estimations are all statistically significant with the exception of βGDP3. Thus, agricultural GHG emissions seem to be connected to economic growth through an EKC relationship.

Table 3: Models 2.1, 2.2 and 2.3 estimation results

Models (Dependent Variable)						
2.1 (ecoEff)		**2.2** (ENERpc)		**2.3** (GhGpc)		
Estimators	**Estimation**	**St.Error**	**Estimation**	**St.Error**	**Estimation**	**St.Error**
βGDP	0.0428 ***	0.0145	1.759 **	0.7169	0.0060 ***	0.00195
βGDP2	-2.62×10^{-9} ***	7.75×10^{-10}	-1.06×10^{7} ***	3.92×10^{-8}	-2.43×10^{-10} **	1.09×10^{-10}
βGDP3	4.30×10^{-14} ***	1.27×10^{-14}	1.89×10^{-12} ***	6.55×10^{-13}	1.65×10^{-15}	1.82×10^{-15}
a	−0.0399	0.0820	−2.6329	4.0450	81.5590 ***	15.490
T	14					
N	27					
WaldTest chi2(1)	8.68		9.75		137.46	
Prob > chi2	0.003		0.021		0.001	
Hausman Test chi2(1)	-		0.01		3.84	
Prob > chi2	-		0.943		0.051	
TurningPoint 1 (000€)	11,326		12,428		12,325	
TurningPoint 2 (000€)	29,294		24,961		-	

Statistical significance: (***) at 0.01 level (**) at 0.05 level. Source: Own elaboration.

As far as the turning points are concerned, these are visualized with Figure 2 which presents the original and fitted values of the dependent variables extracted from the three different models, plotted together with the GDP variable. As can be seen from Figure 2a, Eco-(in)efficiency increases up to 11,326€ and starts decreasing until 29,294€. After this point, there are hints that countries return to more unsustainable practices. Nevertheless, inefficiency at this part of GDP levels is still lower than the respective of the lowest GDP levels. The same trends are observed over the energy use (Figure 2b), although with two remarkable differences. The first regards the turning points, which are closer than those of Eco-(in)efficiency (12,428€ and 24,961€) and the second on the fact that after the second turning point, energy use, in some cases, surpasses the respective use at the lowest GDP levels. Finally, GHG emissions and GDP level relationship seems to match with the characteristics of an EKC curve. Emissions are expanding until the point of 12,325€, and afterwards they present declining trends.

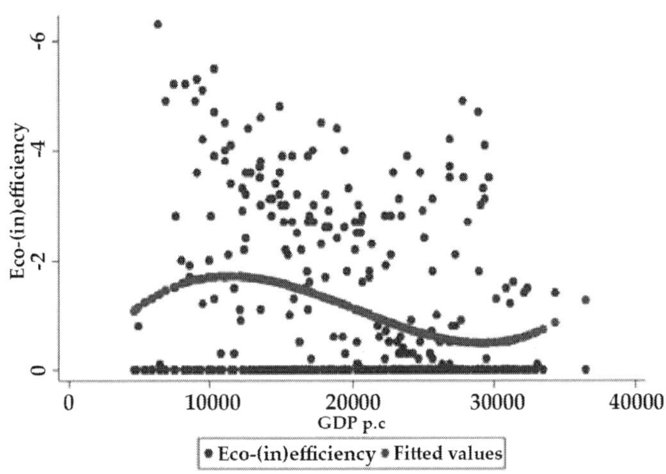

(a)

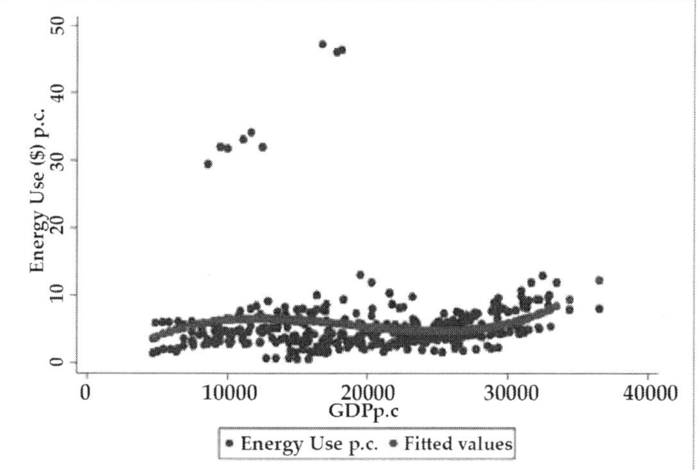

(b)

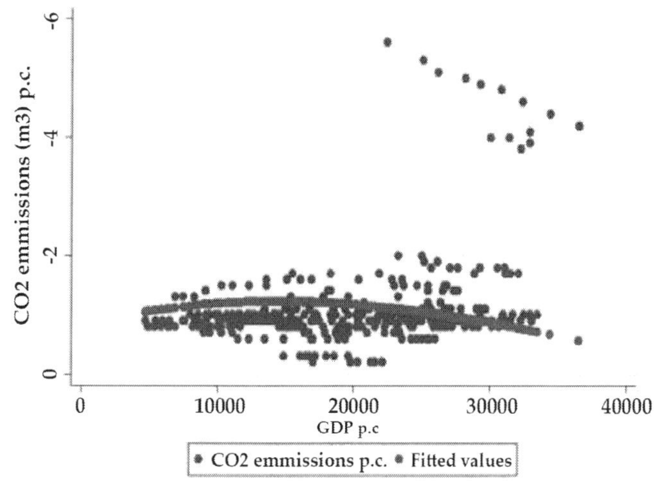

(c)

Figure 2: The Turning Points of Models 2.1, 2.2. and 2.3 (a) The turning points of Eco-(in)efficiency curve (b) The turning points of the energy use curve (c) The turning point of the GHG curve (Source: Own elaboration).

The aforementioned findings provide useful insights into the state and evolution of primary sectors in the EU. A typical analysis focusing only on the GHG emissions would provide hints that the CAP fulfills its targets, at least on the emissions side. As confirmed by the present paper, EKC relationship confirms the general impression that emissions is a matter of economic growth that could be solved together with the global target of EU regional and national economic convergence. Nevertheless, the results of Eco-(in)efficiency and energy use analyses which showed that inefficiency and energy use excesses are also observed in countries with higher income levels, add a question mark for whether EU convergence is only a necessary or/and sufficient conditionfor rendering the EU agricultural sectors more sustainable. The findings of the paper, extend the debate of previous papers [12,16] on the necessity of developing more advanced methods on agricultural sustainability assessments, especially in economies where technological, institutional and scale factors are rather complex.

CONCLUSIONS

This analysis revealed that EU agriculture is quite far from being fully sustainable. The Eco-(in)efficiency index results signifies that the overall EU primary sector could produce 11% more output and less GHG emissions. Inefficiency seems to be higher for the group of countries that accessed the EU in 2004, as for countries such as Slovakia and Lithuania output shortages and GHG emissions excesses, surpass 30%. Moreover, the EKC analysis revealed that despite the fact that GHG emissions seem to be reduced as national incomes increase, a more advanced economic level does not ensure a more sustainable agricultural production, as a whole. This is reflected in the N-Shaped curve linking the Eco-(in)efficiency and energy use levels with the GDP of EU countries during the period of 1999–2012. Thus, a future success of the convergence process within the EU should not be regarded as a decisive factor of agricultural sustainability improvement.

The contradictory results among the three agricultural sustainability dimensions tested by the present paper highlight the challenge of measuring agricultural sustainability in a more holistic and systematic context. To this end, the models of the present paper could be enriched with additional control variables representing particular characteristics of EU countries, such as, their institutional framework, their technological advancement, their specialization in crop and livestock production etc. Moreover, new synthetic indicators for measuring agricultural sustainability could be added to the ones employed by the present paper. In addition, at the policy level, community initiatives such as the European innovation scoreboard should be capitalized and adapted to the specific challenges of the EU agricultural sector. Finally, further improvement of efficiency scores for inefficient countries can be achieved through utilization of good practices being implemented in efficient ones. It is therefore necessary that advanced technology are reachable by farmers, hence more effectively and efficiently utilizing energy resources, while reducing GHG emissions. The continuous training of farmers is also necessary in order to increase their ability to adopt and implement new cultivation and production practices. It has been proven that continuous engagement with new knowledge leads farmers to more appropriate decision-making processes which have a positive impact on efficiency improvements [62]. The new CAP provides considerable degrees of freedom to farmers to organize their cultivation and production plans, and is an appropriate framework for efficiency change and efficiency improvement.

AUTHOR CONTRIBUTIONS

All authors designed the structure of the paper and contributed to its writing; George Vlontzos prepared the data, conducted the literature review and estimated the Eco-(in)efficiency scores; Spyros Niavis and Panos Pardalos developed the framework for Eco-(in)efficiency estimation and tested the EKC curve.

APPENDIX A

Table A1. Annual Eco-(in)efficiency scores of EU agricultural sector (1999–2012).

	Year													
Country	**1999**	**2000**	**2001**	**2002**	**2003**	**2004**	**2005**	**2006**	**2007**	**2008**	**2009**	**2010**	**2011**	**2012**
Belgium	0.06	0.00	0.00	0.07	0.00	0.00	0.01	0.00	0.00	0.00	0.00	0.00	0.00	0.00
Bulgaria	0.00	0.00	0.00	0.00	0.00	0.00	0.00	0.00	0.00	0.00	0.13	0.03	0.03	0.11
Czech Republic	0.00	0.00	0.00	0.00	0.00	0.39	0.45	0.44	0.27	0.27	0.40	0.33	0.25	0.28
Denmark	0.03	0.00	0.00	0.05	0.02	0.08	0.09	0.00	0.00	0.05	0.00	0.00	0.00	0.00
Germany	0.00	0.00	0.00	0.06	0.03	0.00	0.10	0.00	0.00	0.00	0.05	0.00	0.00	0.00
Estonia	0.28	0.00	0.00	0.00	0.21	0.00	0.00	0.10	0.00	0.40	0.48	0.39	0.30	0.29
Ireland	0.07	0.00	0.00	0.00	0.00	0.00	0.15	0.14	0.00	0.00	0.00	0.00	0.00	0.00
Greece	0.00	0.00	0.00	0.00	0.00	0.01	0.00	0.08	0.05	0.05	0.02	0.00	0.05	0.00
Spain	0.02	0.00	0.00	0.00	0.00	0.00	0.00	0.05	0.00	0.00	0.00	0.00	0.00	0.00
France	0.01	0.00	0.00	0.00	0.00	0.00	0.02	0.03	0.00	0.00	0.00	0.00	0.00	0.00
Croatia	0.17	0.12	0.20	0.17	0.15	0.24	0.00	0.00	0.00	0.00	0.00	0.00	0.00	0.00
Italy	0.00	0.00	0.00	0.00	0.00	0.00	0.04	0.00	0.00	0.00	0.00	0.00	0.00	0.00
Cyprus	0.00	0.15	0.23	0.26	0.00	0.00	0.00	0.00	0.00	0.00	0.00	0.00	0.00	0.00
Latvia	0.63	0.00	0.00	0.00	0.36	0.28	0.38	0.36	0.28	0.34	0.44	0.35	0.30	0.22
Lithuania	0.49	0.52	0.52	0.53	0.55	0.40	0.33	0.38	0.21	0.32	0.46	0.39	0.27	0.26
Luxembourg	0.00	0.00	0.00	0.00	0.12	0.00	0.14	0.00	0.00	0.00	0.00	0.00	0.10	0.00
Hungary	0.42	0.39	0.34	0.32	0.36	0.30	0.31	0.32	0.22	0.13	0.30	0.25	0.11	0.16
Malta	0.00	0.00	0.00	0.00	0.00	0.00	0.00	0.00	0.00	0.00	0.00	0.00	0.00	0.00
Netherlands	0.00	0.00	0.00	0.00	0.00	0.00	0.00	0.00	0.00	0.00	0.00	0.00	0.00	0.00
Austria	0.03	0.00	0.00	0.00	0.01	0.00	0.05	0.00	0.00	0.00	0.02	0.00	0.00	0.01
Poland	0.19	0.00	0.00	0.00	0.00	0.00	0.00	0.00	0.00	0.00	0.00	0.00	0.00	0.00
Portugal	0.00	0.00	0.00	0.05	0.00	0.00	0.00	0.00	0.00	0.03	0.00	0.00	0.06	0.00
Romania	0.00	0.08	0.00	0.00	0.01	0.00	0.20	0.00	0.00	0.09	0.00	0.22	0.00	0.11
Slovenia	0.31	0.27	0.27	0.18	0.27	0.06	0.18	0.16	0.00	0.21	0.22	0.25	0.18	0.23
Slovakia	0.49	0.51	0.47	0.45	0.41	0.29	0.37	0.36	0.27	0.17	0.28	0.32	0.24	0.26
Finland	0.30	0.28	0.28	0.28	0.31	0.24	0.31	0.37	0.33	0.35	0.35	0.35	0.30	0.31
Sweden	0.19	0.03	0.09	0.29	0.18	0.08	0.21	0.18	0.12	0.15	0.27	0.13	0.16	0.14
United Kingdom	0.17	0.36	0.39	0.36	0.00	0.42	0.49	0.47	0.41	0.00	0.00	0.00	0.00	0.00

Source: Own elaboration.

REFERENCES

1. Arrow, K.; Bolin, B.; Costanza, R.; Dasgupta, P.; Folke, C.; Holling, C.S.; Jansson, B.-O.; Levin, S.; Maler, K.-G.; Perrings, C.; et al. Economic growth, carrying capacity, and the environment. *Ecol. Econ.* 1995, *268*, 520–521.
2. Han, X.; Chatterjee, L. Impacts of Growth and Structural Change on CO_2 Emissions of Developing Countries. *World Dev.* 1997, *25*, 395–407.
3. Beckerman, W. Economic growth and the environment: Whose growth? whose environment? *World Dev.* 1992, *20*, 481–496.
4. Handl, G. *Declaration of the United Nations Conference on the Human Environment (Stockholm Declaration), 1972 and the Rio Declaration on Environment and Development, 1992*; United Nations Audiovisual Library International Law; Codification Division, Office of Legal Affairs: New York, NY, USA, 2012; pp. 1–11. (In English)
5. Stern, D.I. Environmental Kuznets curve. *Encycl. Energy* 2004, *2*, 1–3.
6. Stern, D.I. The Rise and Fall of the Environmental Kuznets Curve. *World Dev.* 2004, *32*, 1419–1439.
7. Kaika, D.; Zervas, E. The Environmental Kuznets Curve (EKC) theory-Part A: Concept, causes and the CO_2emissions case. *Energy Policy* 2013, *62*, 1392–1402.
8. Stern, D.I. The environmental Kuznets curve after 25 years. *J. Bioecon.* 2017, *19*, 7–28.
9. Pretty, J.; Brett, C.; Gee, D.; Hine, R.; Mason, C.; Morison, J.; Rayment, M.; Van Der Bijl, G.; Dobbs, T. Policy Challenges and Priorities for Internalizing the Externalities of Modern Agriculture. *J. Environ. Plan. Manag.* 2001, *44*, 263–283.
10. Ceddia, M.G.; Sedlacek, S.; Bardsley, N.O.; Gomez-y-Paloma, S. Sustainable agricultural intensification or Jevons paradox? The role of public governance in tropical South America. *Glob. Environ. Chang.* 2013, *23*, 1052–1063.
11. Culas, R.J. Deforestation and the environmental Kuznets curve: An institutional perspective. *Ecol. Econ.* 2007, *61*, 429–437.
12. Managi, S. Are there increasing returns to pollution abatement? Empirical analytics of the Environmental Kuznets Curve in pesticides. *Ecol. Econ.* 2006, *58*, 617–636.

13. Edenhofer, O.; Pichs-Madruga, R.; Sokona, Y.; Kadner, S.; Minx, J.; Brunner, S.; Agrawala, S.; Baiocchi, G.; Bashmakov, I.A.; Blanco, G.; et al. *Climate Change 2014: Mitigation of Climate Change*; Contribution of Working Group III to the Fifth Assessment Report of the Intergovernmental Panel on Climate Change; Intergovernmental Panel on Climate Change: Geneva, Switzerland, 2014.
14. Eurostat Agriculture—Greenhouse Gas Emission Statistics. Available online: http://ec.europa.eu/eurostat/statistics-explained/index.php/Agriculture_-_greenhouse_gas_emission_statistics (accessed on 29 September 2017).
15. Fort, R.; Henrar, M. *Agriculture, Forestry and Fishery Statistics*, 2016th ed.; Eurostat: Luxembourg, 2016; ISBN 978-92-79-63350-8.
16. Yan, Q.; Yin, J.; Baležentis, T.; Makutėnienė, D.; Štreimikienė, D. Energy-related GHG emission in agriculture of the European countries: An application of the Generalized Divisia Index. *J. Clean. Prod.* 2017, *164*, 686–694.
17. Communication from the Commission to the European Council and the European Parliament: An Energy Policy for Europe 2007. Avaliable online: http://eur-lex.europa.eu/legal-content/EN/TXT/?uri=LEGISSUM:l27067 (accessed on 27 November 2017).
18. Christensen, L.A. *Soil, Nutrient, and Water Management Systems Used in U.S. Corn Production*; Agriculture Information Bulletin No. 774; United States Department of Agriculture: Washington, DC, USA, 2002.
19. Goodland, R.; Daly, H. Environmental Sustainability: Universal and Non-Negotiable. *Ecol. Appl.* 1996, *6*, 1002–1017.
20. Färe, R.; Grosskopf, S.; Tyteca, D. An activity analysis model of the environmental performance of firms—Application to fossil-fuel-fired electric utilities. *Ecol. Econ.* 1996, *18*, 161–175.
21. Suri, V.; Chapman, D. Economic growth, trade and energy: Implications for the environmental Kuznets curve. *Ecol. Econ.* 1998, *25*, 195–208.
22. Tamazian, A.; Chousa, J.P.; Vadlamannati, K.C. Does higher economic and financial development lead to environmental degradation: Evidence from BRIC countries. *Energy Policy* 2009, *37*, 246–253.
23. Bohringer, C.; Jochem, P.E.P. Measuring the immeasurable—A survey of sustainability indices. *Ecol. Econ.*2007, *63*, 1–8.
24. Picazo-Tadeo, A.J.; Beltrán-Esteve, M.; Gómez-Limón, J.A. Assessing

eco-efficiency with directional distance functions. *Eur. J. Oper. Res.* 2012, *220*, 798–809.

25. Oude Lansink, A.; Wall, A. Frontier models for evaluating environmental efficiency: An overview. *Econ. Bus. Lett.* 2014, *3*, 43–50.
26. De Koeijer, T.J.; Wossink, G.A.A.; Struik, P.C.; Renkema, J.A. Measuring agricultural sustainability in terms of efficiency: The case of Dutch sugar beet growers. *J. Environ. Manag.* 2002, *66*, 9–17.
27. Farrell, M.J. The Measurement of Productive Efficiency. *J. R. Stat. Soc. Ser. A* 1957, *120*, 253–290.
28. Lauwers, L. Justifying the incorporation of the materials balance principle into frontier-based eco-efficiency models. *Ecol. Econ.* 2009, *68*, 1605–1614.
29. Vlontzos, G.; Niavis, S. Assessing the Evolution of Technical Efficiency of Agriculture in EU Countries: Is There a Role for the Agenda 2000? In *Agricultural Cooperative Management and Policy: New Robust, Reliable and Coherent Modelling Tools*; Zopounidis, C., Kalogeras, N., Mattas, K., van Dijk, G., Baourakis, G., Eds.; Springer International Publishing: Cham, Switzerland, 2014; pp. 339–351. ISBN 978-3-319-06635-6.
30. Vlontzos, G.; Niavis, S.; Manos, B. A DEA approach for estimating the agricultural energy and environmental efficiency of EU countries. *Renew. Sustain. Energy Rev.* 2014, *40*, 91–96.
31. Armundito, E.; Kaneko, S. Environmental Productivities and Carbon Abatement Costs of Manufacturing Sectors. In *Climate Change Policies and Challenges in Indonesia*; Kaneko, S., Kawanishi, M., Eds.; Springer: Tokyo, Japan, 2016; Volume 9, pp. 199–218.
32. Scheel, H. Undesirable outputs in efficiency valuations. *Eur. J. Oper. Res.* 2001, *132*, 400–410.
33. Rodseth, K.L. Environmental efficiency measurement and the materials balance condition reconsidered. *Eur. J. Oper. Res.* 2016, *250*, 342–346.
34. Knox Lovell, C.A.; Pastor, J.T.; Turner, J.A. Measuring macroeconomic performance in the OECD: A comparison of European and non-European countries. *Eur. J. Oper. Res.* 1995, *87*, 507–518.
35. Vlontzos, G.; Pardalos, P.M. Assess and prognosticate green house gas emissions from agricultural production of EU countries, by implementing, DEA Window analysis and artificial neural networks. *Renew. Sustain. Energy Rev.* 2017, *76*, 155–162.

36. Halkos, G.E.; Tzeremes, N.G.; Kourtzidis, S.A. Regional sustainability efficiency index in Europe: An additive two-stage DEA approach. *Oper. Res.* 2015, *15*, 1–23.

37. Chung, Y.H.; Färe, R.; Grosskopf, S. Productivity and Undesirable Outputs: A Directional Distance Function Approach. *J. Environ. Manag.* 1997, *51*, 229–240.

38. Watanabe, M.; Tanaka, K. Efficiency analysis of Chinese industry: A directional distance function approach. *Energy Policy* 2007, *35*, 6323–6331.

39. Suzuki, S.; Nijkamp, P. An evaluation of energy-environment-economic efficiency for EU, APEC and ASEAN countries: Design of a Target-Oriented DFM model with fixed factors in Data Envelopment Analysis. *Energy Policy* 2016, *88*, 100–112.

40. Tyteca, D. Linear programming models for the measurement of environmental performance of firms—Concepts and empirical results. *J. Product. Anal.* 1997, *8*, 183–197.

41. Zaim, O. Measuring environmental performance of state manufacturing through changes in pollution intensities: A DEA framework. *Ecol. Econ.* 2004, *48*, 37–47.

42. Zhou, P.; Ang, B.W.; Wang, H. Energy and CO_2 emission performance in electricity generation: A non-radial directional distance function approach. *Eur. J. Oper. Res.* 2012, *221*, 625–635.

43. Sueyoshi, T.; Goto, M. Data envelopment analysis for environmental assessment: Comparison between public and private ownership in petroleum industry. *Eur. J. Oper. Res.* 2012, *216*, 668–678.

44. Zhou, P.; Ang, B.W.; Poh, K.L. A survey of data envelopment analysis in energy and environmental studies. *Eur. J. Oper. Res.* 2008, *189*, 1–18.

45. Sueyoshi, T.; Yuan, Y.; Goto, M. A literature study for DEA applied to energy and environment. *Energy Econ.*2017, *62*, 104–124.

46. Coelli, T.J.; Rao, D.S. Total factor productivity growth in agriculture: A Malmquist index analysis of 93 countries, 1980–2000. *Agric. Econ.* 2005, *32*, 115–134.

47. Zhu, X.; Lansink, A.O. Impact of CAP subsidies on technical efficiency of crop farms in Germany, the Netherlands and Sweden. *J. Agric. Econ.* 2010, *61*, 545–564.

48. Martinho, V.J.P.D. Efficiency, total factor productivity and returns to

scale in a sustainable perspective: An analysis in the European Union at farm and regional level. *Land Use Policy* 2017, *68*, 232–245.

49. Toma, P.; Miglietta, P.P.; Zurlini, G.; Valente, D.; Petrosillo, I. A non-parametric bootstrap-data envelopment analysis approach for environmental policy planning and management of agricultural efficiency in EU countries. *Ecol. Indic.* 2017, *83*, 132–143.
50. Lauwers, L.; Van Huylenbroeck, G.; Rogiers, G. Technical, economic and environmental efficiency analysis of pig fattening farms. In Proceedings of the 9th European Congress of Agricultural Economists, Warschau, Polen, 24–28 August 1999.
51. Hoang, V.N.; Rao, D.S.P. Measuring and decomposing sustainable efficiency in agricultural production: A cumulative exergy balance approach. *Ecol. Econ.* 2010, *69*, 1765–1776.
52. Tian, D.; Zhao, F.; Mu, W.; Kanianska, R.; Feng, J. Environmental efficiency of chinese open-field grape production: An evaluation using data envelopment analysis and spatial autocorrelation. *Sustainability* 2016, *8*, 1246.
53. Ghali, M.; Latruffe, L.; Daniel, K. Efficient use of energy resources on French farms: An analysis through technical efficiency. *Energies* 2016, *9*, 601.
54. Halkos, G.E.; Tzeremes, N.G. Exploring the existence of Kuznets curve in countries' environmental efficiency using DEA window analysis. *Ecol. Econ.* 2009, *68*, 2168–2176.
55. Zaim, O.; Taskin, F. A Kuznets curve in environmental efficiency: An application on OECD countries. *Environ. Resour. Econ.* 2000, *17*, 21–36.
56. Oggioni, G.; Riccardi, R.; Toninelli, R. Eco-efficiency of the world cement industry: A data envelopment analysis. *Energy Policy* 2011, *39*, 2842–2854.
57. Galeotti, M.; Lanza, A.; Pauli, F. Reassessing the environmental Kuznets curve for CO_2 emissions: A robustness exercise. *Ecol. Econ.* 2006, *57*, 152–163.
58. Hoechle, D. Robust standard errors for panel regressions with cross-sectional dependence. *Stata J.* 2010, *10*, 288–308.
59. Eurostat. Available online: http://appsso.eurostat.ec.europa.eu/nui/show.do?dataset=apro_acs_a&lang=en(accessed on 1 September 2017).

60. Economic Accounts for Agriculture—Values at Current Prices. Available online: http://appsso.eurostat.ec.europa.eu/nui/show.do?dataset=aact_eaa01&lang=en (accessed on 1 September 2017).
61. Eurostat Database. Available online: http://ec.europa.eu/eurostat/data/database (accessed on 1 September 2017).
62. Haug, R. Some leading issues in agricultural extension: A literature review. *J. Agric. Educ. Ext.* 1999, *5*, 263–274.

CHAPTER 6

AN ASSESSMENT OF DIRECT ON-FARM ENERGY USE FOR HIGH VALUE GRAIN CROPS GROWN UNDER DIFFERENT FARMING PRACTICES IN AUSTRALIA

Tek Maraseni, Guangnan Chen, Thomas Banhazi, Jochen Bundschuh , and Talal Yusaf

Institute for Agriculture and the Environment, University of Southern Queensland, Toowoomba, Queensland 4350, Australia

ABSTRACT

Several studies have quantified the energy consumption associated with crop production in various countries. However, these studies have not compared the energy consumption from a broad range of farming practices currently in practice, such as zero tillage, conventional tillage and irrigated farming systems. This study examines direct on-farm energy use for high value grain crops grown under different farming practices in Australia. Grain farming processes are identified and "typical" farming operation data are collected from several sources, including published and unpublished literature, as

Citation: Maraseni, T.; Chen, G.; Banhazi, T.; Bundschuh, J.; Yusaf, T. An Assessment of Direct on-Farm Energy Use for High Value Grain Crops Grown under Different Farming Practices in Australia. *Energies* 2015, *8*, 13033-13046.

well as expert interviews. The direct on-farm energy uses are assessed for 27 scenarios, including three high value grain crops—wheat, barley and sorghum—for three regions (Northern, Southern and Western Australia) under three farming conditions with both dryland (both for conventional and zero-tillage) and irrigated conditions. It is found that energy requirement for farming operations is directly related to the intensity and frequency of farming operations, which in turn is related to tillage practices, soil types, irrigation systems, local climate, and crop types. Among the three studied regions, Western Australia requires less direct on-farm energy for each crop, mainly due to the easily workable sandy soils and adoption of zero tillage systems. In irrigated crops, irrigation energy remains a major contributor to the total on-farm energy demand, accounting for up to 85% of total energy use.

Keywords: grain industry, on-farm energy, irrigation energy, zero tillage, conventional tillage

INTRODUCTION

Despite several efforts, from 2000 to 2010, global greenhouse gas (GHG) emissions grew more quickly (2.2%/year) during that period than in each of the three previous decades (1.3%/year) and reached 49 Gt CO_2e/year in 2010 [1]. Development of new infrastructure in rapidly developing countries, especially in transition countries such as India and China, accounts for much of the increase [2]. This trajectory shows that the avoidance of two degrees of warming proposed by the United Nations Framework Convention on Climate Change (UNFCCC) at Cancun is problematic to reach. In order to meet the two degrees target, the current de-carbonization rate of 1.2%/year needs to be increased to 6.2%/year, every year from now till 2100 [3]. If we stick with current trajectory we will end up at four degrees increase by the end of 2100 [3]. When climate change policy was being developed in the 1990s, developing countries accounted for only 40% of global emissions and this increased to 54.3% in 2004 and is expected to increase to 66% by 2030 [4]. Therefore, every country, whether developed or developing, needs to share emissions reduction responsibility [5].

As a developed country and a major party of UNFCCC, Australia has a significant responsibility for emissions reduction. Currently, Australia shares 1.18% of global GHG emissions, 1.21% of global GDP and 0.33% of the global population [6]. From 2008 to 2013, Australia reduced its annual average carbon intensity by 4.6% [3], and in fact, this was the world record.

This was achieved mainly by structural changes, new technologies, fuel switching and improvements in energy efficiency [7]. However, as noted, this is not enough yet.

Globally, agriculture is one of the major sources of energy consumption and therefore GHG emissions. With more intensive and modernized farming systems, during the period 1990–2005, global GHG emissions from agriculture increased by 14%, at an annual rate of 49 Mt CO_2e/year [8]. In 2010, GHG emissions from agriculture are estimated to be 5.6 Gt CO_2e/year to 6.4 Gt CO_2e/year [9,10,11], *i.e.*, about 11.4% to 13.1% of global emissions. If the production of agricultural inputs and various downstream activities are considered, the agricultural sector contributes a further 3% to 6% of global emissions [12].

Australia is one of the largest GHG emitting countries from the agriculture sector in the world. Its agricultural sector accounts for 15% of national GHG emissions and is the second largest source of emissions [13]. This proportion is significantly higher than those in Central and Eastern Europe, the former Soviet Union and the USA [14]. In order to reduce GHG emissions from the agricultural sector, the previous Australian Government implemented Carbon Farming Initiatives (CFIs). The current Government replaced it with the Direct Action Plan called "Emission Reduction Fund (ERF)" and allocated A$2.55 billion to establish the Fund in the 2014–2015 budgets [15]. ERF allows farmers and other land managers to earn carbon credits by storing carbon or reducing GHG emissions on land. This would allow them to adopt some best management practices and earn carbon credits and reputational benefits at the same time.

The uptake of ERF is likely to be good. As of 31 August 2015, 390 projects are registered and 16.3 million Australian Carbon Credit Units (ACCUs) have been issued to these projects [16]. Australia has set 5% unconditional emission reduction targets for 2020 from 2000 levels, and with this initiative agriculture is expected to meet this target [16]. So far, the majority of the registered projects under the ERF are from sequestration projects, mainly from forestry activities. However, the grain industry in Australia could have a good opportunity in reducing energy consumption and thereby GHG emissions. However, due to limited research the grain industry is not able to target where they have an opportunity to increase energy efficiency and participate in the ERF. This study aims to examine and compare on-farm energy use for three high value grain crops grown under three major farming practices in three agro-ecological zones in Australia. This will help farmers to understand the range of energy uses and also to pinpoint where their energy use is highest and the potential savings achievable.

Several studies have quantified the energy consumption associated with crop production in various countries including: (1) sugar beet production in the UK [17]; (2) arable and outdoor vegetable production in New Zealand [18]; (3) field crops (wheat, cotton, maize, sesame) and vegetables (tomato, melon, watermelon) [19], tomato [20], sugar beet [21], stake-tomato [22] and grape production [23] in Turkey; (4) wheat production in India [24]; and cotton and lucerne production in Australia [25,26,27]. However, these studies have not compared the energy consumption from a broad range of farming practices currently in practice such as zero tillage, conventional tillage and irrigated farming systems. This paper is the first in this direction.

In this study, direct energy used for various on-farm operations such as tillage, fertiliser application, boom spraying, planting, aerial spraying and water pumping is considered. A large amount of indirect energy is also required for: (1) the production, packaging, storing and transportation of various farm inputs such as fertilisers, chemicals (herbicides, insecticides, fungicides and plant regulator), fuels and farm machineries; and (2) post-harvesting operations such as transportation and drying of harvested products. However, energy used for these operations are not considered in this study. Similarly, human energy is also required for on-farm operation but it is insignificant as the production system in Australia is highly mechanised. Therefore, human energy is also not considered in this study.

A BRIEF SNAPSHOT OF THE GRAIN INDUSTRY IN AUSTRALIA AND RATIONALE FOR SELECTING THREE FARMING PRACTICES

The grain industry in Australia operates in all states and makes a major contribution to the state and national economies. Australia produced over 48.5 million tonnes of grain in 2011–2012. However, in 2012–2013, grain production in Australia fell to 38.7 million tonnes [27]. Reduced grain production was a result of increased climatic variability, and decreased production area largely due to rising input prices—including those for energy—and volatility of grain prices [28].

The grain industry in Australia is broadly divided into three agro-ecological regions: (1) the Southern Region that covers south-eastern Australia, including central and southern New South Wales, Victoria,

Tasmania, and south-eastern South Australia. Soils in this area are generally poor (low fertility) with many subsoil constraints, such as salinity, sodicity and toxic levels of some elements; (2) the Northern Region which covers Queensland and northern New South Wales. This region has relatively good soils, but relatively high seasonal rainfall variability and therefore production; and (3) the Western Region which covers Western Australia. This region has poor soils and crop yields largely depend on the winter and spring rainfall [28].

Due to large geographical dispersion, there is a wide range of climatic conditions within each region. In general, rainfall decreases from north to south and from coastal to inland areas. Northern Region has a tropical and subtropical climate, whereas the Southern Region has a temperate climate and the Western Region has a Mediterranean climate.

Depending on the availability and price of water for irrigation, grains in these regions are produced either on irrigated or dryland "rain-fed" conditions. Overall, grain growers are increasingly moving from conventional tillage systems to reduced or zero tillage systems because: (1) continuing cultivation systems result in a loss of soil carbon, and about 75% of Australian agricultural lands have less than 1% soil organic carbon [29]. For example, over a 60 year period, cereal cropping soils of Northern New South Wales and Southern Queensland have lost over 40 t C/ha (146.7 t CO_2e) and 4 t N/ha [30]; (2) the loss of soil carbon adversely affects soil fertility, the soil water holding capacity and plant-available water capacity [31]; and (3) continuous cultivation systems leave soils vulnerable to water and wind erosion, increasing agricultural runoff, degrading soil productivity and releasing GHG by disturbing soils and burning fossil fuels for farm machinery [32,33].

On the other hand, the zero tillage practice reverses these processes by minimizing mechanical soil disturbance, providing permanent soil cover by organic materials and diversifying crop species grown in sequence and/or association [33]. Zero tillage also has adaptation benefits. Soils under zero tillage can hold more moisture and therefore under drought conditions the crops are more resilient and produce more [34]. Therefore, state agriculture departments, regional natural resources management organisations, and local landholder groups generally recommend that farmers move from traditional dryland farming systems to reduced tillage systems and where appropriate towards a zero tillage system [35]. This system is also an eligible activity through the Australian Government ERF, and participating landholders may claim a refundable tax offset of 15% of the purchase price of an eligible no-till seeder. It is likely that a higher proportion of grain production in dryland area will be grown under zero till practices in the future [36]. Therefore, this study aims to examine and compare direct on-farm energy use for all three agro-ecological regions, under both irrigated and dry-land conditions and under dry-land both for conventional and zero-tillage systems (Figure 1).

Now a question may arise why irrigated farming systems are not divided into two categories as in the dry-land farming system. As noted, zero tillage has been largely practiced with rain-fed (dryland) crops as farmers are increasingly realising its value for soil moisture conservation [37]. A global meta-analysis shows that zero tillage performs best under rain-fed conditions in dry climates, either matching or exceeding conventional tillage yields for crops, whereas crop yields declined in irrigated conditions [38]. However, in recent years there are some practices of zero tillage in irrigated crops but they are not common and also the data are not available [36,39]. Therefore, in this study, irrigated condition is not divided into different categories.

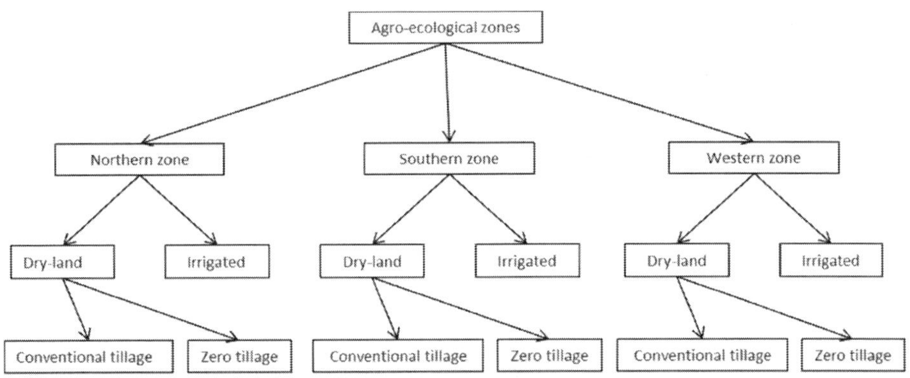

Figure 1: Showing different farming practices in the three agro-ecological regions in Australia.

METHODOLOGY

Selection of Three High Value Grain Crops

As noted, this study focuses on three high value grain crops. In order to determine the high value grain crops, grains yield data for various years, and the prices of these grain crops in seven different time periods (April–June 2012; July–September 2012; October–December 2012; January–March

2013; April–June 2013; July–September 2013; and October–December 2013) were taken from the Australian Bureau of Agricultural and Resource Economics [40]. From these data the five-year average yields and domestic average prices of seven periods were calculated and total values of the grains were estimated (Table 1).

Table 1: Total value of different grain and oilseed crops

Crops	**Five-year Average to 2012–2013**			**Average Domestic Price of 7 Periods * (A$/Tonne)**	**Total Value (Million A$)**
	Area Planted (in 1000 Ha)	**Yield (Tonne/Ha)**	**Total Production in Kilo-Tonne**		
Winter crops					
Wheat	13,518	1.82	24,606	292	**7185** **
Barley	4092	1.96	7909	267	**2112** **
Canola	2226	1.20	2709	548	1485
Chickpeas	490	1.22	586	521	305
Faba beans	154	1.68	264	NA	NA
Field peas	286	1.16	330	366	121
Lentils	155	1.29	212	NA	NA
Lupins	633	1.19	756	NA	NA
Oats	795	1.48	1165	229	267
Summer crops					
Grain sorghum	630	3.27	2076	279	**579** **
Corn (maize)	67	5.94	401	339	136
Sunflower	37	1.28	46	523	24

Note: NA is not available. Production figures could be slightly different to the "product of yield and area planted" figures due to rounding up. * Seven period are: April–June 2012; July–September 2012; October–December 2012; January–March 2013; April–June 2013; July–September 2013; and October–December 2013. ** Selected high value grain crops.

The top three high value grain crops were wheat, barley and canola. However, canola is used for oil and therefore the fourth highest value crop "grain sorghum" is considered instead. Two of the selected grain crops are winter crops (wheat and barley) and the third one is a summer crop (grain sorghum).

Data Collection

Compared to other parts of the world (such as USA, Iran, UK, New Zealand) on-farm energy use data are scarce in Australia. In contrast, carbon footprint data are available for wheat in WA, Victoria and NSW. However, as we are concerned only on on-farm energy use data, these data are of little use.

In this research, a bottom-up approach is taken to estimate the energy uses. Grain farming processes were first identified through literature review and discussed with the relevant experts (see the Acknowledgments). "Typical" farming operation data were then collected from published and unpublished literature for all three grain zones (Northern, Southern and Western) and for three different tillage types (irrigated, dryland-zero tillage and dryland-conventional tillage). Where such data were not available, relevant government organisations and experts were approached and consulted to fill the data gaps. A summary of these field operation data are provided in Table 2, Table 3 and Table 4, which represent the typical farming systems and practices in these areas. These data were used to calculate the on-farm fuel and energy use later. The footnotes in these three tables provide the details about the experts who provided data for the analysis. As these people are renowned experts in the given areas, it was believed that the acquired data are of high quality. However, they should be applied to the given farming conditions and cannot be generalised.

In the Northern Grain region, wheat under conventional tillage system has the highest amount of secondary tillage, followed by barley under the same tillage system. As expected, all crops under zero tillage have lower farming activities than their counterparts in conventional and irrigated systems. However, where irrigation related activities are included, irrigated crops have a higher energy demand than others. In both Northern and Southern Grain Regions (Table 2 and Table 3), it is noted that: (1) surface/furrow irrigation is the dominant irrigation system. However, barley in the Northern Grain Region often uses pressured irrigation; and (2) sorghum needs greater irrigation than the two other crops (wheat and barley). In general, wheat and barley need 2.5 ML of water per hectare whereas sorghum needs 2.93 ML to 7.5 ML per hectare, depending on the soil type and climatic conditions (Table 2 and Table 3). In the case of Western Australia (Western Grain Region), about 90% of grain is produced using zero-till systems. There are neither irrigated grain crops nor dryland sorghum crops. Thus, irrigation is not assumed to take place in Western Australia (Table 4).

Table 2: On-farm operation data for three high value grain crops in Northern Grain Region (example from Darling Downs region)

Farming Operation	Diesel (L Per Ha Per Operation)	Number of Times for Dry Land (Conventional Tillage)			Number of Times for Dry Land (Zero Tillage)			Number of Times for Irrigated Land		
		Wheat	Barley	Sorghum	Wheat	Barley	Sorghum	Wheat	Barley	Sorghum
Primary tillage	18	1	1	0	0	0	0	0	1	1
Secondary tillage	8	3	2	1	0	0	0	0	0	0
Fertiliser application	5	1	0	0	1		1	1	1	0
Boom spraying	2.25	1	2	1	6	6	3	5	2	0
Planting	5	1	1	1	1	1	1	1	1	1
Aerial spray	3	1	0	0	1	1	0	1	0	2
Harvesting (diesel L/ha)		6.6	8.1	8.0	6.6	8.1	8.0	12.9	12.9	11.5
Type of irrigation and amount of water used								2.5 ML (furrow)	2.5 ML (side roll)	2.93 ML (furrow)

Note: Sorghum data are taken from Mr Gary Sandell, Research Engineer, National Centre for Engineering in Agriculture, University of Southern Queensland; all other data are taken from Maraseni and Cockfield [36,39].

Table 3: On-farm operation data for high value grain crops in Southern Grain Region (example from Victoria)

Farming Operation	Diesel (L Per Ha Per Operation)	Number of Times for Dry Land (Conventional Tillage)			Number of Times for Dry Land (Zero Tillage)			Number of Times for Irrigated Land		
		Wheat	Barley	Sorghum	Wheat	Barley	Sorghum	Wheat	Barley	Sorghum
Primary tillage	18	1	1	0	0	0	0	0	0	0
Secondary tillage	8	0–3 (1.5)	0–3 (1.5)	0	0	0	0	0	0	0
Fertiliser application	5	1	1	0	0	0	0	0	0	0
Boom spraying	2.25	1	1	1	0–5 (2.5)	0–5 (2.5)	2	0–5 (2.5)	0–5 (2.5)	0
Planting	5	1	1	1	1	1	1	1	1	1
Aerial spray	3	0	0	0	0	0	2	0	0	0
Harvesting (diesel L/ha) *		6.6	8.1	8.0	6.6	8.1	8.0	12.9	12.9	11.5
Type of irrigation and amount of water used **								2.5 ML(surface)	2.5 ML (surface)	7.5 ML (surface)

Note: Source: * Amount of diesel used for each operation was taken from Maraseni and Cockfield [36,39]. Number of times of different farming operation for dry land (zero till) sorghum was taken from Mr Gary Sandell (Research Engineer, National Centre for Engineering in Agriculture, University of Southern Queensland). This value is for a typical farm in Tamworth. The rest of the data was taken from Darryl Pearl (Project Officer, Productive Soils (Mallee/Wimmera), Rural Recovery Coordinator Mallee, FSV Grains Agriculture Productivity, Department of Environment and Primary Industries, Victoria). The numbers in the brackets are average of given values and they are used for further analysis. ** The amounts of irrigation water for irrigated crops were taken from DPI Victoria [41]. Most of the irrigation in Victoria is surface irrigation (pers comm with Dr S Mushtaq, Deputy Director, International Centre for Applied Climate Sciences, University of Southern Queensland).

Table 4: On-farm operation data for high value grain crops in Western Grain Region (Western Australia)

Farming Operation	Diesel (L Per Ha Per Operation)	Number of Times for Dry Land (Conventional Tillage)			Number of Times for Dry Land (Zero Tillage)			Number of Times for Irrigated Land		
		Wheat	Barley	Sorghum	Wheat	Barley	Sorghum	Wheat	Barley	Sorghum
Primary tillage	5	1	1	NA	0	0	NA	NA	NA	NA
Secondary tillage	0	0	0	NA	0	0	NA	NA	NA	NA
Fertiliser application	0	0	0	NA	0	0	NA	NA	NA	NA
Boom spraying	0.5	3	3	NA	3	3	NA	NA	NA	NA
Planting	4	1	1	NA	1	1	NA	NA	NA	NA
Aerial spray	NA	NA	NA	NA	NA	NA	NA	NA	NA	NA
Harvesting (diesel L/ha)		5	5	NA	5	5	NA	NA	NA	NA
Type of irrigation and amount of water used								NA	NA	NA

Note: Source: Glen Riethmuller, Development Officer, Department of Agriculture and Food, Western Australia, Dryland Research Institute, WA.

Energy Conversion Factors

In all farming operations diesel is used, for which energy conversion factor of 38.6 GJ/kL is used [42]. Similarly, diesel water pumps are common in both Northern and Southern Grain Regions. About 60 MJ of energy is used for per ML water pumped per meter total dynamic head. This means about 0.5 GJ of energy will be used for each megalitre of water under surface irrigation. This value for side roll irrigation is 0.775 GJ/ML [43]. Therefore, energy conversion factors of 0.5 GJ/ML and 0.775 GJ/ML are used for surface irrigation and side roll irrigation systems, respectively.

RESULTS AND DISCUSSION

Fuel Consumption

Table 5, Table 6 and Table 7 present the fuel (diesel) consumption data for three grain crops in Northern, Southern and Western Grain Regions, respectively.

Table 5: Fuel (diesel, L/ha) consumption (except irrigation fuel) for three high value grain crops in Northern Grain Region (From Darling Downs region)

Farming Operation	Fuel Consumption (L/ha) for Dry Land (Conventional Tillage)			Fuel Consumption (L/ha) for Dry Land (Zero Tillage)			Fuel Consumption (L/Ha) for Irrigated Land		
	Wheat	Barley	Sorghum	Wheat	Barley	Sorghum	Wheat	Barley	Sorghum
Primary tillage	18	18	0	0	0	0	0	18	18
Secondary tillage	24	16	8	0	0	0	0	0	0
Fertiliser application	5	0	0	5	0	5	5	5	0
Boom spraying	2.25	4.5	2.25	13.5	13.5	6.75	11.25	4.5	0
Planting	5	5	5	5	5	5	5	5	5
Aerial spray	3	0	0	3	3	0	3	0	6
Harvest	6.6	8.1	8	6.6	8.1	8	12.9	12.9	11.5
Total diesel (L/ha)	63.85	51.6	23.25	33.1	29.6	24.75	37.15	45.4	40.5

Table 6: Fuel (diesel, L/ha) consumption (except irrigation fuel) for three high value grain crops in Southern Grain Region (From Victoria)

Farming Operation	Fuel Consumption (L/ha) for Dry Land (Conventional Tillage)			Fuel Consumption (L/ha) for Dry Land (Zero Tillage)			Fuel Consumption (L/Ha) for Irrigated Land		
	Wheat	Barley	Sorghum	Wheat	Barley	Sorghum	Wheat	Barley	Sorghum
Primary tillage	18.00	18.00	0.00	0.00	0.00	0.00	0.00	0.00	0.00
Secondary tillage	12.00	12.00	0.00	0.00	0.00	0.00	0.00	0.00	0.00
Fertiliser application	5.00	5.00	0.00	0.00	0.00	0.00	0.00	0.00	0.00
Boom spraying	2.25	2.25	2.25	5.63	5.63	4.50	5.63	5.63	0.00
Planting	5.00	5.00	5.00	5.00	5.00	5.00	5.00	5.00	5.00
Aerial spray	0.00	0.00	0.00	0.00	0.00	6.00	0.00	0.00	0.00
Harvest	6.60	8.10	8.00	6.60	8.10	8.00	12.90	12.90	11.50
Total diesel (L/ha)	48.85	50.35	15.25	17.23	18.73	23.50	23.53	23.53	16.50

Table 7: Fuel (diesel, L/ha) consumption (except irrigation fuel) for three high value grain crops in Western Grain Region

Farming Operation	Fuel Consumption (L/ha) for Dry Land (Conventional Tillage)			Fuel Consumption (L/ha) for Dry Land (Zero Tillage)			Fuel Consumption (L/Ha) for Irrigated Land		
	Wheat	Barley	Sorghum	Wheat	Barley	Sorghum	Wheat	Barley	Sorghum
Primary tillage	5.00	5.00	NA	0.00	0.00	NA	NA	NA	NA
Secondary tillage	0.00	0.00	NA	0.00	0.00	NA	NA	NA	NA
Fertiliser application	0.00	0.00	NA	0.00	0.00	NA	NA	NA	NA
Boom spraying	1.50	1.50	NA	1.50	1.50	NA	NA	NA	NA
Planting	4.00	4.00	NA	4.00	4.00	NA	NA	NA	NA
Aerial spray	0.00	0.00	NA	0.00	0.00	NA	NA	NA	NA
Harvest	5.00	5.00	NA	5.00	5.00	NA	NA	NA	NA
Total diesel (L/ha)	15.50	15.50	NA	10.50	10.50	NA	NA	NA	NA

The highest amounts of diesel for farming operations (other than the diesel for irrigation) were required by the wheat and barley farming systems (both under conventional tillage system in the Northern Grain Region), followed by the same crops under a similar tillage system in the Southern Grain Region. Western Region grains required the least amount of diesel. This is mainly due to prevalence of sandy soils and thereby the lower number of farming activities, as sandy soils have large mineral or solid particles and therefore they have more extensive air between the particles. As a result, they are loose and easier to cultivate. For example, on average a primary tillage in Northern and Southern Australian soils requires 18 L of diesel whereas 5 L diesel is enough for the same tillage in the sandy soil in Western Australia.

By getting the average of all three regions, it can be found (Table 5) that wheat (42.7 L/ha) and barley (39.2 L/ha) under the conventional tillage system required the highest amounts of diesel, followed by barley (34.5 L/ha), wheat (30.3 L/ha) and sorghum (28.5 L/ha) under the irrigated farming system. As expected, crops under the zero tillage required the least amount of on-farm diesel usage among the other crops. This finding is in agreement with some other studies [44,45].

Energy Consumption

Figure 2 and Figure 3 and Table 8 present energy consumption data for three grain crops in Northern, Southern and Western Grain Regions. It can be seen that energy consumption from diesel for farming operations (except for irrigation-related diesel) follows similar patterns as diesel consumption. In both Northern and Southern Grain Regions, crops under the zero tillage system required less on-farm energy inputs than the conventional and irrigated systems. In all cultivation types, crops grown in Western Australian region require the smallest amount of energy when compared to their counterparts in other regions, as zero tillage systems are practiced. This finding is in agreement with the results of Baillie [46] and Maraseni and Cockfield [36]. Baillie [46] compared energy use from three scenarios on Keytah (irrigated cotton and grains farming operation west of Moree in Northern NSW) and found that reduced and zero tillage operations could result in 12% and 24% energy savings, respectively. Similarly, Maraseni and Cockfield [36] conducted research in the Darling Downs District and reported that the fossil fuel-related emissions from wheat, durum, barley and chickpea cultivation under a zero tillage system is much lower than that of other cultivation systems.

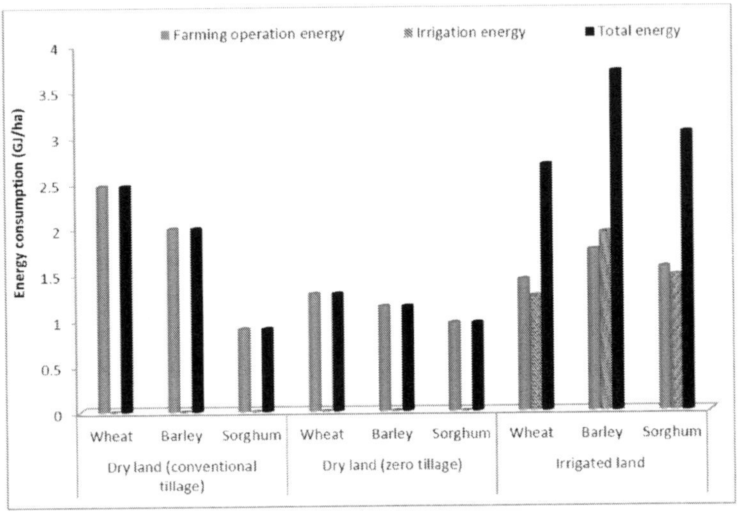

Figure 2: Energy consumption (GJ/ha) for three high value grain crops in Northern Grain Region.

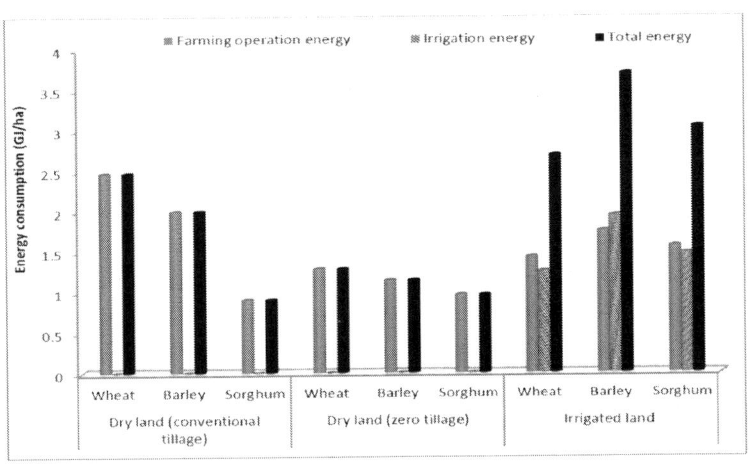

Figure 3: Energy consumption (GJ/ha) for three high value grain crops in Southern Grain Region.

Table 8: Energy consumption (GJ/ha) for three high value grain crops in Western Grain Region

Farming Operation	Fuel Consumption (L/ha) for Dry Land (Conventional Tillage)			Fuel Consumption (L/ha) for Dry Land (Zero Tillage)			Fuel Consumption (L/Ha) for Irrigated Land		
	Wheat	Barley	Sorghum	Wheat	Barley	Sorghum	Wheat	Barley	Sorghum
Farming operation energy	0.60	0.60	NA	0.41	0.41	NA	NA	NA	NA

It can be seen from Figure 2 and Figure 3 that when irrigation is practiced, it required the highest amounts of diesel fuel energy for all three crops. Australia wide, the highest amount of energy was required for sorghum crops (4.4 GJ/ha) grown under irrigated systems in the Southern Grain Region, followed by barley (3.7 GJ/ha) and sorghum (3.0 GJ/ha) grown under irrigated systems in the Northern Grain Region. Irrigation-related energy also accounted for a higher proportion of total on-farm direct energy use for all grains in Southern Grain Region than that of their counterparts in the Northern Grain Region.

In both the Southern and Northern Regions, both wheat and barley required the same amount of irrigation water (2.5 ML/ha), but sorghum required a higher amount of irrigation water, especially in the Southern Grain Region (7.5 ML/ha). The most common irrigation system in both regions was surface irrigation. However, pressured irrigation was the most common irrigation system for barley in the Northern Grain Region. Because a pressured irrigation system requires a higher amount of energy than surface irrigation systems, barley in the Northern Grain Region thus consumed higher amounts of energy than wheat and barley in the Southern Grain Region. Therefore, sorghum in Southern Grain Region required the highest amount of irrigation related energy (3.8 GJ/ha), followed by barley in the Northern Grain Region (1.9 GJ/ha).

How Our Results Compare with Other Studies

Recent international literature on energy use by the arable cropping industry is relatively limited. Pellizzi *et al.*[47] found that in Europe, for wheat-like cereals, about 2.5–4.3 GJ/ha of direct energy is used. Similarly, Safa *et al.*[48] found that about 6.5 GJ/ha and 3.2 GJ/ha of direct energy (fuel and electricity) is used for irrigated and dryland wheat crops in New Zealand, respectively. Pellizzi *et al.*'s estimate is similar to our results from dryland and irrigated farming systems in the Northern Grain Region. However, New Zealand estimates are higher than our results, especially in case of the irrigated wheat cropping system. This is mainly due to use of energy-intensive irrigation systems (*i.e.*, gun, centre pivot and rotary rainers) in New Zealand, whereas in our case it was gravity-fed surface and furrow irrigation systems.

Back in Australia, direct energy use for the production of wheat and barley was investigated by Khan *et al.* [49], based on the farm survey data in Coleambally Irrigation Areas (CIA) and Murrumbidgee Irrigation Area

(MIA) of New South Wales. Their estimates (5.8 GJ/ha for wheat and 5.7 GJ/ha for barley) are slightly higher than our estimates, mainly due to use of energy-intensive pressurised irrigation systems and water pumping from greater depths. Similarly, Sandell *et al.* [50] investigated the energy saving opportunities for various farming enterprises in Western Australia. The average on-farm energy use was 0.83 GJ/ha. Biswas *et al.* [51] found this can be as low as 0.35 GJ/ha in Southwestern Australia. Our estimates are in agreement with Biswas *et al.* [51].

CONCLUSIONS AND RECOMMENDATIONS

This study has assessed on-farm energy usages for three high value grain crops grown under three major farming practices in three agro-ecological zones of Australia. As expected, it has been found that fuel input in grain production in dryland systems is less than in irrigated systems. The highest amount of energy is required for sorghum crops (4.4 GJ/ha) grown under irrigated systems in the Southern Grain Region, followed by barley (3.7 GJ/ha) and sorghum (3.0 GJ/ha) grown under irrigation in the Northern Grain Region. Under dryland conditions, crops under zero tillage require less energy per hectare for each crop than conventional tillage. Among the three regions, Western Australia requires less energy for each crop, mainly due to its easily workable sandy soils. The lowest energy requirements (0.4 GJ/ha) are for wheat and barley grown in dryland by zero tillage methods in Western Australia. This data provides significant insights for energy use by different crops in different farming systems.

In irrigated crops, irrigation energy has been identified as a major contributor (47%–86%) of total energy use. It is noted that farmers are now increasingly utilising pressurised irrigation systems powered by electricity. Therefore, identifying strategies that are both water and energy efficient would be a matter of priority for further research.

Energy consumption for on-farm cropping activities depends on several factors such as tillage practices, irrigation type, water source, depth of ground water and soil type. Australia being a diverse continent, these factors not only vary between the different grain regions, but also vary within a grain region. However, in this study, only the most common ("average farm") attributes are considered for each region, which may not necessarily reflect specific areas within the region. Therefore, more research across a larger number of sites is recommended in order to determine if the results found in this project are sufficiently indicative of the given regions.

Furthermore, in order to improve the sustainability of food production, a complete life cycle analysis may be needed. This is because energy is not only consumed for the direct on-farm operations such as cultivation, fertilising, irrigating and harvesting activities, but also indirectly for production, storage and transportation of several other farm inputs such as machinery, fertiliser, herbicides, insecticides, fungicides, and plant growth regulators, *etc.*[52,53]. The first part, also called direct energy, is covered by this study but the second part, called indirect energy, is not covered.

ACKNOWLEDGMENTS

The authors would like to acknowledge the support of the RIRDC for funding this project. The authors would also like to thank the contributions and useful discussions of many people, among them particularly include Craig Baillie, Gary Sandell, Shahbaz Mushtaq, Darryl Pearl, Geoff Cockfield and Glen Riethmuller. Their detailed comments and feedback have significantly improved the quality and relevance of this work, and this is greatly appreciated by the whole project team.

AUTHOR CONTRIBUTIONS

Tek Maraseni, Guangnan Chen, Thomas Banhazi and Jochen Bundschuh designed research; Tek Maraseni collected and analysed the data; all authors wrote the manuscript and approved the final version.

REFERENCES

1. Intergovernmental Panel on Climate Change (IPCC). *Technical Summary of Climate Change 2014: Mitigation of Climate Change, Working Group III Contribution to the IPCC Fifth Assessment Report (AR5)*; IPCC: New York, NY, USA, 2014.
2. Liu, L.C.; Wu, G.; Wang, J.N.; Wei, Y.M. China's carbon emissions from urban and rural households during 1992–2007. *J. Clean Prod.* 2011, *19*, 1754–1762.
3. PricewaterhouseCoopers International Limited (PWC). *Two Degrees of Separation: Ambition and Reality, Low Carbon Economy Index 2014*; PWC: London, UK, 2014.
4. Department of the Prime Minister and Cabinet. *Australia's Climate Change Policy Our Economy, Our Environment, Our Future*; One National Circuit: Barton, Australia, 2007; p. 46.
5. Maraseni, T.N.; Maroulis, J.; Cockfield, G. An analysis of Australia's carbon pollution reduction scheme. *Int. J. Environ. Stud.* 2009, *66*, 591–603.
6. Climate Change Performance Index. Climate Change Performance Index Component. 2014. Available online: https://germanwatch.org/en/download/8601.pdf (accessed on 11 August 2014).
7. Climate Change Authority. *Final Reports on Australia's Future Emissions Reduction Targets*; Climate Change Authority: Canberra, Australia, 2015; p. 10.
8. United States Environmental Protection Agency (US-EPA). *Global Anthropogenic Non-CO_2 Greenhouse Gas Emissions: 1990–2020*; US-EPA: Washington, DC, USA, 2006.
9. Ceschia, E.; Béziat, P.; Dejoux, J.F.; Aubinet, M.; Bernhofer, C.; Bodson, B.; Buchmann, N.; Carrara, A.; Cellier, P.; Tommasi, D.I.; *et al.* Management effects on net ecosystem carbon and GHG budgets at European crop sites. *Agric. Ecosyst. Environ.* 2010, *139*, 363–383.
10. FAOSTAT. FAOSTAT Database. Food and Agriculture Organization of the United Nations. 2013. Available online: http://faostat.fao.org/ (accessed on 12 August 2015).
11. Tubiello, F.; Salvatore, M.; Rossi, S.; Ferrara, A. Analysis of global emissions, carbon intensity and efficiency of food production. *Energia Ambient. Innov. Anno* 2013, *4–5*, 87–93.
12. Vermuelen, S.J.; Campbell, B.M.; Ingram, J.S.I. Climate change and

food systems. *Ann. Rev. Environ. Resour.* 2012, *37*, 195–222.

13. Department of Climate Change and Energy Efficiency (DCCEE). *Australia's Emissions Projections*; DCCEE: Canberra, Australia, 2012.
14. Smith, P.; Martino, D.; Cai, Z.; Gwary, D.; Janzen, H.H.; Kumar, P.; McCarl, B.; Ogle, S.; O'Mara, F.; Rice, C.; *et al.* Greenhouse gas mitigation in agriculture. *Phil. Trans. R. Soc. B* 2008, *363*, 789–813.
15. Australian Government. *Emissions Reduction Fund Overview*; Commonwealth of Australia: Canberra, Australia, 2015; p. 2.
16. Department of Environment. *Carbon Farming Initiatives News*; Government of Australia: Canberra, Australia, 2015.
17. Tzilivakis, J.; Warner, D.J.; May, M.; Lewis, K.A.; Jaggard, K. An assessment of the energy inputs and greenhouse gas emissions in sugar beet (Beta vulgaris) production in the UK. *Agric. Syst.* 2005, *85*, 101–119.
18. Barber, A. *Seven Case Study Farms: Total Energy and Carbon Indicators for New Zealand Arable and Outdoor Vegetable Production*; AgriLINK: Kumeu, New Zealand, 2004.
19. Canakci, M.; Topakci, M.; Akinci, I.; Ozmerzi, A. Energy use pattern of some field crops and vegetable production: Case study for Antalya Region, Turkey. *Energy Convers. Manag.* 2005, *46*, 655–666.
20. Hatirli, S.A.; Ozkan, B.; Fert, C. Energy inputs and crop yield relationship in greenhouse tomato production. *Renew. Energy* 2006, *31*, 427–438.
21. Erdal, G.; Esungun, K.; Erdal, H.; Gunduz, O. Energy use and economical analysis of sugar beet production in Tokat province of Turkey. *Energy* 2007, *32*, 35–41.
22. Esungun, K.; Erdal, G.; Gunduz, O.; Erdal, H. An economic analysis and energy use in stake-tomato production in Tokat province of Turkey. *Renew. Energy* 2007, *32*, 1873–1881.
23. Ozkan, B.; Fert, C.; Karadeniz, C.F. Energy and cost analysis for greenhouse and open-field grape production. *Energy* 2007, *32*, 1500–1504.
24. Singh, H.; Singh, A.K.; Kushwaha, H.L.; Singh, A. Energy consumption pattern of wheat production in India. *Energy* 2007, *32*, 1848–1854.
25. Maraseni, T.N.; Mushtaq, S.; Reardon-Smith, K. Climate change, water security and the need for integrated policy development: The

case of on-farm infrastructure investment in the Australian irrigation sector. *Environ. Res. Lett.* 2012, *7*, 1–12.

26. Mushtaq, S.; Maraseni, T.N.; Reardon-Smith, K. Climate change and water security: Estimating the greenhouse gas costs of achieving water security through investments in modern irrigation technology. *Agric. Syst.* 2013, *117*, 78–89.
27. Australian Bureau of Agricultural and Resource Economics and Sciences (ABARE). *Australian Crop Reports and Agricultural Commodities (Formerly Australian Commodities) Reports*; Australian Government: Canberra, Ausrtalia, 2013.
28. Grain Research and Development Corporation (GRDC). *Annual Report 2012–2013*; Grain Research and Development Corporation: Canberra, Australia, 2013.
29. Rssell, K. Zero waste Australia. A Letter Submitted to the Garnaut Climate Change Review, Melbourne. 2008. Available online: http://www.zerowasteaustralia.org (accessed on 17 March 2014).
30. Dalal, R.; Wang, W. EEnergy Use and Carbon Emissions in NNSW/SQLD Farming Systems—Are We Ahead or Behind? 2010. Available online: http://www.grdc.com.au/Research-and-Development/GRDC (accessed on 18 March 2014).
31. Pattanayak, S.K.; Mccarl, B.A.; Sommer, A.J.; Murray, B.C.; Bondelid, T.; Gillig, D.; Deangelo, B. Water quality co-effects of greenhouse gas mitigation in US agriculture. *Clim. Chang.* 2005, *71*, 341–372.
32. Llewellyn, R.S.; D'emden, F.H. *Adoption of No-Till Cropping Practices in Australian Grain Growing Regions*; Grain Research and Development Corporation: Canberra, Australia, 2010.
33. FAO. Conservation Agriculture. Agriculture and Consumer Protection Department. 2013. Available online: http://www.fao.org/ag/ca/ (accessed on 18 March 2014).
34. FAO. Advancement and Impact of Conservation Agriculture/No-Till Technology Adoption in Kazakhstan. 2012. Available online: http://www.fao.org/ag/ca/doc/Importance_Zero_Tillage_Northern_Kazakhstan.pdf(accessed on 18 March 2014).
35. Australian Greenhouse Office (AGO). *Reducing GHG Emissions from Australian Agriculture: The Role of Benchmarking in Driving Best Management Practice*; Australian Greenhouse Office: Canberra, Australia, 2006.

36. Maraseni, T.N.; Cockfield, G. Does the adoption of zero tillage reduce greenhouse gas emissions? An assessment for the grains industry in Australia. *Agric. Syst.* 2011, *104*, 451–458.
37. Friedrich, T.; Derpsch, R.; Kassam, A. Overview of the Global Spread of Conservation Agriculture: Field Actions Science Reports, Special Issue 6; 2012. Available online: http://factsreports.revues.org/1941(accessed on 4 November 2015).
38. Pittelkow, C.M.; Linquist, B.A.; Lundy, M.E.; Liang, X.; van Groenigen, K.J.; Lee, J.; van Gestel, N.; Six, J.; Venterea, R.T.; van Kessel, C. When does no-till yield more? A global meta-analysis. *Field Crop. Res.* 2015, *183*, 156–168.
39. Maraseni, T.N.; Cockfield, G. Including the costs of water and greenhouse gas emissions in a reassessment of the profitability of irrigation. *Agric. Water Manag.* 2012, *103*, 25–32.
40. Australian Bureau of Agricultural and Resource Economics (ABARE). *Australian Crop Report*; No.169; Department of Agriculture, Australian Government: Canberra, Australia, 2014.
41. Department of Primary Industries (DPI). *Victoria. Northern Victoria Irrigated Cropping Gross Margins 2009–2010*; Department of Primary Industries: Victoria, Australia, 2010; p. 42.
42. Department of Climate Change (DCC). *National Greenhouse Gas Accounts (NGA) Factors*; Department of Climate Change: Canberra, Australia, 2009.
43. Jacob, S. *Comparison of Life Cycle Energy Consumption of Alternative Irrigation Systems*; University of Southern Queensland: Queensland, Australia, 2006.
44. Tullberg, J.N. Traffic effects on tillage energy. *J. Agric. Eng. Res.* 2000, *75*, 375–382.
45. McLaughlin, N.B.; Drury, C.F.; Reynolds, W.D.; Yang, X.M.; Li, Y.X.; Welacky, T.W.; Stewart, G. Energy inputs for conservation and conventional primary tillage implements in a clay loam soil. *Trans. ASABE* 2008, *51*, 1153–1163.
46. Baillie, C. *Energy and Carbon Accounting Case Study on Keytah, a Project Report for the Cotton Research and Development Corporation (CRDC)*; National Centre for Engineering in Agriculture University of Southern Queensland: Toowoomba, Australia, 2009.
47. Pellizzi, G.; Cavalchini, A.G.; Lazzari, M. *Energy Savings in*

Agricultural Machinery and Mechanization; Elsevier Science Publishing Co.: New York, NY, USA, 1998.

48. Safa, M.; Samarasinghe, S.; Mohssen, A. Field study of energy consumption in wheat production in Canterbury, New Zealand. *Energy Convers. Manag.* 2011, *52*, 2526–2532.
49. Khan, S.; Khan, M.A.; Latif, N. Energy requirements and economic analysis of wheat, rice and barley production in Australia. *Soil Environ.* 2010, *29*, 61–68.
50. Sandell, G.R.; Szabo, P.M.; Baillie, C.P.; Woodhouse, N.P.; Schmidt, E. *North-East Farming Futures R & D Support: Increasing the Knowledge and Adoption of Energy Saving Initiatives Amongst Farming Enterprises in Western Australia*; National Centre for Engineering in Agriculture: Toowoomba, Australia, 2013.
51. Biswas, W.K.; Barton, L.; Carter, D. Global warming potential of wheat production in Western Australia: A life cycle assessment. *Water Environ. J.* 2008, *22*, 6–16.
52. Maraseni, T.N.; Cockfield, G.; Maroulis, J. An assessment of greenhouse gas emissions from the Australian vegetables industry. *J. Environ. Sci. Health B* 2010, *45*, 578–588.
53. Maraseni, T.N.; Cockfield, G.; Maroulis, J. An assessment of greenhouse gas emissions: Implications for the Australian cotton industry. *J. Agric. Sci.* 2010, *148*, 501–510.

CHAPTER 7

DIESEL CONSUMPTION OF AGRICULTURE IN CHINA

Nan Li, Hailin Mu, Huanan Li , and Shusen Gui

Key Laboratory of Ocean Energy Utilization and Energy Conservation of Ministry of Education, Dalian University of Technology, Dalian 116024, China

ABSTRACT

As agricultural mechanization accelerates the development of agriculture in China, to control the growth of the resulting energy consumption of mechanized agriculture without negatively affecting economic development has become a major challenge. A systematic analysis of the factors (total power, unit diesel consumption, *etc.*) influencing diesel consumption using the SECA model, combined with simulations on agricultural diesel flows in China between 1996 and 2010 is performed in this work. Seven agricultural subsectors, fifteen categories of agricultural machinery and five farm operations are considered. The results show that farming and transportation are the two largest diesel consumers, accounting for 86.23% of the total diesel consumption in agriculture in 2010. Technological progress has led to a decrease in the unit diesel consumption and an increase in the unit productivity of all machinery, and there is still much potential for future progress. Additionally, the annual average working hours have decreased

Citation: Li, N.; Mu, H.; Li, H.; Gui, S. Diesel Consumption of Agriculture in China. *Energies* 2012, *5*, 5126-5149.

rapidly for most agricultural machinery, thereby influencing the development of mechanized agriculture.

Keywords: Diesel, agriculture, mechanization

INTRODUCTION

The limited supply of traditional fossil fuels and the associated consumption limitations needed for dealing with global climate change have considerably restricted economic development in recent years. This has raised the concern of energy analysts and policy makers regarding the adverse effects of energy overuse. A series of policy measures aimed at reducing energy consumption have been implemented in order to meet the compulsory targets stated in the China government's Eleventh Five-Year Plan (2006–2010). Agriculture development in China is not an exception. Calculated at constant prices, the average elasticity of energy consumption in agriculture declined from 2.33 (2001–2005) to 0.29 (2006–2010). The rapid growth of energy consumption in agriculture was thus restrained without negatively affecting economic development.

With a gradually increasing level of mechanization, agricultural energy consumption in China has increased from 36.88 million tons of coal equivalent (Mtce) in 1996 to 64.77 Mtce in 2010, which translates to an annual increase of 4.10%. Energy plays a critical role in the development of agriculture as it does in the manufacturing, construction and service industry. This has motivated many researchers to focus on agricultural energy issues, and analysis of energy and exergy efficiency in the agricultural sector has become a research hotspot. Two energy resources, namely diesel for tractors and electricity for pumps are usually the research topics in this area. Such analysis has been applied in Saudi Arabia [1], Turkey [2], Jordan [3], Iran [4] and Malaysia [5]. The relationship between energy inputs and agricultural production outputs is another research hotspot. These energy inputs usually include direct and indirect energy, *i.e.*, human and animal labor, machinery, electricity, diesel oil, fertilizers, seeds, *etc.* Rijal and Bansal [6] examined the total energy input and output of subsistence agriculture in the rural areas of Nepal. Ozkan and Akcaoz [7] estimated the input-output ratio in the Turkish agricultural sector for the period of 1975–2000, where their output is composed of 36 agricultural commodities. On the other hand, agricultural output in Hatirli and Ozkan [8] comprises 104 agricultural commodities. Alam and Alam [9] evaluated the impact of energy input on agricultural production output in Bangladesh from 1980 to 2000. There are also researches that provided meaningful econometrics methods. Uri [10] quantified the relationship between the energy price and the use of conservation tillage via Granger causality over the period of 1963–1997. A regression analysis of the relationship between energy use and agricultural productivity was done by Karkacier and Gokalp Goktolga [11]. Using co-integration and error correction analysis, Türkekul and Unakıtan [12] estimated the long- and short-run relationship among energy consumption, agricultural GDP, and energy prices from 1970 to 2008 in Turkey's agriculture.

Based on a bottom-up modeling approach, the model named Save Production simulated the development of energy use in the Dutch industry and agriculture [13]. Baruah and Bora [14] assessed the energy demands in the state of Assam, India. In that study, they considered four strategic scenarios of mechanization that incorporated some proven technologies. Nevertheless, only a few simulation and forecast models were established to study the energy demand of agriculture and little attention has been paid to the relationship between energy consumption and end-use machinery in agriculture, especially in China. In the Twelfth Five-Year Plan, the Chinese government has planned to reduce the energy consumption *per* unit GDP by 16% during this five year horizon. A special model, Simulation and Analysis of Energy Consumption for Agriculture (SECA), is designed to answer all kinds of questions on how energy was consumed in different agricultural sectors to achieve the new goal. Furthermore it also serves as the foundation for the agricultural energy demand forecasting model. In this study, we identify the factors (total power, unit diesel consumption, *etc.*) influencing diesel consumption in China and simulate the diesel flows of agriculture of China in detail. Based on availability, our dataset spans the period of 1996–2010.

MODEL SPECIFICATION AND DATA

Overall Structure

This paper uses a generalized definition of the word agriculture to include farming (i.e., agriculture in narrow definition), fishery, forestry, animal husbandry and services supporting agriculture. Fifteen kinds of the agricultural machinery from seven agricultural subsectors are considered. Figure 1 shows the overall structure of SECA. In the Distribution Module, it is assumed that the change in the number of agricultural machinery with different rated powers follows a certain curved distribution. Simulation results of the curved distribution are processed by the Operation Module and transferred into the Unit Consumption Module and the Productivity Module which generate the weighted average of unit consumption and that of unit productivity, respectively. Then the two weighted averages are input into the Main Module along with workload statistics, machinery capacity and load factors, which are obtained from the database in the Operation Module. Finally, technological progress and other effects are considered in the Correction Module to narrow the gap between the empirical statistics and the model calculations. The following subsections describe each module in detail.

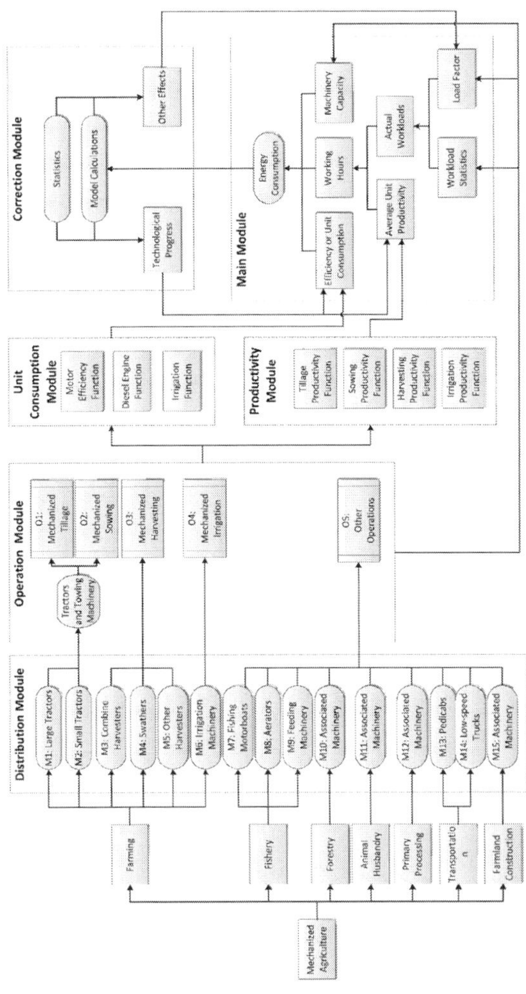

Figure 1: Overall structure of SECA.

Operation Module

In SECA, all the agricultural machinery are categorized into five farming related operations ($o1$~$o5$, Table 1). These operations include mechanized tillage, mechanized sowing, mechanized harvesting, mechanized irrigation and other operations. They are assumed as follows:

- Mechanized tillage ($o1$) and mechanized sowing ($o2$) are performed by tractors and associated towing farm machinery. Both the large and medium-sized tractors and the small ones are included.
- Mechanized harvesting ($o3$) is performed by combine harvesters, swathers and other harvesters.
- Mechanized irrigation ($o4$) is performed by irrigation machinery.
- Other operations ($o5$) are performed by rest agricultural machinery, the workload statistics of which are not available.

Table 1: The relationships between operations and machinery

Operations	**Machinery**	**Relationships**
$o1$: Mechanized tillage	$m1$: Large and medium-sized tractors $m2$: Small tractors (and associated towing machinery)	$T_{o1,y} = \sum_{m=m1}^{m2} T_{m,y}$
$o2$: Mechanized sowing	$m1$: Large and medium-sized tractors $m2$: Small tractors (and associated towing machinery)	$T_{o2,y} = \sum_{m=m1}^{m2} T_{m,y}$
$o3$: Mechanized harvesting	$m3$: Combine harvesters $m4$: Swathers $m5$: Other harvesters	$T_{o3,y} = \sum_{m=m3}^{m5} T_{m,y}$
$o4$: Mechanized irrigation	$m6$: Irrigation machinery	$T_{o4,y} = \sum_{m=m6}^{m6} T_{m,y}$
$o5$: Other operation	$m7$~$m15$: Rest machinery	$T_{o5,y} = \sum_{m=m7}^{m15} T_{m,y}$

Note: T is a generalized symbol which can be replaced by P, L, l_i, C, *etc.*

Main Module

Two kinds of the energy carriers, diesel and electricity, are considered in SECA. The energy required to perform the selected farming operation is estimated using the equation below:

$$C_y = \sum_o C_{o,y} = \sum_o P_{o,y} \times t_{o,y} \times \overline{uc}_{o,y} \tag{1}$$

wherein the subscript o indicates the farm operations mentioned in the Operation Module ($o = o1, o2, \ldots, o5$); the subscript y indicates the year ($y = 1996, 1997, \ldots, 2010$); C refers to the quantity of energy consumption; P is the total power of agricultural machinery; t refers to the average annual working hours of unit machinery; $\overline{uc}$ is the weighted average of unit energy consumption (for diesel) or motor efficiency (for electricity).

The workload statistics for the four categories of agricultural operations, *viz.* $o1$, $o2$, $o3$ and $o4$, can be obtained from specific statistics. Their average annual working hours of unit machinery $t_{o,y}$ can be determined from the following equation:

$$t_{o,y} = \frac{AW_{o,y}}{\overline{up}_{o,y}} = \frac{WS_{o,y} \times lf_{o,y} \times oec_y}{\overline{up}_{o,y}} \tag{2}$$

wherein AW is the actual workloads of the farming operation; $\overline{up}$ indicates the weighted average of unit productivity; WS refers to the workload statistics; lf is the load factor, which is used for describing the actual work intensity and is assumed to be proportional to the machinery power in unit area; oec is the other effect coefficient obtained from the Correction Module.

Because workload statistics are not available for "other operations" ($o5$), in this study we assume that its average annual working hours of unit machinery t_{o5} is constant. That constant is determined from the following equation:

$$t_{o5} = \frac{\sum_y [(P_{o5,y} \times \overline{uc}_{o5,y}) \times \Delta C_y]}{\sum_y (P_{o5,y} \times \overline{uc}_{o5,y})^2} = \frac{\sum_y [(P_{o5,y} \times \overline{uc}_{o5,y}) \times (C_{s,y} - \sum_{o=o1}^{o4} C_{o,y})]}{\sum_y (P_{o5,y} \times \overline{uc}_{o5,y})^2} \tag{3}$$

wherein $C_{s,y}$ indicates the diesel consumption statistics in year y.

Distribution Module

The mechanization of agriculture is a process of replacing human and animal with agricultural machinery powered by either diesel or electricity. The rated power is an important factor affecting the device performance of agricultural machinery. The rated power of the agricultural machinery is negatively correlated with its unit consumption (or motor efficiency) and is positively correlated with its unit productivity.

Smaller agricultural machinery are more widely used than larger equipment in China due to reasons such as the high cost of larger machinery and the small amount of arable land *per capita*. It is assumed that the change in the number $l_{i,m,y}$ of agricultural machinery (except tractors) with the representative power $p_{i,m,y}$ of the interval i follows an exponential distribution as shown in Figure 2(a). The equation is expressed below:

$$l_{i,m,y} = a_{m,y} \times e^{b_{m,y} \times p_{i,m,y}} \quad (4)$$

wherein m indicates machinery category mentioned in the Distribution Module ($m = m3, m4, \ldots, m15$); $a_{m,y}$ and $b_{m,y}$ are the undetermined parameters.

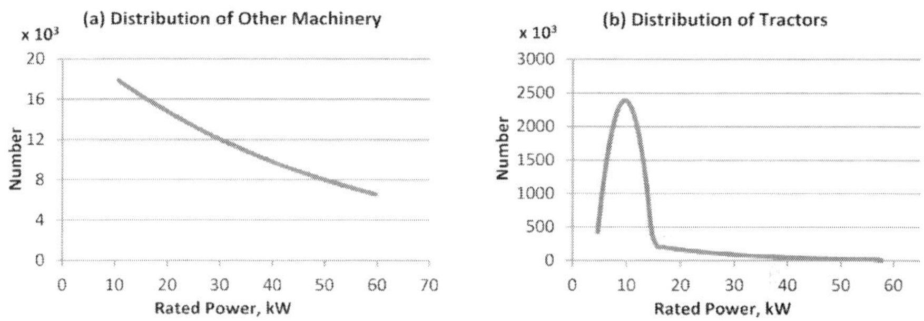

Figure 2: Distribution curves of the agricultural machinery with rated power.

In the Distribution Module, the most common power range ($p_{beginning,m,y}$, $p_{end,m,y}$) of machinery is selected on the basis of experience and divided into n intervals. The midpoint of interval i as its representative power $p_{i,m,y}$ can be obtained using the following relationship:

$$p_{i,m,y} = p_{beginning,m,y} + \frac{(2i-1) \times (p_{end,m,y} - p_{beginning,m,y})}{2n} \quad (5)$$

The total number of machinery $L_{m,y,c}$ and the total power $P_{m,y,c}$ can be obtained from statistics, so another two equations can be established:

$$L_{m,y} = \sum_{i} l_{i,m,y} \quad (6)$$

$$P_{m,y} = \sum_i (p_{i,m,y} \times l_{i,m,y}) \quad (7)$$

Equations (4)–(7) should be solved simultaneously to estimate parameters $a_{m,y}$ and $b_{m,y}$. Then the number $l_{i,m,y}$ of the agricultural machinery (excluding tractors) with the representative power $p_{i,m,y}$ in the interval i can be determined.

Tractors are the most common agricultural machinery and play an important role in mechanized agriculture in China. The number and total power of tractors increased from 9.86 million and 1.08×10^8 kW in 1996 to 21.78 million and 2.84×10^8 kW in 2010, with annual increases of 5.82% and 7.16%, respectively. The rated power of the main tractor models in Chinese market is between 8 and 12 kW, while in China's Department of Statistics, the rated power for tractors starts from 2.2 kW, so in SECA, it is assumed that the number of large and medium-sized tractors, referring to tractors with power ratings greater than or equal to 14.7 kW, still follows an exponential distribution. On the other hand, the number of small tractors, referring to tractors less than 14.7 kW and greater than 2.2 kW, follows a parabolic distribution [Figure 2(b)]:

$$l_{i,m1,y} = a_{m1,y} \times e^{b_{m1,y} \times p_{i,m1,y}} \quad (8)$$

$$l_{i,m2,y} = a_{m2,y} \times (p_{i,m2,y} - b_{m2,y})^2 + c_{m2,y} \quad (9)$$

wherein $c_{m,y}$ is the undetermined parameter.

Another equation is established as follow in order to ensure the continuity of Equations (8) and (9):

$$l_{n+1,m2,y} = l_{1,m1,y} \quad (10)$$

Equations (5)–(10) should be solved simultaneously to estimate parameter $a_{m,y}$, $b_{m,y}$ and $c_{m,y}$. Then the number of tractors with the representative power in the interval i can be determined.

Unit Consumption Module and Productivity Module

Unit consumption (or motor efficiency) and productivity of agricultural machinery are both functions of the machinery's rated power. The simulation results from the Distribution Module are processed by the Operation Module and then output into the Unit Consumption Module and the Productivity Module which generate the weighted average of unit consumption and that of productivity. The data describing the relationship between the input and output of the functions are collected from the relevant national standards of China and product manuals of those agricultural machinery [15,16]. They are fitted by the least square method (Figure 3 and Figure 4).

The fitting equation of the unit consumption is as follows:

$$uc_{i,o,y} = \begin{cases} 0.0082 \times p_{i,o,y}^2 - 2.0807 \times p_{i,o,y} + 398.21 & (diesel; o \neq o4) \\ 534.32 \times p_{i,o,y}^{-0.136} & (diesel; o = o4) \\ 3.6572 \times \ln(p_{i,o,y}) + 77.516 & (electricity) \end{cases} \quad (11)$$

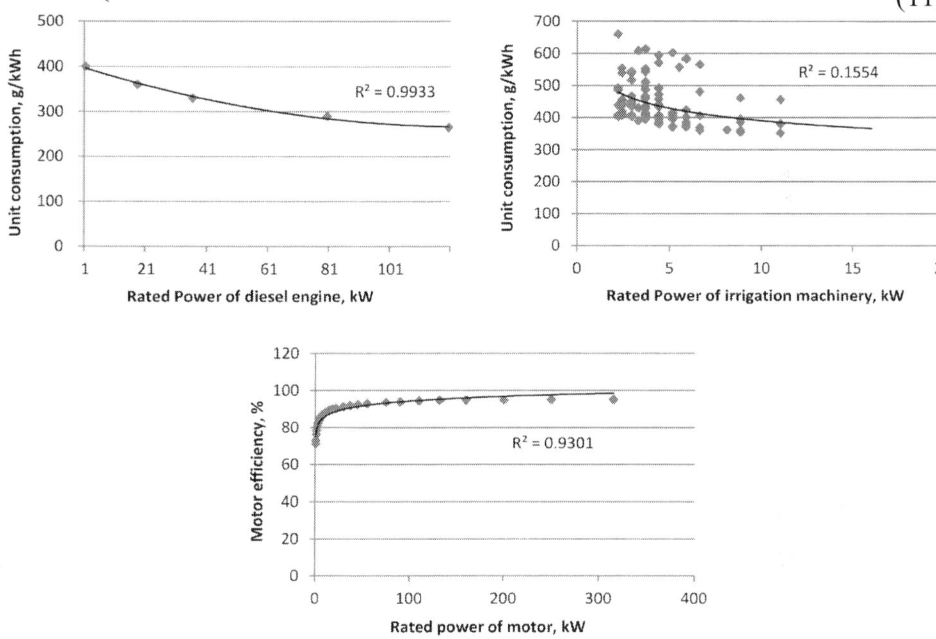

Figure 3: Fitting curves of the unit consumption.

The unit productivity for each operation is estimated using the following equations:

$$up_{i,o,y} = \begin{cases} 0.9903 \times p_{i,o1,y}^2 + 26.72 \times p_{i,o1,y} + 379.12 & (o = o1) \\ 2068.6 \times e^{0.0235 \times p_{i,o2,y}} & (o = o2) \\ 0.1875 \times P_{i,o3,y}^2 + 71.6 \times P_{i,o3,y} + 386.6 & (o = o3) \\ 8.9867 \times p_{i,o4,y} + 10.23 & (o = o4) \end{cases} \quad (12)$$

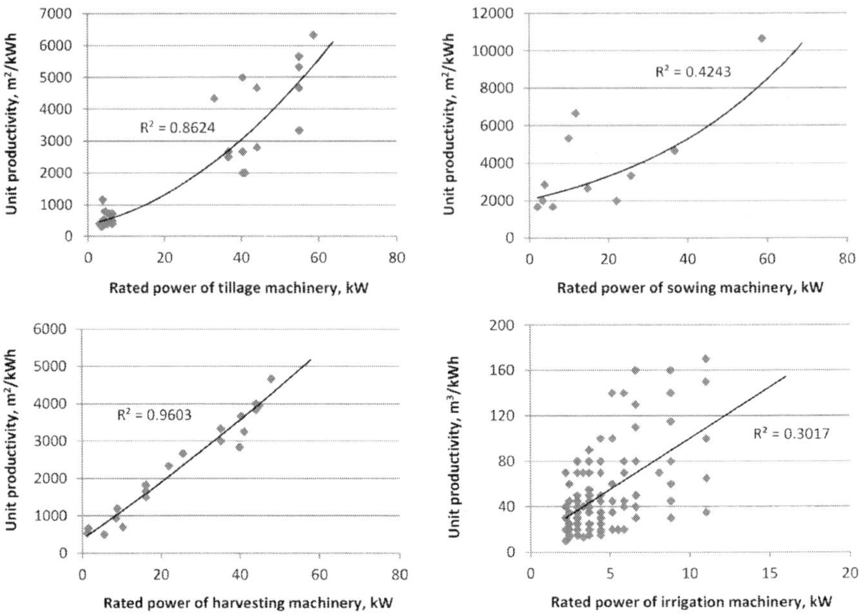

Figure 4: Fitting curves of the unit productivity.

As the output in the Unit Consumption Module and the Productivity, the weighted averages of unit consumption and productivity are obtained according to the following relationships:

$$\overline{uc}_{o,y} = \frac{\sum_i (l_{i,o,y} \times uc_{i,o,y})}{L_{o,y}} \times tpc_{uc,y} \tag{13}$$

$$\overline{up}_{o,y} = \frac{\sum_i (l_{i,o,y} \times up_{i,o,y})}{L_{o,y}} \times tpc_{up,y} \tag{14}$$

wherein tpc_{uc} and tpc_{up} are the technological progress coefficients obtained from the Correction Module.

Correction Module

Technological progress and other effects are considered in the Correction Module to narrow the gap between the statistics and the model calculations in the Main Module. The correction coefficient is the correction on the basic assumptions in the other modules of SECA. These assumptions which are either explicit or implicit include:

- In the Unit Consumption Module, the weighted average of unit consumption changes without being affected by the technological progress.
- In the Productivity Module, the weighted average of unit productivity changes without being affected by the technological progress.
- In the Main Module, the load factor is proportional to the machinery power in unit area.
- In the Main Module, the average annual working hours of the other operation is constant the value of which remains unchanged over the year.

The correction coefficient *coc* can be calculated with the following relationship:

$$coc_y = \frac{C_{s,y}}{C_y} \quad (15)$$

The correction coefficient *coc* can be divided into two parts: technological progress coefficient *tpc* and the other effect coefficient *oec*. The relationship between *coc*, *tpc* and *oec* is assumed to following equation:

$$tpc_y = oec_y = coc_y^{\frac{1}{2}} \quad (16)$$

Technological progress is typically accompanied with the reduction of the unit consumption and the increase of the unit productivity. The relationship between *tpc*, tpc_{uc} and tpc_{up} is assumed according to Equation (17):

$$tpc_{uc,y} = tpc_{up,y} = tpc_y^{\frac{1}{2}} \quad (17)$$

The other effect coefficient *oec* mainly works on the load factor *lf* mentioned in the Main Module. The other effect coefficient *oec* means that the load factor is not proportional to machinery power in unit area any longer. In short, we can conclude that technological progress leads to the reduction of average unit consumption and changes in working hours result from the changes in unit productivity and load factor.

DATA SOURCES

The data related to the agricultural land, machinery and energy consumption in this paper are mainly obtained from the China Statistical Yearbook [17], China Energy Statistical Yearbook [18], China Rural Statistical Yearbook [19] and China Agriculture Statistical Report [20]. The first three data resources are published by the National Bureau of Statistics of China and the last one is published by the Ministry of Agriculture of China.

The data related to land and water resources in this paper are given by the China Land and Resources Bulletin [21] and the China Water Resources Bulletin [22]. The former is published by the Ministry of Land and Resources of China, and its data, especially the farmland area data, are more reliable than the other data sources [23]. The latter is published by the Ministry of Water Resources of China.

Table A1 in the Appendix presents the workload statistics of four operations: *i.e.*, mechanized tillage, mechanized sowing, mechanized harvesting and mechanized irrigation. Considering that parts of the land are repeatedly cultivated in a year, a re-seeding coefficient (the ratio of sowing area to tillage area) is introduced to correct the workload statistic of the mechanized tillage when the data are input into the model. Table A2 presents the number and the total power of the agricultural machinery mentioned in this study.

RESULTS AND DISCUSSION

Changes of Correction Coefficient

Figure 5 presents the changes of the correction coefficient. The correction coefficient has declined from 1.37 in 1996 to 0.84 in 2010, with an average annual decline of 3.46%. The decline in the correction coefficient proves that technological progress has been affecting the unit diesel consumption and the unit productivity of the agricultural machinery positively. One also can find that the load factor has not been growing as expected.

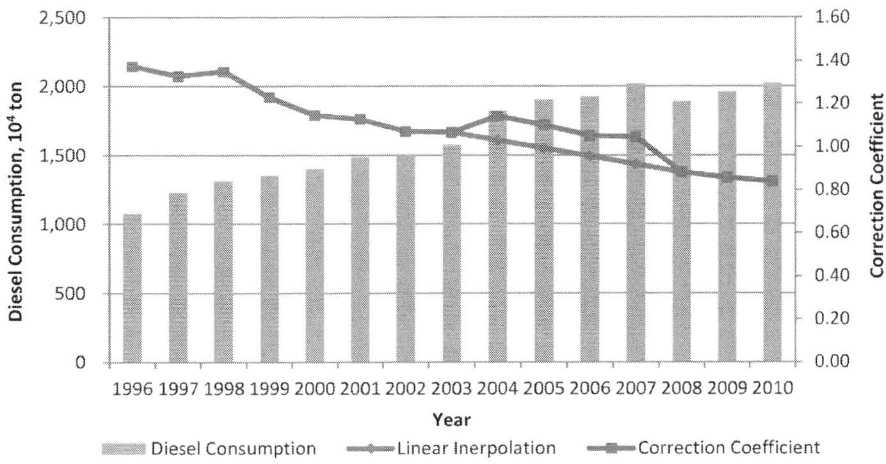

Figure 5: Correction coefficient and diesel consumption statistics.

The curve of the correction coefficient shows a significant linear downward trend. However, there are step changes in the correction coefficient corresponding to the step changes in the diesel consumption from 2004 to 2007. The step changes of the energy consumption statistics are widespread in most sectors during the Eleventh Five-Year period including the diesel consumption in agriculture. According to careful analysis, it is believed that the step changes cannot reflect the real energy consumption, and they probably result from either changes in statistical methodology or artificially adjusted energy consumption numbers. The latter is more likely the main reason due to the existence of the compulsory target for reducing energy consumption stated in the Eleventh Five-Year Plan of China government, so the correction coefficients from 2004 to 2007 are corrected using linear interpolation.

Changes of Unit Consumption

Table 2 provides the unit diesel consumption of agricultural machinery. The unit diesel consumption generally maintained a steady downward trend from 1996 to 2010.

- For tractors, the unit diesel consumption declined from 407.96 g/kWh in 1996 to 356.93 g/kWh in 2010, an average annual decline of 0.95%.
- For harvesters, it declined from 423.67 g/kWh to 352.70 g/kWh, an annual decline of 1.30%.
- For irrigation machinery, it declined from 450.99 g/kWh to

401.25 g/kWh, an annual decline of 0.83%.

- For primary processing machinery, it declined from 411.91 g/kWh to 364.70 g/kWh, an annual decline of 0.87%.
- For animal husbandry machinery, it declined from 426.16 g/kWh to 375.13 g/kWh, an annual decline of 0.91%.
- For fishery machinery, it declined from 383.17 g/kWh to 328.06 g/kWh, an annual decline of 1.10%.
- For forestry machinery, it declined from 414.97 g/kWh to 370.71 g/kWh, an annual decline of 0.80%.
- For transportation machinery, it declined from 408.68 g/kWh to 352.89 g/kWh, an annual decline of 1.04%.
- For farmland construction machinery, it declined from 338.56 g/kWh to 297.35 g/kWh, an annual decline of 0.92%.

Figure 6 shows the changes in the unit diesel consumption compared with that in the previous year. The red part shows the changes in the unit diesel consumption caused by technological progress. The blue part displays the changes in the unit diesel consumption caused by changes in machinery quantity at different intervals (structural influence). It is obvious that the effect of the technological progress plays a major role in almost all agricultural machinery. It can be predicted that the trend will not change in the foreseeable future. However, the effect of technological progress is no longer significant for some machinery with low unit diesel consumption, such as farmland construction machinery.

Another way to reduce unit diesel consumption is to adjust the structure of the distribution of agricultural machinery. Results show that most of the agricultural machinery, such as tractors, harvesters, animal husbandry machinery, fishery machinery, transportation machinery and farmland construction machinery are becoming larger in size and lower in unit diesel consumption with the development of the agricultural economy. This is not the case for irrigation machinery, primary processing machinery and forestry machinery.

Table 2: Unit diesel consumption (g/kWh) of the agricultural machinery

Year	Tractors	Harvesters	Machinery						
			Irrigation	Primary Processing	Animal Husbandry	Fishery	Forestry	Transportation	Farmland Construction
1996	407.96	423.67	450.99	411.91	426.16	383.17	414.97	408.68	338.56
1997	405.21	419.81	446.92	408.40	422.47	374.80	412.29	404.72	344.16
1998	406.88	420.81	449.53	410.07	424.14	373.89	414.53	405.75	348.48
1999	397.07	409.81	439.71	400.63	414.33	362.89	405.37	395.98	339.26
2000	389.93	402.37	432.35	393.59	406.36	358.04	397.80	388.52	330.26
2001	388.29	399.76	430.27	392.30	405.35	357.19	398.05	386.61	325.88
2002	383.16	394.04	425.47	386.99	400.21	353.13	391.89	381.17	318.80
2003	382.76	391.78	425.62	386.70	399.85	350.62	394.08	380.37	315.59
2004	379.09	386.31	421.85	383.29	396.51	349.70	386.81	376.77	311.79
2005	375.32	380.44	418.10	379.67	392.81	342.10	385.62	372.60	307.53
2006	371.08	373.16	414.74	376.20	389.43	340.16	382.12	369.09	318.42
2007	367.09	367.66	411.02	372.58	385.57	336.61	378.09	365.04	312.34
2008	362.76	360.42	406.74	368.98	380.35	332.78	373.67	359.09	306.54
2009	359.45	355.74	403.54	366.50	377.30	330.13	372.37	355.66	301.60
2010	356.93	352.70	401.25	364.70	375.13	328.06	370.71	352.89	297.35

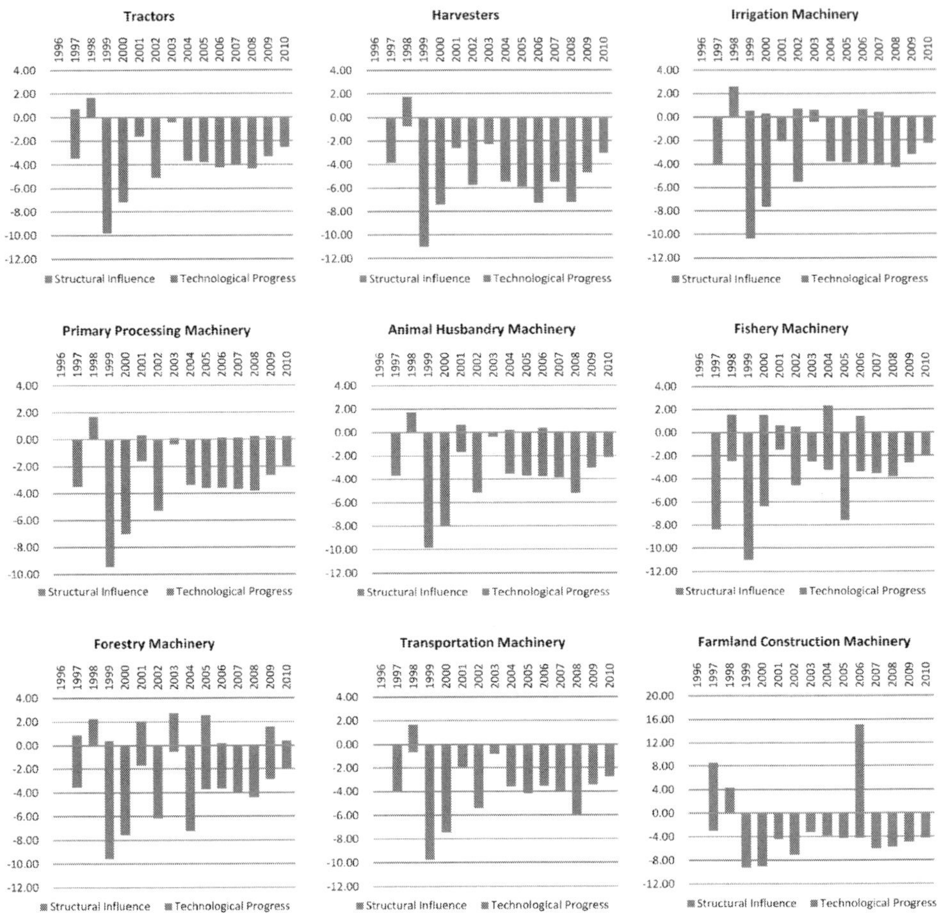

Figure 6: Changes in unit diesel consumption compared with that in the previous year.

Changes of Unit Productivity

Table 3 shows the unit productivity of the four mechanized operations. The unit productivity generally maintained a steady upward trend from 1996 to 2010:

- The unit productivity of the mechanized tillage increased from 0.08 ha/h in 1996 to 0.10 ha/h in 2010, an average annual increase of 1.86%.
- The unit productivity of the mechanized sowing increased from 0.25 ha/h to 0.30 ha/h, an average annual increase of 1.27%.
- The unit productivity of the mechanized harvesting increased

from 0.06 ha/h to 0.19 ha/h, an average annual increase of 8.13%, which is the largest growth rate in these four operations.

- The unit productivity of the mechanized irrigation increased from 74.41 m^3/h to 79.87 m^3/h, an average annual increase of 0.51%.

Table 3: Unit productivity of the four operations

Year	Mechanized Tillage (ha/h)	Mechanized Sowing (ha/h)	Mechanized Harvesting (ha/h)	Mechanized Irrigation (m^3/h)
1996	0.08	0.25	0.06	74.41
1997	0.08	0.25	0.06	75.35
1998	0.08	0.25	0.07	74.01
1999	0.08	0.26	0.07	75.03
2000	0.08	0.26	0.07	75.95
2001	0.08	0.26	0.08	76.74
2002	0.08	0.27	0.08	76.77
2003	0.08	0.27	0.09	76.05
2004	0.09	0.27	0.10	76.75
2005	0.09	0.28	0.12	77.33
2006	0.09	0.28	0.13	77.17
2007	0.09	0.29	0.15	77.40
2008	0.09	0.29	0.17	78.30
2009	0.10	0.30	0.18	79.28
2010	0.10	0.30	0.19	79.87

Figure 7 shows the changes in unit productivity compared with that in the previous year. The red part indicates changes in unit productivity caused by technological progress. The blue part indicates changes in unit productivity caused by changes in machinery quantity at different intervals (structural influence).

Basically, technological progress has a positive influence on changes in the unit productivity of four operations. Improvement of unit productivity in mechanized harvesting is less than that in the other operations.

Structural adjustments to the machinery number caused an increase in the unit productivity of mechanized tillage, mechanized sowing and mechanized harvesting. Moreover, it also enables the unit productivity of the mechanized irrigation to decrease gradually with average rated power.

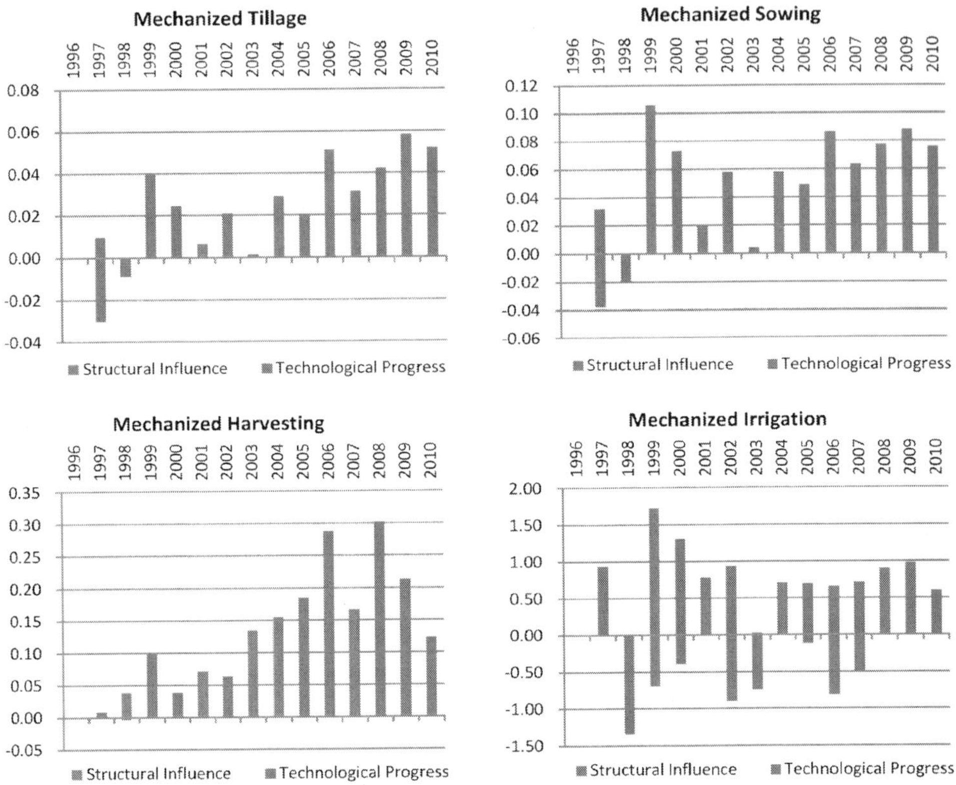

Figure 7: Changes in the unit productivity compared with that in the previous year.

Changes of Working Hours

Table 4 shows the average annual working hours of agricultural machinery. Almost all average annual working hours of the agricultural machinery show a decreasing trend from 1996 to 2010, except for harvesters.

- The average annual working hours of the tractors declined from 56.33 h in 1996 to 41.00 h in 2010, an average annual decline of 2.24%.
- The average annual working hours of the harvesters increased from 87.19 h to 105.65 h, an average annual increase of 1.38%.
- The average annual working hours of the irrigation machinery declined from 217.24 h to 109.06 h, an average annual decline of 4.80%.
- The average annual working hours of the other machinery

declined from 165.70 h to 114.43 h, an average annual decline of 2.74%.

Figure 8 shows the changes in annual average working hours compared with that in the previous year. The changes in average annual working hours could be caused by changes in actual workloads and changes in the unit productivity. In Figure 8, they are indicated by the red part and the blue part, respectively.

There is no doubt that an increase in unit productivity leads to a reduction in working hours. This is the case for all kinds of agricultural machinery. However, its effect is negligible compared with the effect caused by changes of actual workloads.

Table 4: Average annual working hours (h) of agricultural machinery

Year	Tractors	Harvesters	Irrigation Machinery	Other Machinery
1996	56.33	87.19	217.24	165.70
1997	54.31	85.94	227.71	161.50
1998	55.07	91.48	195.59	163.48
1999	51.88	92.38	176.55	152.45
2000	49.58	89.13	158.42	144.64
2001	49.07	92.53	153.42	142.90
2002	47.49	91.69	143.85	137.53
2003	47.42	98.06	134.68	137.14
2004	46.29	101.17	134.90	133.55
2005	45.40	103.76	129.25	129.93
2006	44.45	107.22	127.96	126.28
2007	43.33	106.57	120.87	122.59
2008	42.38	107.16	115.53	118.86
2009	41.60	106.56	113.22	116.30
2010	41.00	105.65	109.06	114.43

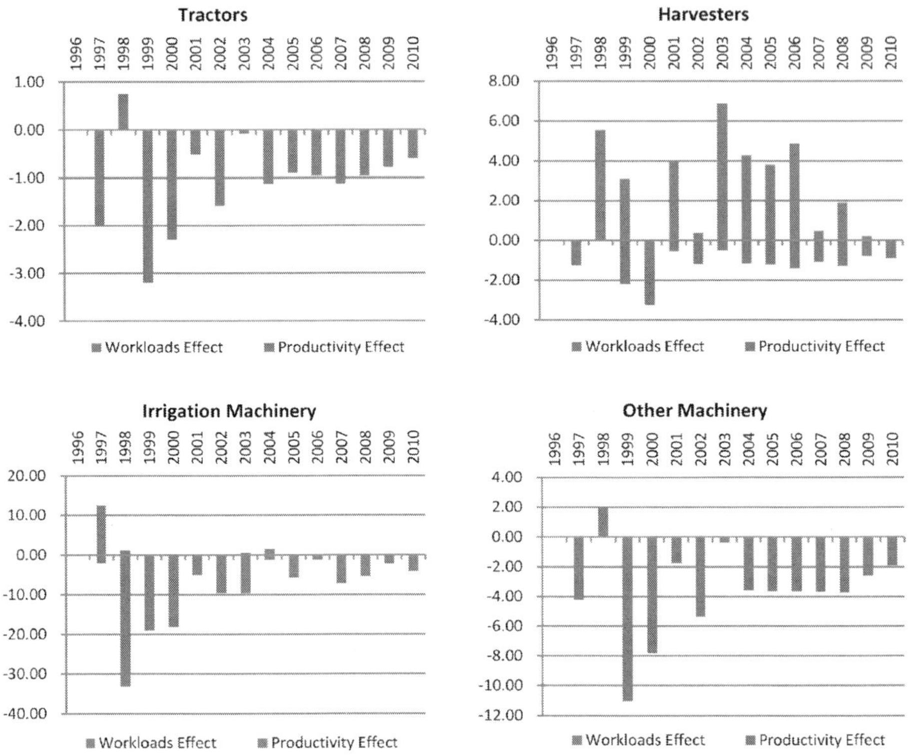

Figure 8: Changes in the annual average working hours compared with that in the previous year.

Due to the effect of actual workloads, working hours of the tractors, irrigation machinery and other machinery continue to decline from 1996 to 2010. The harvesting machinery is the only one whose working hours have increased. This means that the total power of the agricultural machinery (excluding harvesters) grew faster than the actual workloads. Subsidies policy for purchasing agricultural machines upon 2004 results in the massive growth of the agricultural machinery at the expense of the waste of partial production capacity. Furthermore, economic life of most agricultural machinery does not exceed 15 years in China and a large number of scrapped agricultural machinery need to be recycle every year. Chinese energy policy maker should pay attention to this problem. Some materials such as aluminum and steel, are easily recyclable and thus their post-consumer recycling takes much less energy than production of finished materials from virgin feedstocks [24,25].

In addition, the average annual working hours of agricultural machinery powered by electricity is about 500–600 h in China. It is significantly higher than the working hours of machinery powered by diesel. It is believed that instability of the diesel supply and rising prices are the reasons for the low utilization rate of diesel machinery.

Diesel Flows in Agriculture

Table 5 shows diesel consumption in different sectors of agriculture in China from 1996 to 2010.

- Diesel consumption in four farming operations increased from 652.31 × 10^4 ton in 1996 to 884.01×10^4 ton in 2010, with an annual increase of 2.19%. In these four operations, the mechanized harvesting has the largest annual growth rate (*i.e.*, 19.39%) of the diesel consumption. The annual growth rates of mechanized tillage and mechanized sowing are 3.68% and 4.28%, respectively, which are slightly higher than the average level among all farming operations. Slight negative growth in the diesel consumption of mechanized irrigation was witnessed in the past thirteen years.
- Diesel consumption by primary processing increased from 107.88 × 10^4 ton in 1996 to 111.78 × 10^4 ton in 2010, with an annual increase of 0.25%.
- Diesel consumption by animal husbandry increased from 8.07 × 10^4 ton in 1996 to 17.87 × 10^4 ton in 2010, with an annual increase of 5.84%.
- Diesel consumption by fishery increased from 54.86 × 10^4 ton in 1996 to 68.35 × 10^4 ton in 2010, with an annual increase of 1.58%.
- Diesel consumption by forestry increased from 0.31 × 10^4 ton in 1996 to 3.95 × 10^4 ton in 2010, with an annual increase of 20.07%.
- Diesel consumption by transportation increased from 238.74 × 10^4 ton in 1996 to 860.53 × 10^4 ton in 2010, with an annual increase of 9.59%.
- Diesel consumption by farmland construction increased from 13.94 × 10^4 ton in 1996 to 76.61 × 10^4 ton in 2010, with an annual increase of 12.94%.

Table 5: Diesel consumption (10^4 ton) in different sectors of agriculture in China

Year	Farming					Primary Processing	Animal Husbandry	Fishery	Forestry	Transportation	Farmland Construction
	Total	Mechanized Tillage	Mechanized Sowing	Mechanized Harvesting	Mechanized Irrigation						
1996	652.31	214.53	33.76	13.61	390.40	107.88	8.07	54.86	0.31	238.74	13.94
1997	713.39	225.42	34.92	16.22	436.83	110.28	8.14	65.40	0.23	317.59	14.37
1998	699.85	245.02	37.87	21.38	395.58	117.83	9.49	71.79	0.21	399.54	16.00
1999	692.02	245.79	38.22	24.94	383.07	118.24	10.40	69.33	0.21	447.28	16.82
2000	672.33	248.06	38.63	27.25	358.39	120.66	11.23	69.32	0.71	508.50	22.26
2001	688.59	249.97	38.95	31.32	368.36	124.17	11.26	70.42	0.74	561.88	28.25
2002	668.75	248.18	38.76	34.91	346.90	123.14	12.09	67.08	0.98	602.70	32.76
2003	663.40	255.74	39.95	47.13	320.59	126.51	12.74	68.56	1.24	663.72	38.43
2004	693.71	266.36	41.98	55.06	330.32	120.78	12.89	64.65	1.59	705.20	42.67
2005	716.95	278.87	44.12	67.89	326.07	123.50	13.90	61.17	1.71	750.62	49.00
2006	747.16	290.89	46.86	83.09	326.32	121.20	13.70	64.36	1.91	743.74	57.19
2007	752.66	298.75	48.51	93.26	312.14	119.74	13.98	66.24	2.28	758.98	62.37
2008	803.33	327.90	53.90	113.20	308.33	116.04	17.27	66.43	2.66	810.93	71.23
2009	847.41	341.90	57.33	137.69	310.49	113.58	17.53	67.14	3.22	835.99	75.03
2010	884.01	355.64	60.69	162.68	304.99	111.78	17.87	68.35	3.95	860.53	76.61

Figure 9 shows the diesel flows of the agriculture in China. It is obvious that farming and transportation are the two largest diesel consumers, while other sectors account for a negligible share. From 1996 to 2010, diesel consumption by farming grew smoothly while diesel consumption by transportation grew sharply. Transportation consumed nearly half of the total diesel in 2010 and its share can be expected to continue to grow in the future.

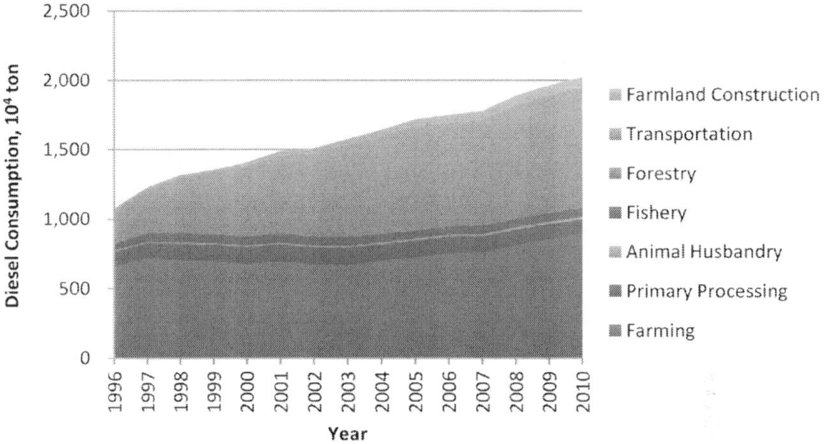

Figure 9: Diesel flows of the agriculture in China.

Diesel Consumption Intensity

Table 6: Diesel consumption intensity (kg/ha) of the operations

Year	Mechanized Tillage	Mechanized Sowing	Mechanized Harvesting	Mechanized Irrigation
1996	33.47	10.46	7.49	598.99
1997	34.16	10.19	7.90	607.62
1998	34.12	9.90	9.16	586.66
1999	32.66	9.51	9.75	567.48
2000	32.53	9.61	10.22	533.29
2001	32.84	9.49	11.66	552.11
2002	32.89	9.34	12.77	517.31
2003	33.89	9.79	17.19	476.54
2004	33.35	9.47	18.05	496.21
2005	33.00	9.36	19.73	483.05
2006	33.74	9.38	21.70	476.45
2007	33.11	9.20	22.11	443.55
2008	33.06	9.16	23.89	420.89
2009	32.72	8.82	25.83	416.73

2010	31.90	8.80	27.27	400.19

Table 6 shows diesel consumption intensity in different operations from 1996 to 2010. Little change occurred to the diesel consumption intensity of mechanized tillage and its value remained at around 33 kg/ha. The diesel consumption intensity of mechanized sowing declined slightly from 10.46 kg/ha in 1996 to 8.80 kg/ha. By an annual growth rate of 9.99%, the diesel consumption intensity of mechanized harvesting increased rapidly from 7.49 kg/ha in 1996 to 27.27 kg/ha in 2010. Because of the popularity of irrigation machinery powered by electricity, the diesel consumption intensity of mechanized irrigation which is far more than that of other operations declined from 598.99 kg/ha in 1996 to 400.19 kg/ha in 2010, with an annual decrease of 2.75%.

CONCLUSIONS

This study is a fundamental research for establishing the agricultural energy demand forecasting model. The simulation results demonstrate that the methodology used in this study is proper and accurate. The conclusions and the relevant policy recommendations are summarized as follows:

- For agriculture in China, farming and transportation are the two largest diesel consumers, accounting for 86.23% of the total diesel consumption in agriculture in 2010, while the other sectors account for a negligible share. Differing from the farming in this respect, more attention should be paid to the fast growth of the diesel consumption in the transportation in the forecasting model.
- Technological progress positively affected unit diesel consumption and the unit productivity of all machinery from 1996 to 2010. However, there is great potential in reducing unit diesel consumption and increasing unit productivity. The Chinese government should continue to promote technological progress and to improve in the field of mechanized agriculture.
- With the development of the agricultural economy, most of the agricultural machinery becomes larger and larger in size, more diesel fuel efficient and productive. However, irrigation machinery has proved to be an exception. Diesel consumption in mechanized agriculture can be reduced by preventing the miniaturization trend of irrigation machinery and raising the proportion of the medium-sized and large-scale agricultural machinery.

- The annual average working hours of the agricultural machinery (except harvesters) continue to decline from 1996 to 2010. Subsidies policy for purchasing agricultural machines upon 2004 leads to the massive growth of the agricultural machinery at the expense of the waste of the partial production capacity. This means that machinery sits idle in the yard for most of the time. Although this may not directly affect diesel fuel consumption, it directs resources to the manufacturing of agricultural machinery and increases the cost of the agricultural production.
- The annual average working hours of the agricultural machinery powered by diesel are about 40–120 h which is much fewer than that of agricultural machinery powered by electricity (*i.e.*, 500–600 h). With an adequate power supply and feasible techniques, it is effective to save energy and improve utilization by replacing diesel machinery with electricity machinery.

ACKNOWLEDGMENTS

The authors gratefully acknowledge the financial support from the National Natural Science Foundation of China (No. 70873013 & No. 71273039).

APPENDIX A

Table A1: Workload statistics of the four operations

Year	Tillage		Sowing		Harvesting		Irrigation	
	Total area	Mechanized area	Total area	Mechanized area	Total area	Mechanized area	Total water consumption	Mechanized Irrigation
	(10^4 ha)	(10^4 ha)	(10^4 ha)	(10^4 ha)	(10^4 ha)	(10^4 ha)	(10^8 m^3)	(10^8 m^3)
1996	13003.92	5518.45	15242.33	3258.81	15249.54	1834.52	3900.00	2342.41
1997	12993.33	5791.05	15398.33	3564.71	15391.90	2134.86	3919.72	2418.95
1998	12966.70	6005.29	15548.61	3835.84	15544.40	2342.54	3766.46	2319.94
1999	12920.00	6192.42	15605.44	3993.43	15605.11	2542.07	3869.00	2389.00
2000	12826.70	6208.78	15629.57	3990.23	14483.03	2644.60	3783.54	2315.24
2001	12761.58	6165.28	15572.85	4058.29	14750.77	2653.66	3825.73	2326.23
2002	12653.30	6129.30	15467.98	4120.67	14813.48	2715.31	3736.18	2248.65
2003	12340.00	6094.36	15243.13	4071.44	14385.33	2736.09	3432.81	2091.34
2004	12243.92	6359.31	15352.72	4426.19	14583.48	3045.03	3585.70	2159.53
2005	12208.27	6630.05	15548.80	4710.17	14652.49	3437.47	3580.00	2173.60
2006	12177.59	6900.79	15214.90	4994.15	14707.82	3829.92	3664.45	2207.93
2007	12173.52	7171.54	15330.03	5278.13	14753.18	4222.36	3599.51	2188.50
2008	12171.60	7658.37	15793.86	5897.43	15224.12	4748.40	3663.46	2239.40
2009	12198.50	8049.79	15864.75	6509.31	15373.82	5340.87	3723.11	2287.77
2010	12214.58	8502.57	16068.99	6916.09	15581.02	5984.67	3689.10	2266.91

Table A2: Number and total power of agricultural machinery

Year	Small tractors		Large and medium-sized tractors		Combine harvesters		Swathers		Other harvesters	
	Number (10^4)	Total power (10^4 kW)	Number (10^4)	Total power (10^4 kW)	Number (10^4)	Total power (10^4 kW)	Number (10^4)	Total power (10^4 kW)	Number (10^4)	Total power (10^4 kW)
1996	918.92	8385.20	67.08	2415.10	9.64	301.50	78.09	30.46	10.19	12.25
1997	1048.48	9337.20	68.91	2486.50	14.13	370.60	93.00	37.20	11.21	13.88
1998	1122.06	10031.50	72.52	2587.90	18.26	467.50	103.08	42.26	12.33	15.88
1999	1200.25	11008.90	78.42	2772.80	22.60	570.50	100.03	42.01	13.56	18.34
2000	1264.37	11663.90	97.45	3161.10	26.26	660.90	110.69	47.81	14.92	21.38
2001	1305.08	12257.90	82.99	2901.70	28.29	744.50	105.61	57.54	18.21	28.18
2002	1339.39	12695.00	91.17	3073.40	31.01	855.70	112.35	57.97	18.79	32.00
2003	1377.71	13060.20	98.06	3229.80	36.50	1085.80	95.08	45.46	42.04	74.45
2004	1454.93	13855.40	111.86	3713.10	41.05	1265.60	87.91	44.30	41.90	81.61
2005	1526.89	14660.90	139.60	4293.50	48.04	1549.90	82.51	47.52	47.82	112.18
2006	1567.90	15229.10	171.82	5245.30	56.56	1896.20	68.25	45.97	50.45	132.66
2007	1619.11	15729.20	206.27	6101.10	63.38	2194.50	60.89	40.25	58.79	143.17
2008	1722.41	16647.70	299.52	8186.50	74.35	2707.40	49.21	46.31	69.37	189.21
2009	1750.90	16922.70	351.58	9772.60	85.84	3364.30	48.93	47.46	83.44	233.63
2010	1785.79	17278.39	392.17	11166.99	99.21	4042.30	49.85	48.85	101.29	288.68

Year	Irrigation machinery		Primary processing machinery		Animal husbandry machinery		Fishing motorboats		Forestry machinery		Pedicabs		Low-speed trucks		Farmland construction machinery	
	Number	Total power	Number	Total power	Number	Total power	Number	Total power	Number	Total power	Number	Total power	Number	Total power	Number	Total power
	(10^4)	(10^4kW)	(10^4)	(10^4kW)	(10^4)	(10^4kW)	(10^4)	(10^4kW)	(10^4)	(10^4kW)	(10^4)	(10^4kW)	(10^4)	(10^4kW)	(10^4)	(10^4kW)
1996	509.29	3984.90	176.81	1580.62	52.19	114.28	35.89	864.00	0.59	4.44	288.77	2482.73	43.76	1040.48	4.52	248.45
1997	546.12	4292.40	187.04	1672.03	53.88	119.35	39.88	1080.50	0.49	3.43	388.04	3412.58	60.05	1446.40	5.37	258.61
1998	581.61	4499.00	196.60	1757.55	61.10	136.92	41.13	1174.40	0.44	3.00	462.77	4160.86	76.34	1862.41	6.12	280.84
1999	644.95	4934.60	216.56	1935.98	72.62	164.59	41.74	1253.10	0.52	3.44	557.34	5120.84	92.62	2288.50	6.95	325.29
2000	688.12	5232.60	236.15	2119.58	72.58	191.00	45.99	1338.60	1.78	12.18	660.68	6219.97	112.07	2828.87	9.44	465.94
2001	728.57	5580.00	250.99	2215.04	83.36	194.35	48.01	1379.60	2.19	12.97	730.21	6996.78	124.36	3173.99	11.68	606.59
2002	750.61	5667.90	257.10	2313.68	94.27	219.58	48.57	1381.20	2.79	18.09	800.20	7830.44	142.25	3667.05	13.72	747.23
2003	749.57	5592.80	266.16	2385.51	100.12	232.38	47.81	1425.90	4.41	22.92	866.79	8622.85	155.66	4100.84	15.57	887.88
2004	777.54	5804.20	263.24	2359.40	109.07	243.47	48.69	1384.20	4.27	30.42	944.45	9608.98	169.77	4405.82	17.69	1024.84
2005	809.91	6034.00	277.15	2503.47	119.61	272.32	43.96	1376.10	5.73	33.99	1002.39	10418.32	189.84	5086.09	20.71	1226.35
2006	836.35	6148.80	284.45	2551.20	132.29	278.58	49.21	1498.30	6.72	39.32	1026.87	10777.86	202.85	5179.50	30.26	1422.34
2007	861.50	6282.80	293.83	2621.56	139.24	295.70	52.48	1605.30	8.16	48.92	1049.29	11205.87	235.96	5754.34	32.87	1628.80
2008	898.39	6561.70	300.41	2645.75	137.47	382.05	54.45	1679.49	9.47	59.51	1065.97	11807.90	233.86	7190.90	37.68	1954.85
2009	924.92	6795.50	306.28	2664.62	136.30	399.52	56.40	1748.73	13.36	73.97	1078.71	12034.09	255.51	8176.45	39.35	2139.06
2010	946.25	6969.20	311.46	2678.58	138.76	416.29	58.36	1820.82	17.36	92.88	1087.09	12341.52	271.80	8969.50	39.64	2251.53

REFERENCES

1. Dincer, I.; Hussain, M.; Al-Zaharnah, I. Energy and exergy utilization in agricultural sector of Saudi Arabia. *Energy Policy* 2005, *33*, 1461–1467.
2. Utlu, Z.; Hepbasli, A. Assessment of the energy and exergy utilization efficiencies in the Turkish agricultural sector. *Int. J. Energy Res.* 2006, *30*, 659–670.
3. Al-Ghandoor, A.; Jaber, J.O. Analysis of energy and exergy utilisation of Jordan's agricultural sector. *Int. J. Exergy* 2009, *6*, 491–508.
4. Avara, A.; Karami, M. Energy and exergy efficiencies in agricultural and utility sectors of Iran compared with other countries. In Proceedings of the 2010 2nd International Conference on Mechanical and Electrical Technology (ICMET), Singapore, 10–12 September 2010; pp. 6–10.
5. Ahamed, J.U.; Saidur, R.; Masjuki, H.H.; Mekhilef, S.; Ali, M.B.; Furqon, M.H. An application of energy and exergy analysis in agricultural sector of Malaysia. *Energy Policy* 2011, *39*, 7922–7929.
6. Rijal, K.; Bansal, N.; Grover, P. Energy in subsistence agriculture: a case study of Nepal. *Int. J. Energy Res.* 1991, *15*, 109–122.
7. Ozkan, B.; Akcaoz, H.; Fert, C. Energy input-output analysis in Turkish agriculture. *Renew. Energy* 2004, *29*, 39–51.
8. Hatirli, S.A.; Ozkan, B.; Fert, C. An econometric analysis of energy input–output in Turkish agriculture. *Renew. Sustain. Energy Rev.* 2005, *9*, 608–623.
9. Alam, M.; Alam, M.; Islam, K. Energy flow in agriculture: Bangladesh. *Am. J. Environ. Sci.* 2005, *1*, 213–220.
10. Uri, N.D. Energy and the use of conservation tillage in US agriculture. *Energy Sources* 1999, *21*, 757–771.
11. Karkacier, O.; Gokalp Goktolga, Z.; Cicek, A. A regression analysis of the effect of energy use in agriculture. *Energy Policy* 2006, *34*, 3796–3800.
12. Türkekul, B.; Unakıtan, G. A co-integration analysis of the price and income elasticities of energy demand in Turkish agriculture. *Energy Policy* 2011, *39*, 2416–2423.
13. Danièls, B.; Van Dril, A. Save production: A bottom-up energy model for Dutch industry and agriculture. *Energy Economics* 2007, *29*, 847–867.

14. Baruah, D.C.; Bora, G.C. Energy demand forecast for mechanized agriculture in rural India. *Energy Policy*2008, *36*, 2628–2636.
15. Feng, L.J.; Shang, X.; Xing, Z. Survey analysis of wheel tractor fuel consumption [in Chinese]. *Tractor Farm Transp.* 2007, *34*, 1–3.
16. GB/T15370. *General Requirement of Agricultural Wheeled Tractors and Crawler Tractors*; General Administration of Quality Supervision, Inspection and Quarantine of China, Standardization Administration of China: Beijing, China, 2004.
17. CSY. *China Statistical Yearbook*; National Bureau of Statistics of China, National Development and Reform Commission: Beijing, China, 1997–2011.
18. CESY. *China Energy Statistical Yearbook*; National Bureau of Statistics of China, National Development and Reform Commission: Beijing, China, 1997–2011.
19. CRSY. *China Rural Statistical Yearbook*; National Bureau of Statistics of China: Beijing, China, 1997–2011.
20. CASR. *China Agriculture Statistical Report*; Ministry of Agriculture of China: Beijing, China, 1996–2010.
21. CLRB. *China Land and Resources Bulletin*; Ministry of Land and Resources of China: Beijing, China, 1996–2010.
22. CWRB. *China Water Resources Bulletin*; Ministry of Water Resources of China: Beijing, China, 1996–2010.
23. Renpu, B. Statistical error analysis of operating level of Chinese agricultural mechanization. *Modern Agric. Equip.* 2005, *Z2*, 72–76.
24. Cheremnykh, E.; Gori, F. Exergy and Extended Exergy Cost Assessment of a Commercial Truck. In Proceedings of the ASME 2010 International Mechanical Engineering Congress and Exposition (IMECE2010), Vancouver, Canada, 12–18 November 2010; Volume 5, pp. 461–468.
25. Seckin, C.; Sciubba, E.; Bayulken, A.R. An application of the extended exergy accounting method to the Turkish society, year 2006. *Energy* 2012, *40*, 151–163.

CHAPTER 8

AN OPTIMIZATION METHOD FOR LOCAL CONSUMPTION OF PHOTOVOLTAIC POWER IN A FACILITY AGRICULTURE MICRO ENERGY NETWORK

Yuzhu Wang [1], Huanna Niu [1], Lu Yang [1], Weizhou Wang [2], and Fuchao Liu [2]

[1]College of Information and Electrical Engineering China Agricultural University, Beijing 100083, China

[2]State Grid Gansu Electric Power Research Institute, Lanzhou 730000, China

ABSTRACT

In order to solve the problem of optimal dispatching of photovoltaic power for local consumption to the greatest degree in a photovoltaic greenhouse, this paper proposes a multiform energy optimal dispatching model and a solution algorithm. First, an input-output power model is established for energy storages which are reservoir, biogas digester, and block wall with phase-change thermal storage. Based on it, multiform energy storages play a bridging role of energy transfer in optimal energy dispatching. Subsequently, an optimal energy dispatching model is proposed with the

Citation: Wang, Y.; Niu, H.; Yang, L.; Wang, W.; Liu, F. An Optimization Method for Local Consumption of Photovoltaic Power in a Facility Agriculture Micro Energy Network. *Energies* 2018, *11*, 1503.

objective of minimizing the sum of the squares of the difference between the loads and the photovoltaic generation in dispatching periods. Control variables are working state quantities of the time-shiftable loads and input-output state quantities of energy storages in dispatching periods. Finally, a genetic algorithm with matrix binary coding is used to solve the energy optimal dispatching model. Simulation results of a practical photovoltaic greenhouse facility agricultural micro energy network system in three typical weather conditions showed that the method could fully utilize the energy transfer function of the multiform energy storage and the time-shiftable characteristics of the agricultural load to achieve the maximum effect of increasing the local consumption of the photovoltaic power.

Keywords: local consumption, photovoltaic power, optimization, micro energy grid

INTRODUCTION

Under the support of the Chinese government's "Photovoltaic Poverty Alleviation" and other policies [1,2,3], especially in western China, due to abundant solar energy resources and high altitude terrain, building distributed photovoltaic power generation on idle land in rural areas has become an effective way to increase farmers' income and solve the problem of agricultural electricity use. However, the existing rural grids in the western region are simple and cannot meet the security and stability requirements for power grids when using tie lines for distributed power to grid power delivery. On the other hand, the main network has already saturated its capacity for absorbing new energy. Therefore, only if coordinated with carrying out local consumption of photovoltaic power in accordance with local conditions, can the photovoltaic poverty alleviation projects be well implemented. Based on this, the photovoltaic greenhouse has become a model for rural areas in western China to use the local consumption technology of photovoltaic power for implementing photovoltaic poverty alleviation. Photovoltaic power generation has been widely used in the vast rural areas of China. The photovoltaic greenhouse is a typical application of photovoltaic power generation combined with facility agriculture. A facility agriculture micro energy network (FAMEN) system is formed based on modern agricultural planting. This facility-based agricultural micro energy network system combines multi-energy complementary control technology with modern agricultural cultivation technology, taking full advantage of solar energy, biomass energy, and new agricultural load. It consumes the photovoltaic peak output by optimal dispatch control, completes the process of "collecting on-site, storing on-site, and using on-site" of solar energy, and achieves the goal of promoting the growth of crops in photovoltaic greenhouses. The key to the successful implementation of this technology is how to optimize the multiform energy including energy storages, and agricultural load in the facility's agricultural micro energy network system to realize the local consumption of photovoltaic power.

As an extension of the energy internet, the concept of a micro energy network [4,5,6] with multiform energy has provided new ideas for solving the issue of local consumption of photovoltaic power. At present, research on micro energy networks in the literature at home and abroad are mostly based on the discussion of concept and architecture [7,8,9,10]. There is no report on the local consumption of photovoltaic power in a micro energy network through optimal dispatching of multiform energy. Much research has been conducted for the problem of energy optimal dispatching of microgrids containing photovoltaic power, which mainly focuses on energy optimization from the aspects of economic operation [11,12,13,14], peak clipping and valley filling [15,16,17,18], and demand-side response [19,20,21] at home and abroad. However, the investment and maintenance costs of electricity storage in microgrids increase dramatically with the increase of battery installation capacity, which makes the configuration of energy storage in the microgrid more conservative and the cost and maintenance requirement too high. In addition, the energy-optimized dispatching strategy of microgrids, which aims at economy, is ultimately achieved through the use of the energy space-time shifting function of energy storage. Therefore, the frequent charge and discharge operations of battery storage cannot be avoided, and the energy storage life is greatly affected by frequent charging and discharging. Moreover, the battery-based energy storage form is single in the microgrid, and the limitation of demand-side response technology equipment makes the load dispatching not yet reach a substantial application. The two factors responsible for this are that the power fluctuation in the microgrid has a great impact on the upper-level power grid, and the microgrid system has a low photovoltaic absorption capability.

Compared to the currently widely studied microgrid systems, the facility agriculture micro energy network system has a variety of energy types of energy storage, and most of the load has time-shifting characteristics. The photovoltaic power's maximal local consumption also makes it impossible for the system to use the economic cost as the goal to carry out power exchange with the main network. Therefore, the conventional microgrid power optimization dispatching method is no longer applicable in facility agriculture micro energy networks. In view of the above problems, this paper first analyzes the typical structure of photovoltaic micro energy grids in greenhouse facilities, and then proposes a multimodal energy input-output model. Then, an energy optimization dispatching model is established, aiming at minimizing the sum of the square of the difference between the load energy and the photovoltaic (PV)-generated energy in dispatching periods, so as to achieve the goal of maximum local consumption of photovoltaic power. A genetic algorithm with matrix binary coding is used to solve the problem. Finally, an example of an actual photovoltaic greenhouse facility agricultural micro energy network system is used to verify the effectiveness of the energy optimization model and algorithm.

TYPICAL STRUCTURE OF PHOTOVOLTAIC GREENHOUSE FACILITY AGRICULTURAL MICRO ENERGY NETWORK SYSTEM

A typical facility agricultural micro energy network with a photovoltaic greenhouse is externally connected to the distribution network. The interior of it is mainly composed of a photovoltaic power generator, reservoir, biogas digester, block wall with phase-change thermal storage, and various types of loads. The schematic diagram of the structure is shown in Figure 1.

According to the type of energy that the load belongs to and its time-shiftable characteristics, the load can be divided into six categories. They are time-shiftable electrical load, non-time-shiftable electric load, time-shiftable heat load, non-time-shiftable heat load, time-shiftable potential energy load, and non-time-shiftable potential energy load. Table 1 shows the classification of various types of loads and their characteristics in the typical photovoltaic greenhouse facility agricultural micro energy network system.

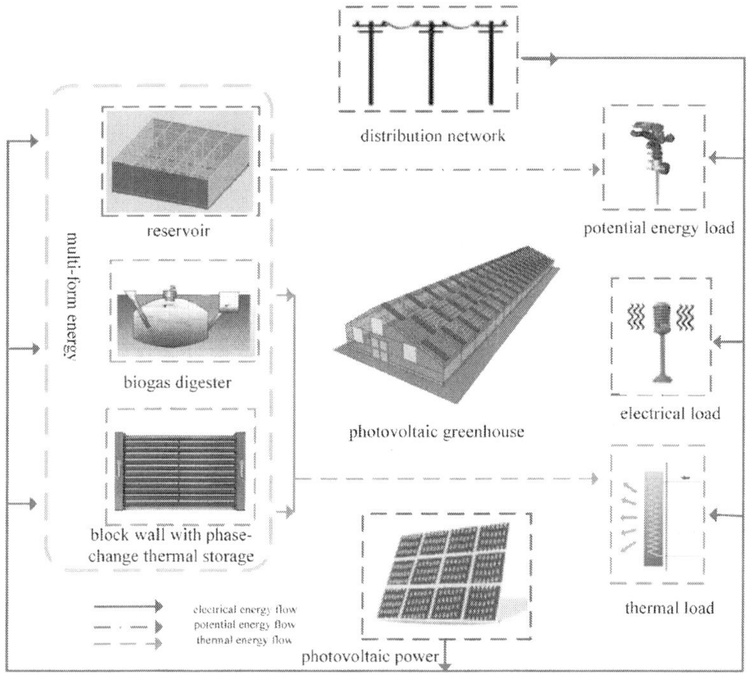

Figure 1: Schematic diagram of the typical structure of facility agriculture micro energy network with photovoltaic greenhouse.

Table 1: Load characteristics classification in photovoltaic greenhouse facilities agricultural micro energy networks

Load Names	Uses	Load Characteristics
LED plant growth lighting	Increase production	Time-shiftable electric load
Plasma nitrogen fixation and water treatment	Increase production and prevent diseases	Time-shiftable electric load
Biogas heat pump	Biogas production	Time-shiftable heat load
Far infrared heating	Insulation	Time-shiftable heat load
Greenhouse heat pump storage system of phase-change	Insulation	Time-shiftable heat load
Physical insecticide	Organic agriculture	Time-shiftable electric load
Irrigation watering	Organic agriculture	Time-shifted potential energy load
Space electric field	Purify and increase production	Non-time-shiftable electric load
Rolling shutter motor	Insulation	Non-time-shiftable electric load

Shed lighting	Daily work	Non-time-shiftable electric load
Smart agricultural monitoring and control system	Monitoring control	Non-time-shiftable electric load
Plant nutrient recycling	Sterilization	Non-time-shiftable electric load
Sound waves encourage	Increase production	Time-shiftable electric load
Ventilator	ventilation	Non-time-shiftable electric load

Energy stores in the facility agriculture micro energy network with photovoltaic greenhouse exist in many forms, including reservoir, biogas digester, and block wall. They play a bridging role of energy transfer in optimal dispatching of photovoltaic power. Energy storage of the reservoir utilizes pumped storage to convert electrical energy into potential energy for water storage, providing energy for plant irrigation when required; energy storage of the biogas digester utilizes bio-gas heat pump heating to promote biogas production, converting the chemical energy of biomass wastes and electrical energy into biomass chemical energy of biogas, and directly heating biogas to provide heat for greenhouse warming when needed; energy storage of the block wall with phase-change thermal storage uses phase change heat storage heat pumps to convert electrical energy into heat energy storage, and provide heat to heat up the greenhouses when needed.

In order to more vividly express the significance of the optimization goal to improve energy efficiency, the exergy analysis method was used to analyse the photovoltaic greenhouses' agricultural micro energy network system. The exergy analysis diagram is shown in Figure 2.

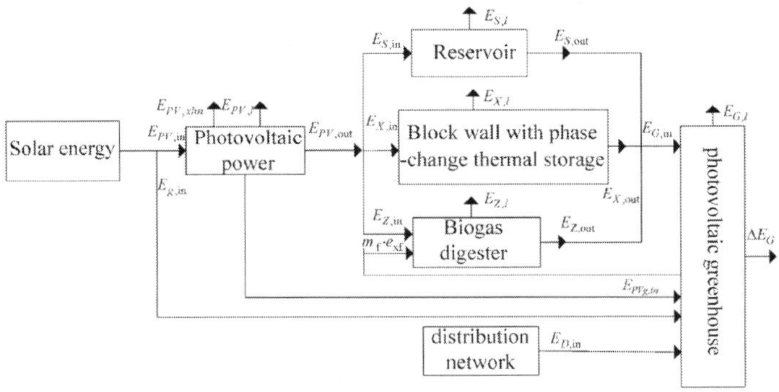

Figure 2: Photovoltaic greenhouse facility agricultural micro energy network exergy analysis diagram.

The exergy equilibrium equation of photovoltaic greenhouse micro energy network system is:

$$E_{PV,in} + E_{g,in} + E_{D,in} - E_{loss} = \Delta E_G \tag{1}$$

Among them, $E_{PV,in}$ is the solar exergy input to the photovoltaic power source, so that the photovoltaic power source works to produce electrical energy; $E_{g,in}$ is the direct exergy input of the solar energy to the plants in the greenhouse for photosynthesis; $E_{D,in}$ is the exergy input of the distribution network to the electrical loads in the greenhouse which help promote the growth of plants; ΔE_G is the exergy increase of the photovoltaic greenhouse micro energy network; E_{loss} is the total exergy loss of the micro energy network, which consists of exergy loss in each part of the micro energy network, as shown in Equation (2).

$$E_{loss} = E_{PV,xlin} + E_{PV,l} + E_{S,l} + E_{X,l} + E_{Z,l} + E_{G,l} \tag{2}$$

Among them, $E_{PV,xlin}$ is the unavoidable exergy loss when the photovoltaic power source carries out energy conversion; $E_{PV,l}$ is the exergy loss caused by the abandonment of the photovoltaic power source; $E_{S,l}$ is the exergy loss when the energy storage of the reservoir is converted; $E_{X,l}$ is the exergy loss in the process of energy conversion in a block wall with phase-change thermal energy storage; $E_{Z,l}$ is the exergy loss in the process of energy conversion in biogas digester energy storage; $E_{G,l}$ is the exergy loss of the various types of loads in the greenhouse.

Reducing exergy loss can improve the exergy efficiency of the micro energy network system. In the exergy loss of various parts of the micro energy network system, the exergy loss caused by the abandonment of the photovoltaic power source has the greatest optimization space. Through reasonable optimization of the exergy input of photovoltaic power to the electric loads in the greenhouse $E_{PV,gin}$, the exergy input of the photovoltaic power to the energy storage of reservoir $E_{S,in}$, the exergy input of the photovoltaic power to the energy storage of block wall with phase-change thermal storage $E_{X,in}$, and the exergy input of the photovoltaic power to the energy storage of biogas digester $E_{Z,in}$, the exergy loss caused by the abandonment of the photovoltaic power $E_{PV,l}$ can be reduced, and the exergy efficiency of the micro energy network system can be improved. In other words, by fully utilizing the energy transfer function of multiform energy storage and time-shiftable loads, it is possible to increase the local consumption of the photovoltaic power, thereby achieving the effect of improving the exergy efficiency of the micro energy network system.

MULTIFORM ENERGY STORAGE INPUT-OUTPUT POWER MODEL

Reservoirs, biogas digesters, and block walls with phase-change thermal storage are among the common multiform energy storages in facility agriculture micro energy networks with photovoltaic greenhouses. They can be used as the energy transfer function to realize the goal of local consumption of photovoltaic power, serving as intermediate bridges between photovoltaic power and time-shiftable agricultural loads. That is, they consume electrical energy as electrical load, and store it as other forms of energy during peak PV output period, providing energy for agricultural load directly as other forms of energy during periods of low PV output. Similar to electricity storage, multiform energy stores are regarded as loads when energy is input, and they are used as energy sources when energy is output. Based on this, the input-output power model of three types of storage energy is established as follows:

(1) Reservoir energy input-output power model

The reservoir in the facility agriculture micro energy network can use pumped storage to convert electrical energy into potential energy for water storage for plant watering. Based on this, the model of the energy storage input-output of the reservoir is established as follows.

$$\begin{cases} P_{SI} = P_W \\ P_{SO} = P_{SI} t_{SImax} \eta_{SI} / t_{SOmax} \end{cases} \quad (3)$$

In the formula, P_{SI} is the input power of the reservoir, kW; P_w is the electric pump power of the electric load, kW; P_{SO} is the output power when the reservoir is used for watering the plants, that is, the energy supply power when the reservoir is used as an energy source, kW; t_{SImax} is the maximum input time of the reservoir, that is, the time required by the electric water pump to fill the reservoir from empty to full water capacity, h; t_{SOmax} is the maximum output time of the reservoir, that is, the time required for complete discharge of the full reservoir for watering plants, h; η_{SI} is the energy storage efficiency of the reservoir.

(2) Reservoir energy input-output power model

On one hand, a biogas digester in the facility agriculture micro energy network can utilize an electric heat pump to promote biogas production, thereby converting the chemical energy biomass waste and electric energy into bio-chemical energy of biogas; on the other hand, when the greenhouse requires heating, biogas can be directly burned to provide heat. Based on this, we established a biogas digester energy storage input-output power model as follows.

$$\begin{cases} P_{ZI} = P_Z \\ P_{ZO} = P_{ZI} t_{ZImax} \eta_{ZI} / t_{ZOmax} \end{cases} \quad (4)$$

In the formula, P_{ZI} is the input power of the biogas digester, kW; P_z is the heat pump power for the biogas production, kW; P_{ZO} is the output power of the biogas digester, kW; t_{ZImax} is the maximum input time of the biogas digester, that is, the time required to use the digester heat pump to promote the gas storage from empty gas storage to full gas storage during biogas production, h; η_{ZI} is the energy storage efficiency of the digester; $t_{ZOmaxis}$ the maximum output time of the digester, that is, the time required by the biogas digester to heat up the greenhouse from the full gas storage capacity to empty gas storage capacity, h.

(3) Block wall with phase-change thermal storage energy input-output power model

A block wall with phase-change thermal storage in the facility agriculture micro energy network can use phase-change heat storage heat pumps to convert electrical energy into heat energy. On the other hand, it can be used as a heat energy source to heat greenhouses. The input-output power model of the block wall with phase-change thermal storage is established as follows.

$$\begin{cases} P_{XI} = P_X \\ P_{XO} = P_{ZI} t_{ZImax} \eta_{ZI} / t_{XOmax} \end{cases} \quad (5)$$

In the formula: P_{XI} is the input power of the block wall with phase-change thermal storage, kW; P_x is the phase change heat storage heat pump power, kW; P_{XO} is the output power of the block wall with phase-change thermal storage, kW. t_{ZImax} is the maximum input time of the block wall with phase-change thermal storage, that is, the time required to store heat from 0 to full heat storage when the phase change heat storage pylon is used by the heat pump, h; η_{ZI} is the energy storage efficiency of the block wall with phase-change thermal storage; t_{XOmax} is the maximum output time of the block wall with phase-change thermal storage, that is, the time required to heat up the greenhouse from full heat storage capacity to 0 heat storage capacity, only using the block wall with phase-change thermal storage

OPTIMAL ENERGY DISPATCHING MODEL FOR FACILITY AGRICULTURE MICRO ENERGY NETWORK SYSTEMS WITH PHOTOVOLTAIC GREENHOUSES

Objective Function

In a facility agriculture micro energy network system with a photovoltaic greenhouse, the local multiform loads in the system are the main targets for system operation optimization. The objective function for this optimal dispatching problem is described in terms of the square of the difference between the load energy and the photovoltaic power generation in the dispatching period. It can make the energy consumption curve of the micro energy network system and the photovoltaic output curve the most similar in the dispatching period. The objective function is as follows.

$$\min \sum_{t=1}^{T} (L_t - L_{\mathrm{pv}t})^2 \tag{6}$$

In the formula, T is the number of the dispatching periods in one cycle, L_t is the using energy of loads after optimization of t period, kWh; L_{pvt} is the predicted value of photovoltaic power in t, period kWh.

Lt consists of the non-time-shiftable load, time-shiftable load, and input/output energy of energy storage in the t period.

$$L_t = L_{\mathrm{dx}t} + L_{\mathrm{move}t} + L_{\mathrm{c}t} \tag{7}$$

In the formula, L_{dxt} is the total energy of the non-time-shiftable load in t period, kWh, L_{movet} is the energy of time-shiftable agricultural electricity, heat, and potential energy loads in t period, kWh, as shown in Formula (8); L_{ct} is the input and output energy of energy storage in t period, kWh, as shown in Equation (9).

$$L_{\mathrm{move}t} = \sum_{i=1}^{n} P_{\mathrm{e}i} x_{\mathrm{e}i,t} \Delta t + \sum_{i=1}^{m} P_{\mathrm{p}i} x_{\mathrm{p}i,t} \Delta t + \sum_{i=1}^{v} P_{\mathrm{q}i} x_{\mathrm{q}i,t} \Delta t \tag{8}$$

In the formula, n is the number of time-shiftable agriculture electric loads; m is the number of time-shiftable agricultural potential loads; v is the number of time-shiftable agricultural heat loads; P_{ei} is the power of the i-th time-shiftable agricultural electric load kW; P_{pi} is the power of the i-th type of time-shiftable agricultural potential load, kW; P_{qi} is the power of the i-th type of time-shiftable agricultural heat load, respectively, kW; $x_{ei,t}$ is the working state quantity of the i-th type of time-shiftable agriculture electric energy load in the t-th dispatching period; $x_{pi,t}$ is the working state quantity of the i-th type of time-shiftable agricultural potential load in the t-th dispatching period; $x_{qi,t}$ is the working state quantity of the i-th type of time-shiftable agricultural thermal load in the t-th dispatching period; Δt is the interval time between each dispatching period, h.

$$L_{ct} = (P_{SI}x_{SIt} - P_{SO}x_{SOt} + P_{ZI}x_{ZIt} - P_{ZO}x_{ZOt} + P_{XI}x_{XIt} - P_{XO}x_{XOt})\Delta t \tag{9}$$

In the formulas: x_{SIt}, x_{ZIt}, and x_{XIt} respectively represent the input state quantity of energy storage in the reservoir, biogas digester, and block wall with phase-change thermal storage in the t-th dispatching period; x_{SOt}, x_{ZOt}, and x_{XOt} represent the output state quantities of energy storage in the reservoir, biogas digester, and block wall with phase-change thermal storage in the t-th dispatching period, respectively. The input and output power of each energy storage is calculated according to Equations (3)–(5).

Constraints

(1) Energy balance constraint

$$\begin{aligned}
&\sum_{t=1}^{T}(L_{dx,et} + L_{move,et}) + S'_e - S_e^0 - \sum L_e = 0\\
&\sum_{t=1}^{T}\left(L_{dx,pt} + L_{move,pt}\right) + S'_p - S_p^0 - \sum L_p = 0\\
&\sum_{t=1}^{T}\left(L_{dx,qt} + L_{move,qt}\right) + S'_q - S_q^0 - \sum L_q = 0
\end{aligned} \tag{10}$$

In the formulas, $L_{dx,et}$, $L_{dx,pt}$, and $L_{dx,qt}$ are respectively the sum of non-time-shiftable agricultural electrical loads, the non-time-shiftable potential loads and the non-time-shiftable heat loads in the dispatching period t, kWh; $L_{move,et}$, $L_{move,pt}$, and $L_{move,qt}$ are respectively the sum of time-shiftable agricultural electric loads, the time-shiftable potential loads, and the time-shiftable heat loads in the dispatching period t, kWh. S'_e, S'_p, and S_q^0 correspond to the stored electrical, potential, and heat energy of the storage devices at the end of the current dispatching cycle, kWh; S_e^0, S_p^0, and S_q^0 correspond to the stored electrical, potential, and heat energy of the storage devices at the beginning of the current dispatching cycle, kWh; $\sum L_e, \sum L_p$, and $\sum L_q$ correspond to the total using energy by the electric loads, the potential loads, and the heat loads, respectively, in the current dispatching cycle, kWh.

(2) Multiform energy storage space constraint

$$Q^0{}_S + \sum_{t=1}^{T} (x_{SIt}P_{SI}\eta_{SI} - x_{SOt}P_{SO})\Delta t \le A_S$$
$$Q^0{}_Z + \sum_{t=1}^{T} (x_{ZIt}P_{ZI}\eta_{SI} - x_{ZOt}P_{ZO})\Delta t \le A_Z$$
$$Q^0{}_X + \sum_{t=1}^{T} (x_{XIt}P_{XI}\eta_{XI} - x_{XOt}P_{XO})\Delta t \le A_X \quad (11)$$

In the formula, $Q^0{}_S$, $Q^0{}_Z$, and $Q^0{}_X$ represent the energy initially stored in the reservoir, the biogas digester, and block wall with phase-change thermal storage at the beginning of the current dispatching cycle, kWh; A_S, A_Z, and A_X indicate the energy storage space of the reservoir, biogas digester, and block wall with phase-change thermal storage, respectively.

(3) Time-shiftable load continuity work status constraint

Some time-shiftable loads must work for several dispatching periods in succession due to the limitations of the production process. For example, the physical insecticide load must work continuously for 2 h to achieve the desired effect. Therefore, the following constraint is established for time-shiftable loads that require continuous operation:

$$x_{lt} \cdot x_{lt+1} \cdots x_{lt+k-1} = 1 \quad (i = 1, 2, \cdots T \quad \text{and } l \in \mathbf{G}) \quad (12)$$

In the formula, x_{lt}, x_{lt+1}, and x_{lt+k-1} represent the working state quantities of the time-shiftable load l in the t, $t + 1$, and $t + k - 1$ dispatching periods respectively. G is a set for time-shiftable loads of continuous work.

(4) Time-shiftable load non-working time constraint

Some time-shiftable loads cannot work at certain times due to the limitations of the production process. For example, watering irrigation load is difficult from 0:00 to 10:00 and 19:00 to 24:00, considering the influences on plant growth. Therefore, an inoperable period state constraint is established for such a time-shiftable load.

$$x_{lt} = 0 \quad (t = \mathbf{N} \quad \text{and } l \in \mathbf{F}) \quad (13)$$

In the formula, x_{lt} represents the work state quantity of the time-shiftable load l in the t-th dispatching period; F is the set of restrictive time-shiftable loads of the working period; N is the complete set of the dispatching periods that the time-shiftable load l cannot work in.

(5) Input and output status of energy storage mutually exclusive constraint

It is not possible for all types of energy storage to be in the input and output states at the same time. The energy input/output state quantity must satisfy the mutual exclusion constraint:

$$\begin{aligned} x_{\mathrm{SI}t}x_{\mathrm{SO}t} &= 0 \\ x_{\mathrm{ZI}t}x_{\mathrm{ZO}t} &= 0 \quad (t = 1,2,\cdots T) \\ x_{\mathrm{XI}t}x_{\mathrm{XO}t} &= 0 \end{aligned} \tag{14}$$

In the formulas: x_{SIt}, x_{ZIt}, and x_{XIt} respectively represent the input state quantity of energy storage in the reservoir, biogas digester and block wall with phase-change thermal storage in the *t*-th dispatching period; x_{SOt}, x_{ZOt}, and x_{XOt} represent the output state quantities of energy storage in the reservoir, biogas digester, and block wall with phase-change thermal storage in the *t*-th dispatching period, respectively.

SOLUTION OF THE OPTIMAL DISPATCHING MODEL BASED ON A GENETIC ALGORITHM WITH MATRIX BINARY CODING

Matrix Binary Coding of Control Variables

The control variables of this dispatching problem are: working state quantities of the time-shiftable electric loads, heat loads, potential loads, and input and output state quantities of the reservoir, biogas digester, and block wall with phase-change thermal storage energy in dispatching periods. In this work, the genetic algorithm adopts the grouping matrix binary coding method and decomposes control variables into two relatively independent sub-chromosomal codes. One is various types of time-shiftable loads in each dispatching period of work state quantities matrix sub-chromosome L as shown in Formula (14), the other is each form of energy storage input and output state quantities in each dispatching period matrix sub-chromosome Q, as shown in Formula (15). The matrix Cformed by the control variables is represented as:

$$\mathbf{C} = [\ \mathbf{L}\ \vdots\ \mathbf{Q}\ ] \tag{15}$$

$$\mathbf{L}=\begin{bmatrix} x_{el,1} & \cdots & x_{ei,1} & \cdots & x_{en,1} & x_{ql,1} & \cdots & x_{qi,1} & \cdots & x_{qm,1} & x_{pl,1} & \cdots & x_{pi,1} & \cdots & x_{pv,1} \\ \vdots & & \vdots & & \vdots & \vdots & & \vdots & & \vdots & \vdots & & \vdots & & \vdots \\ x_{el,t} & \cdots & x_{ei,t} & \cdots & x_{en,t} & x_{ql,t} & \cdots & x_{qi,t} & \cdots & x_{qm,t} & x_{pl,t} & \cdots & x_{pi,t} & \cdots & x_{pv,t} \\ \vdots & & \vdots & & \vdots & \vdots & & \vdots & & \vdots & \vdots & & \vdots & & \vdots \\ x_{el,T} & \cdots & x_{ei,T} & \cdots & x_{en,T} & x_{ql,T} & \cdots & x_{qi,T} & \cdots & x_{qm,T} & x_{pl,T} & \cdots & x_{pi,T} & \cdots & x_{pv,T} \end{bmatrix} \tag{16}$$

In the formula, $x_{ei,t}$ is the work state quantity of the *i*-th type time-shiftable agricultural electric load in the *t*-th dispatching period; $x_{pi,t}$ is the work state quantity of the *i*-th type time-shiftable agricultural potential load in the *t*-th dispatching period; $x_{qi,t}$ is the work state quantity of the *i*-th type time-shiftable agricultural heat load in the *t*-th dispatching period. A status of 0 means it is inactive, and 1 means it is in work.

$$\mathbf{Q}=\begin{bmatrix} x_{SI1} & x_{SO1} & x_{ZI1} & x_{ZO1} & x_{XI1} & x_{XO1} \\ \vdots & \vdots & \vdots & \vdots & \vdots & \vdots \\ x_{SIt} & x_{SOt} & x_{ZIt} & x_{ZOt} & x_{XIt} & x_{XOt} \\ \vdots & \vdots & \vdots & \vdots & \vdots & \vdots \\ x_{SIT} & x_{SOT} & x_{ZIT} & x_{ZOT} & x_{XIT} & x_{XOT} \end{bmatrix} \tag{17}$$

In the formula: x_{SIt}, x_{ZIt}, and x_{XIt} respectively represent the input state quantity of energy storage in the reservoir, biogas digester, and block wall with phase-change thermal storage in the *t*-th dispatching period. The state value is 0 if it is in a non-operating state and 1 if it is in an input state. x_{SOt}, x_{ZOt}, and x_{XOt} represent the output state quantities of energy storage in the reservoir, biogas digester, and block wall with phase-change thermal storage in the *t*-th dispatching period, respectively. The state value is 0 if it is in a non-operating state and 1 if it is in an output state.

Genetic Algorithm Solution Process

In this paper, a genetic algorithm with matrix binary coding is used to solve the established optimal dispatching model. The solution process is shown in Figure 2.

Step 1: Randomly generate the working state quantities of each type of time-shiftable load in each period within the corresponding constraint conditions to form a matrix L; randomly generate the state quantities of the input and output states of each form of energy storage within the corresponding constraint conditions to form a matrix Q, and according to (15), form an initial individual of the binary coded control variable matrix. Then, repeat N times to form the initial population of the control variables matrix, then the population size is *N*.

Step 2: Set the individual fitness function as the reciprocal of the objective function (6) of this work.

Step 3: Genetic Operators and Genetic Operations

? crossover

i. Formation of sub-chromosome L after crossover

The two parent sub-chromosomes selected for crossover are represented as:

$$\begin{aligned} F_1 &= [L_1^{F1}, L_2^{F1}, \ldots, L_i^{F1}, \ldots, L_M^{F1}] \\ F_2 &= [L_1^{F2}, L_2^{F2}, \ldots, L_i^{F2}, \ldots, L_M^{F2}] \end{aligned} \tag{18}$$

where: M represents the number of load types.

The crossover operation is performed on a column of the two parent sub-chromosomes, and the resulting new daughter chromosomes are denoted as F_1, F_2. The specific crossover operation process is:

(a) Generate a random crossover position j between 1 and M.

(b) Perform a crossover on the two column vectors at crossover position j to generate the new sub-chromosomes S1, S2:

$$\begin{aligned} S_1 &= [L_1^{F1}, L_2^{F1}, \ldots, L_{j-1}^{F1}, L_j^{F2}, L_{j+1}^{F2}, \ldots, L_M^{F2}] \\ S_2 &= [L_1^{F2}, L_2^{F2}, \ldots, L_{j-1}^{F2}, L_j^{F1}, L_{j+1}^{F1}, \ldots, L_M^{F1}] \end{aligned} \tag{19}$$

ii. Formation of sub-chromosome Q after crossover

The two parent sub-chromosomes selected for crossover are represented as:

$$\begin{aligned} B_1 &= [Q_{11}^{F1}, Q_{12}^{F1}, \ldots, Q_{i1}^{F1}, Q_{i2}^{F1} \ldots, Q_{m1}^{F1}, Q_{m2}^{F1}] \\ B_2 &= [Q_{11}^{F2}, Q_{12}^{F2}, \ldots, Q_{i1}^{F2}, Q_{i2}^{F2}, \ldots, Q_{m1}^{F2}, Q_{m2}^{F2}] \end{aligned} \tag{20}$$

where m represents the number of energy storage species.

The crossover operation is performed on two columns of the two parent daughter chromosomes corresponding to the input and output state quantities of energy storage. The resulting daughter chromosomes are denoted as D1 D_1 and D_2. The specific crossover operation process is:

(a) Generate a random crossover position j between 1 and m.

(b) Perform a crossover operation on the two column vectors at crossover position j to generate the the new sub-chromosomes D_1 and D_2:

$$\begin{aligned} D_1 &= [Q_{11}^{F1}, Q_{12}^{F1}, \ldots, Q_{(j-1)2}^{F1}, Q_{j1}^{F2}, Q_{j2}^{F2}, \ldots, Q_{m1}^{F2}, Q_{m2}^{F2}] \\ D_2 &= [Q_{11}^{F2}, Q_{12}^{F2}, \ldots, Q_{(j-1)2}^{F2}, Q_{j1}^{F1}, Q_{j2}^{F1}, \ldots, Q_{m1}^{F2}, Q_{m2}^{F1}] \end{aligned} \tag{21}$$

? variation

Whether to mutate is determined by the mutation probability, Pm. Randomly select the mutation bit to perform the mutation operation as shown in Equation (22). Individuals that have undergone mutation operations perform feasibility tests based on constraints. Retain viable variant individuals that meet the constraints.

$$\boldsymbol{M} = [X_1, \boldsymbol{X}_2, \ldots, \boxed{\boldsymbol{X}_i}, \ldots] \quad \uparrow \; 1 - X_i \tag{22}$$

? copy

Set the flag vector to flag if the constraint condition is satisfied. Kick out individuals who do not satisfy the constraints, calculate and store each individual fitness value within the current population, and select the better individuals to copy to the next generation.

Step 4: Repeat steps 2 through 4 until the termination condition is met.

EXAMPLES

Basic Information

This paper uses the simulation of an actual photovoltaic greenhouse facility agricultural micro energy network in western rural China as an example. The system's structure is shown in Figure 1. The system is a demonstration project of 'Research and Demonstration of Key Technologies for Local Consumption Dissipation of Distributed Power Sources Based on Time-Shiftable Agricultural Loads'. The capacity parameters of photovoltaic power and multiform energy storage have been addressed by another sub-topic research of the project which is on the optimization of agricultural micro energy network system configuration based on the actual situation in the area. The micro energy network system has 280 kWp photovoltaic power installed; the energy space of the reservoir is 200 kWh; the energy space of the biogas digester is 5400 kWh; the energy space of the block wall with phase-change thermal storage is 300 kWh.

According to the actual situation in the region, the produced energy conditions of photovoltaic greenhouse under three typical scenarios of a winter sunny day, cloudy day, and rainy day are used as examples to verify the effectiveness of the models and algorithms constructed in this work. The dispatching cycle is one day and the dispatching period is one hour. The output curves of photovoltaic power under three typical weather conditions are shown in Figure 3.

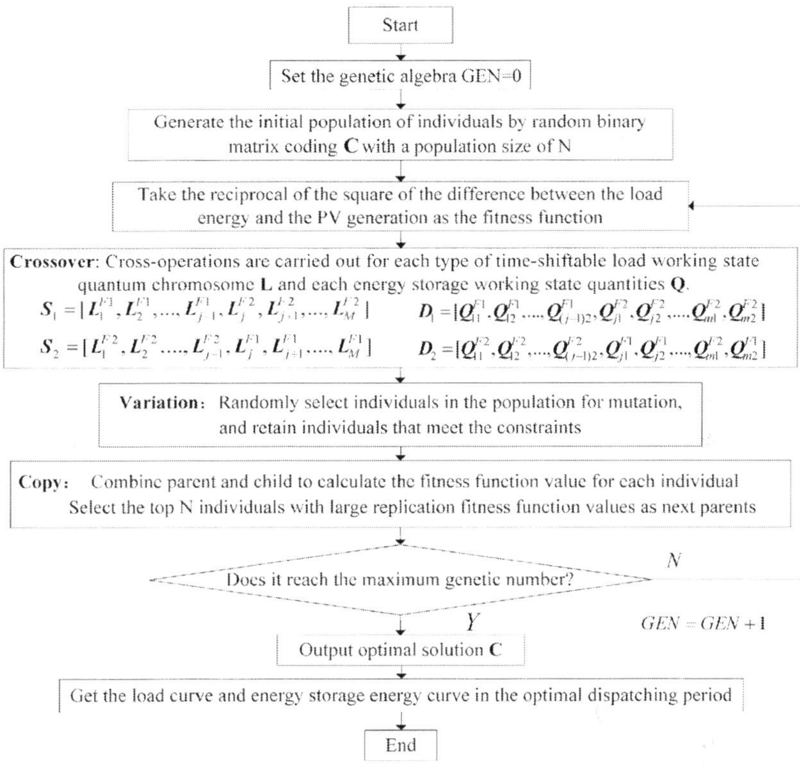

Figure 3: Genetic algorithm with matrix binary coding solution flowchart.

As can be seen from Figure 4, the photovoltaic output on a sunny day is significantly greater than on the cloudy and rainy days. On the sunny winter day, peak hours for photovoltaic power during the day are generated from 11:00 to 13:00. In the winter cloudy days, photovoltaic power output also reached a peak of one day near 12 o'clock. In the winter rainy day, the photovoltaic output fluctuates about a relatively low value with no apparent peak period.

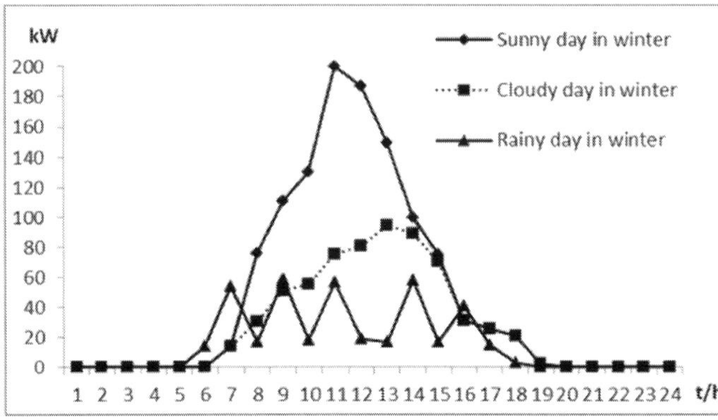

Figure 4: Photovoltaic output curves for three typical weather days.

Analysis of Examples

(1) Sunny day in winter

The model and algorithm established in this work we used for optimization calculation. On a sunny winter day, the loads curve before and after the optimization of the photovoltaic greenhouse facility agricultural micro energy network is shown in Figure 5. After optimization, the stored energy curve of each form of energy storage is shown in Figure 6. The comparison of the time-shiftable load and energy storage working time before and after optimization is shown in Table 2.

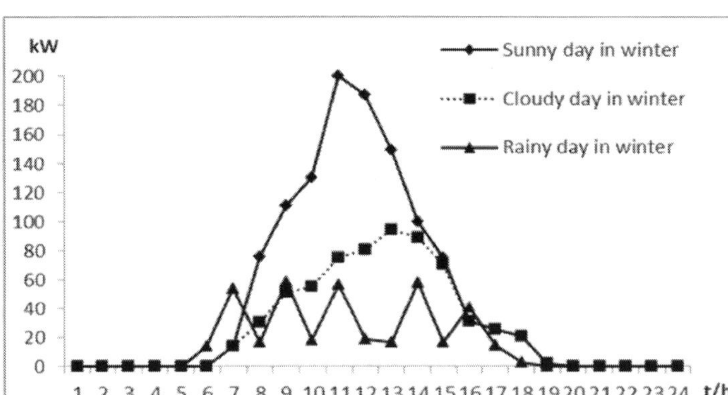

Figure 5: Comparison curve of loads before and after optimization on a sunny winter day.

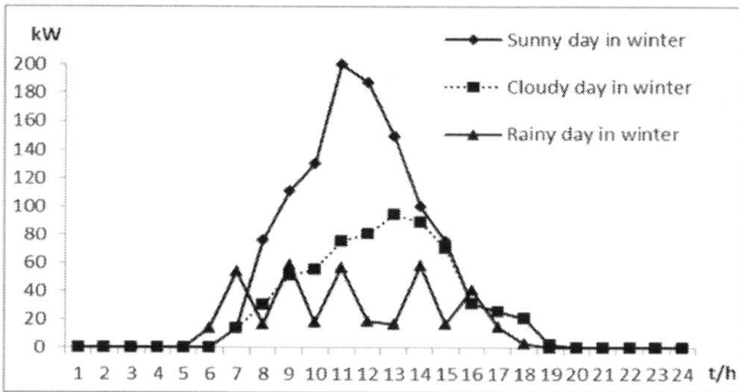

Figure 6: The stored energy curve of energy stores on a winter sunny day.

Table 2: Timetable of loads and energy storages working time before and after optimization on a sunny winter day.

Load	**Power kW**	**Workable Hours**	**Work Time before Optimization**	**Work Time after Optimization**
ventilator	20	0–23	8; 9; 12–17	10–16
LED growth lighting	50	6–20	6–11; 14	8; 9; 11–15
Plasma nitrogen fixation and water treatment	20	0–23	12; 14	7; 11
Plant nutrient recycling	2	0–23	10–12	12–14
Sound waves encourage	3	0 -23	7–9; 14–19	8–16
Physical insecticide	1.5	0–23	12; 13	11; 12
Far infrared heating	40	0–23	0–5; 20–23	0–5, 24 phase change heat storage heat pump works to provide heat; 20–23 biogas combustion provides heat
Irrigation watering	20	10 -17	14–17	14–17 Water Storage Irrigation Water Potential Energy
Reservoir pump input	30	0–23	--	10–13
Biogas pump heat pump input	60	0–23	--	9–11; 13

Phase change heat storage heat pump input	50	0–23	--	8–12

As can be seen from Figure 5 and Figure 6 and Table 2, the load curve of the micro energy network system before the optimization in the winter sunny day is very different from the PV output curve. During the daytime from 8:00 to 15:00, PV output is much greater than the load. There is no photovoltaic output during the night, but the demand for electricity at night is greater due to the need to heat the greenhouse through far-infrared heating. After optimization, the load curve is close to the PV output curve. In the 8:00–15:00 time period, during which the PV output is relatively high during the day, the reservoir, biogas digester, and block wall with phase-change thermal storage are used to store the electric energy as other forms of energy, thereby increasing the electric load during PV peak output period. On the other hand, the working time of time-shiftable loads, such as sound wave encouragement, Light Emitting Diode (LED) growth lighting, ventilators, etc., are also adjusted to this time period as a supplementary adjustment of the electric load. The two together maximize the consumption of photovoltaic power during daytime hours. In the absence of photovoltaic output during the night, the biogas digester, and block wall with phase-change thermal storage are used together to increase the temperature of the greenhouse to replace the far-infrared heating as the electric load in order to reduce the electric load at night.

Before the optimization, the micro energy network purchased 623.71 kWh from the distribution network and consumed 491.30 kWh of photovoltaic power. After optimization, the micro energy network purchased 26.45 kWh from the distribution network and consumed 1077.71 kWh of photovoltaic power, increasing the amount of photovoltaic energy consumption by 586.41 kWh.

It can be seen that on a sunny winter day, the local optimization method of photovoltaic power supply provided by the facility agriculture micro energy network proposed in this work can fully utilize the energy transfer function of multiform energy storage and the time-shiftable characteristics of time-shiftable agricultural load to achieve the optimal local consumption of photovoltaic power.

(2) Rainy day in winter

The model and algorithm established in this paper were used for the optimization calculation. The loads curve before and after the optimization of the photovoltaic greenhouse facility agricultural micro energy network on a rainy winter day is shown in Figure 7. The stored energy curve of each form of energy storage after optimization is shown in Figure 8. The comparison of working time before and after the time-shiftable load optimization is shown in Table 3.

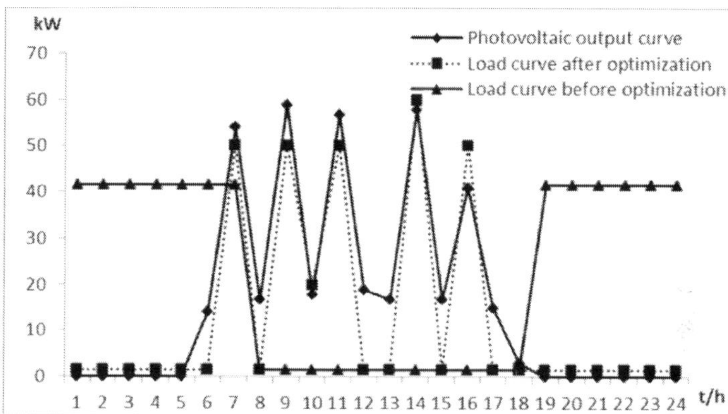

Figure 7: Comparison curve of load before and after optimization on a rainy winter day.

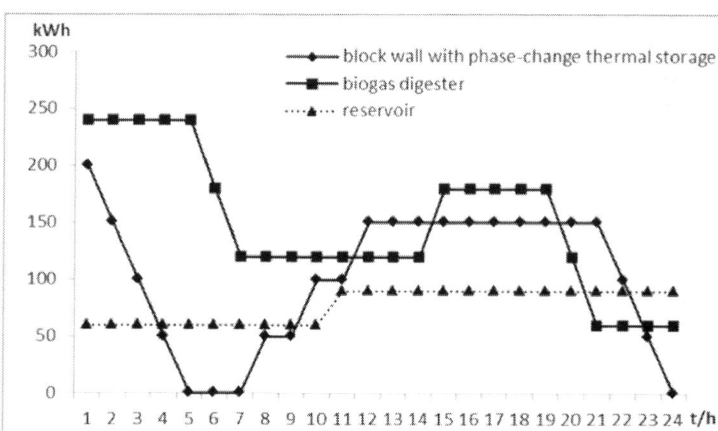

Figure 8: The stored energy curve of energy stores on a winter rainy day.

Table 3: Energy storage working time before and after optimization on a rainy winter day

Load	Power kW	Work-able Hours	Work Time before Optimization	Work Time after Optimization
Far infrared heating	40	0–23	0–6; 19–23	0–4, 21–23 phase change heat storage heat pump works to provide heat; 5–6, 19–20 biogas combustion provides heat
Reservoir pump input	30	0–23	--	10
Biogas pump heat pump input	60	0–23	--	14
Phase change heat storage heat pump	50	0–23	--	7; 9; 11; 16

From Figure 7 and Figure 8 and Table 3, it can be found that on a cloudy rainy day in winter, the load curve of the micro energy network system before the optimization is different from the PV output curve: during the period of 7:00 to 14:00, the photovoltaic output fluctuates to a certain degree, but the load using energy is very small and there is no time-shiftable load; in the night, there is no PV output, but due to the need to increase the temperature of the greenhouse through far-infrared heating, the demand for electricity is increased. After optimization, the load curve is close to the PV output curve: in the 7:00–14:00 time period, the storage reservoir, biogas digester, and block wall with phase-change thermal storage are used to store the electric energy as other forms of energy, thereby increasing the electric load during PV peak output period. In the absence of photovoltaic output during the night, the biogas digester and block wall with phase-change thermal storage are used together to increase the temperature of the greenhouse to replace the far-infrared heating in order to reduce the electrical load at night.

Before the optimization, the micro energy network purchased 483.92 kWh from the distribution network and consumed 72.08 kWh of photovoltaic power. After optimization, the micro energy network purchased 29.86 kWh from the distribution network and consumed 277.14 kWh of photovoltaic power, increasing the amount of photovoltaic energy consumption by 205.06 kWh.

It can be seen that on a rainy winter day, the local optimization method of photovoltaic power supply provided by the facility agriculture micro energy network proposed in this work can fully utilize the energy transfer function of multiform energy storage to achieve the optimal local consumption of photovoltaic power.

(3) Cloudy day in winter

In order to analyze the influence of the final state of each form of energy storage on the next dispatching period in the previous dispatching cycle, the scenario simulation under the cloudy winter day is set to two conditions: (a) perform the simulation calculation of the cloudy day after the above-described '(1) Sunny day in winter' dispatch, when the initial state of the energy storage of each form on this cloudy day is the final state of energy storage after the optimal dispatching of the sunny winter day; (b) perform the simulation calculation of the cloudy day after the above-described '(2) Rainy day in winter' dispatch, when the initial state of the energy storage of each form on this cloudy day is the final state of energy storage after the optimal dispatching of the rainy winter day.

(a) Cloudy day in winter after sunny day

On a cloudy winter day after sunny day dispatching, the loads curve before and after the optimization of the photovoltaic greenhouse facility agricultural micro energy network is shown in Figure 9. After optimization, the stored energy curve of each form of energy storage is shown in Figure 10. The comparison of the time-shiftable load and energy storage working time before and after optimization is shown in Table 4.

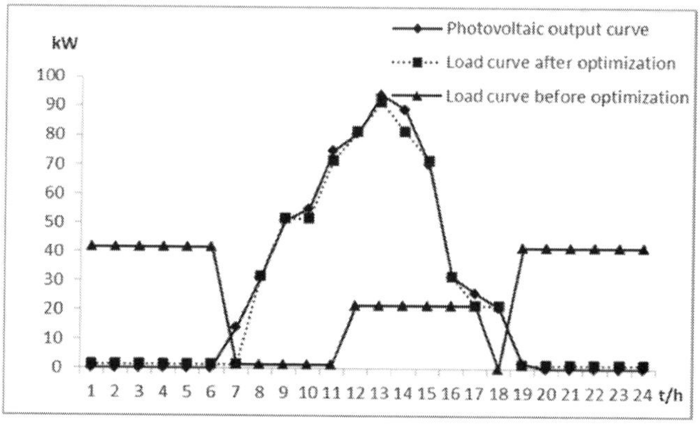

Figure 9: Comparison curve of load before and after optimization on a cloudy winter day after a sunny day.

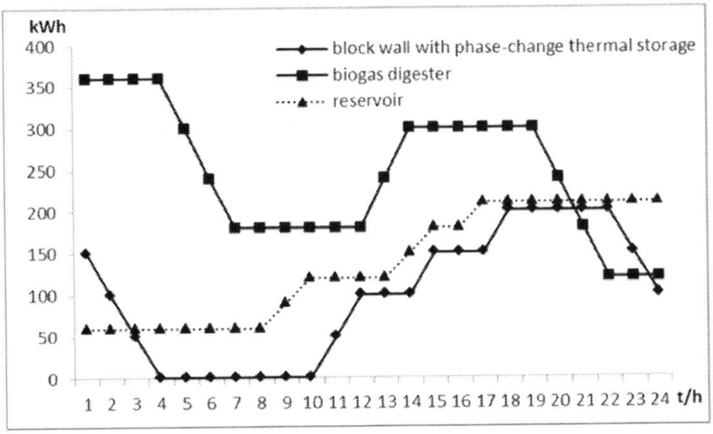

Figure 10: The stored energy curve of energy stores on a winter cloudy day after a sunny day.

Table 4: Timetable of loads and energy stores working time before and after optimization on a cloudy winter day after a sunny day

Load	Power kW	Workable Hours	Work time before Optimization	Work Time after Optimization
Ventilator	20	0–23	12–17	9; 11; 12; 15; 17; 18
Far infrared heating	40	0–23	0–6; 19–23	0–3; 22; 23 phase change heat storage heat pump works to provide heat; 4–6; 19–21; biogas combustion provides heat
Reservoir pump input	30	0–23	--	8; 9; 13; 14; 16
Biogas pump heat pump input	60	0–23	--	12; 13
Phase change heat storage heat pump input	50	0–23	--	10; 11; 14; 15

As can be seen from Figure 9 and Figure 10, and Table 4, the load curve of the micro energy network system before the optimization on the winter cloudy day is very different from the PV output curve. During the daytime from 7:00 to 16:00, PV output is much greater than the load. There is no photovoltaic output during the night, but the demand for electricity at night is greater due to the need to heat the greenhouse through far-infrared heating. After optimization, the load curve is close to the PV output curve. In the 7:00–16:00 time period, during which the PV output is relatively high during the day, the reservoir, biogas digester, and block wall with phase-change thermal storage are used to store the electric energy as other forms of energy, thereby increasing the electrical load during PV peak output period. On the other hand, the working time of the ventilator is also adjusted to this time period as a supplementary adjustment of the electric load. The two together maximize the consumption of photovoltaic power during daytime hours. In the absence of photovoltaic output during the night, the biogas digester and block wall with phase-change thermal storage are used together to increase the temperature of the greenhouse to replace the far-infrared heating in order to reduce the electric load at night.

Before the optimization, the micro energy network purchased 495.91 kWh from the distribution network and consumed 57.10 kWh of photovoltaic power. After optimization, the micro energy network purchased 13.25 kWh from the distribution network and consumed 551.80 kWh of photovoltaic power, increasing the amount of photovoltaic energy consumption by 494.70 kWh.

(b) Cloudy day in winter after rainy day

The loads curve before and after the optimization of the photovoltaic greenhouse facility agricultural micro energy network on a cloudy winter day after rainy day dispatching is shown in Figure 11. The stored energy curve of each form of energy storage after optimization is shown in Figure 12. The comparison of the time-shiftable load and energy storage working time before and after optimization is shown in Table 5.

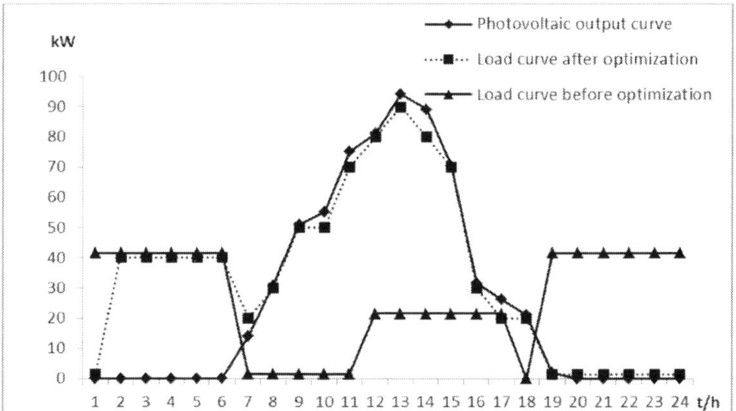

Figure 11: Comparison curve of load before and after optimization on a cloudy winter day after a rainy day.

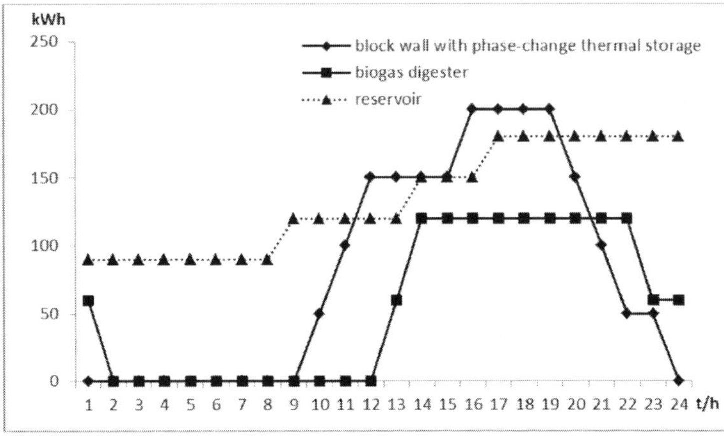

Figure 12: The stored energy curves of energy stores on a winter cloudy day after a rainy day.

Table 5: Timetable of loads and energy stores working time before and after optimization on a cloudy winter day after a rainy day

Load	Power kW	Workable Hours	Work time before Optimization	Work Time after Optimization
Ventilator	20	0–23	12–17	7; 11; 12; 15; 17; 18
Far infrared heating	40	0–23	0–6; 19–23	0; 19–21; 23 phase change heat storage heat pump works to provide heat; 22 biogas combustion provides heat
Reservoir pump input	30	0–23	--	8; 13; 16
Biogas pump heat pump input	60	0–23	--	12; 13
Phase change heat storage heat pump input	50	0–23	--	9–11; 15

After optimization, the micro energy network purchased 214.9 kWh from the distribution network and consumed 582.40 kWh of photovoltaic power, increasing the amount of photovoltaic energy consumption by 525.40 kWh.

Similarly, it can be seen from Figure 11 and Figure 12 and Table 5 that the goal of maximizing the PV output can be achieved after being optimized under the cloudy day after the rainy day. In addition, comparing the optimal results of the cloudy day scenario (a) after a sunny day and (b) after the rainy day, it can be found that, due to the fact that in scenario (a), all types of energy stores have stored more energy at the end of the previous dispatching period (sunny day), the energy storages of biogas digester and block wall with phase-change thermal storage release heat for warming up the greenhouses between 0:00 and 7:00. In scenario (b), at the end of the previous dispatching period (rainy day), all types of stored energy have less energy and cannot meet the heating demand between 0:00 and 7:00. The temperature of the greenhouse can only be increased by means of far-infrared lighting, so the electricity purchased from the distribution network under scenario (b) (214.9 kWh) greatly increases compared to that under scenario (a) (13.25 kWh). Since the photovoltaic power supply basically has no output during this period, it has little effect on the amount of photovoltaic consumption, making little difference in the total amount of photovoltaic consumption between the two scenarios. From the analysis of the above two scenarios, it can be seen that the last state of the energy stores from the last dispatching period has an effect reflected mostly in the initial periods of the next dispatching period. However, in both scenarios, the optimization method proposed in this paper can be used to guide the dispatching system to make full use of the energy transfer function of multiform energy stores and the time-shiftable characteristics of agricultural loads, so as to maximize the local consumption of photovoltaic power.

(4) Analysis of photovoltaic power consumption

As shown in Table 6, after optimization, the photovoltaic power consumption on a sunny day reached 96.1%, an increase of 52.4%. On rainy days and cloudy days under two conditions, the percentage of photovoltaic power consumption reached 86.0%, 90.7%, and 71.5%. Compared with a rainy day, the optimized PV power consumption and consumption ratio are higher than on sunny days and cloudy days because more time-shiftable loads allow greater optimization space.

Table 6: Local photovoltaic power consumption results after optimization

Scenarios	Consumption before Optimization	Percentage of Consumption before Optimization	Consumption after Optimization	Percentage of Consumption after Optimization	Percentage of Increase in Consumption before and after Optimization
Sunny day	491.3 kWh	43.7%	1077.7 kWh	96.1%	52.4%
Rainy day	72.1 kWh	18.6%	277.1 kWh	71.5%	52.9%
Cloudy day after sunny day	57.1 kWh	8.9%	551.8 kWh	86.0%	77.1%
Cloudy day after rainy day	57.1 kWh	8.9%	582.4 kWh	90.7%	81.8%

CONCLUSIONS

This work proposes a model for input and output of multiform energy storage power for photovoltaic energy greenhouses. Based on this, an energy optimization dispatching model for microgrid systems is established, which aims to minimize the sum of the squared difference between load energy and photovoltaic generation in various dispatching periods. Finally, the genetic algorithm with matrix binary coding for state quantities of the time-shiftable loads and each form of energy store input/output is used to solve the dispatching problem.

Simulation examples in three typical weather days show that the proposed optimization model and algorithm can fully utilize the energy transfer function of multiform energy storage and the characteristics of time-shiftable agricultural loads to achieve the purpose of optimal local consumption of photovoltaic power. The final state of each form of energy storage in the previous dispatching cycle has an effect on the next dispatching cycle, with a focus on the start periods of the day. Due to the limitation of the load in different weather conditions, the examples of the sunny day and cloudy day are more optimized than those of rainy days which have greater photovoltaic power consumption.

AUTHOR CONTRIBUTIONS

Y.W. and H.N. performed the data analyses and wrote the manuscript; L.Y. helped perform the analysis with constructive discussions; W.W. and F.L. contributed significantly to analysis and manuscript preparation.

ACKNOWLEDGMENTS

This work was supported as a project in science and technology of the State Grid Corporation of China, No. 5227221600KQ.

REFERENCES

1. National Energy Administration. The State Council Leading Group Office of Poverty Alleviation and Development. In *Engineering Work Plan about Implementing PV Poverty Alleviation*; National Energy Administration: Beijing, China, 2014.
2. Wang, X. Tension and Digestion in the Process of Precise Poverty Alleviation Policy in Contiguous. Master's Thesis, Shangdong University, Jinan, China, 2017.
3. Xu, B. Research on the Implementation of the Policy of Precise Poverty Alleviation in Jining, Shangdong. Master's Thesis, Northeast Agricultural University, Harbin, China, 2017.
4. Cho, H.; Smith, A.D.; Mago, P. Heating cooling heating and power: A review of performance improvement and optimization. *Energy* 2014, *136*, 168–185.
5. Rafiee Sandgani, M.; Sirouspour, S. Priority-based Microgrid Energy Management in a Network Environment. *IEEE Trans. Sustain. Energy* 2018, *9*, 980–990.
6. Peacock, A.D.; Newborough, M. Impact of micro-combined heat-and-power system on energy flows in the UK electricity supply industry. *Energy* 2006, *31*, 1468–1482.
7. Werth, A.; Kitamura, N.; Tanaka, K. Conceptual study for open energy systems: Distributed energy network using interconnected dc nanogridsWerth. *IEEE Trans. Smart Grid* 2015, *6*, 1621–1630.
8. Vesterlund, M.; Toffolo, A.; Dahl, J. Simulation and analysis of a meshed district heating network. *Energy Convers. Manag.* 2016, *122*, 63–73.
9. Vesterlund, M.; Dahl, J. A method for the simulation and opti-mization of district heating systems with meshed networks. *Energy Convers. Manag.* 2015, *89*, 555–567.
10. Tao, Z.; Fuxing, Z.; Yan, Z. Study on energy management system of energy internet. *Power Syst. Technol.*2016, *40*, 146–155.
11. Tianqi, L.; Donglin, J. Economic operation of microgrid based on operation mode optimization of energy storage unit. *Power Syst. Technol.* 2012, *36*, 45–50.
12. Yulong, H.; Chunjuan, J.; Yuankui, F. Research on capacity allocation and economy of islanded microgrid. In Proceedings of the 2017 IEEE Transportation Electrification Conference and Expo, Asia-Pacific

(ITEC Asia-Pacific), Harbin, China, 7–10 August 2017.

13. Hao, X.; Wei, P.; Yanhong, Y.; Zhiping, Q.; Li, K. Energy storage capacity optimization for microgrid considering battery life and economic operation. *High Volt. Eng.* 2015, *41*, 3256–3265.
14. Dan, X.; Qiang, D.; Yi, P. Study on optimizing capacity of storage battery in microgrid system based on economic dispatch. *Power Syst. Prot. Control* 2011, *39*, 55–59.
15. Xueting, Z.; Tianqi, L.; Qian, L. A dynamic peak load regulation margin based coordinated optimal dispatching under grid-connection of wind farm. *Power Syst. Technol.* 2015, *39*, 1685–1690.
16. Dany, G. Power reserve in interconnected systems with high wind power production. *IEEE Porto Power Tech Proc.* 2001, *4*, 10–13.
17. Tiejiang, Y.; Qin, C.; Yibulayin, T.; Yiyan, L. Optimized economic and environment-friendly dispatching modeling for large scale wind power integration. *Proc. CESS* 2010, *30*, 7–13.
18. Dongfeng, Y.; Suquan, Z.; Feng, B. Analysis on peak load regulation capability of power grid integrated with wind farms in valley load period. *Power Syst. Technol.* 2014, *38*, 1446–1451.
19. Carpinelli, G.; Mottola, F.; Proto, D. Optimal scheduling of a microgrid with demand response resources. *IET Gener. Transm. Distrib.* 2014, *8*, 1891–1899.
20. Xiang, Y.; Liu, J.; Liu, Y. Robust energy management of microgrid with uncertain renewable generation and load. *IEEE Trans. Smart Grid* 2016, *7*, 1034–1043.
21. Mehdizadeh, A.; Taghizadegan, N. Robust optimisation approach for bidding strategy of renewable generation-based microgrid under demand side management. *IET Renew. Power Gener.* 2017, *11*, 1446–1455.

CHAPTER 9

DO THE DIFFERENT EXERGY ACCOUNTING METHODOLOGIES PROVIDE CONSISTENT OR CONTRADICTORY RESULTS? A CASE STUDY WITH THE PORTUGUESE AGRICULTURAL, FORESTRY AND FISHERIES SECTOR

Ricardo Manso, Tânia Sousa , and Tiago Domingos

Marine, Environment and Technology Centre—MARETEC, Department of Mechanical Engineering, Instituto Superior Técnico, Universidade de Lisboa, Avenida Rovisco Pais, 1, 1049-001 Lisbon, Portugal

ABSTRACT

Three exergy accounting approaches are used to evaluate exergy efficiency: the Energy Resources Exergy Accounting (EREA), the Natural Resources' Exergy Accounting (NREA) and the Extended Exergy Accounting (EEA).

Citation: Manso, R.; Sousa, T.; Domingos, T. Do the Different Exergy Accounting Methodologies Provide Consistent or Contradictory Results? A Case Study with the Portuguese Agricultural, Forestry and Fisheries Sector. *Energies*2017, *10*, 1219.

To test the consistency of the results provided by these methodologies, we apply them to evaluate the Portuguese agricultural, forestry and fisheries (AFF) sector, from 2000 to 2012. EREA shows an increase of 30% in the efficiency of the Portuguese AFF sector, while NREA and EEA methodologies increases of 27% and 43%, respectively. Although the results are consistent for the AFF sector, the same does not happen in the fisheries subsector, whose exergetic efficiency increases 14% with the EREA but decreases 42% with the NREA approach. The ratio of output to useful exergy reveals that a better thermodynamic efficiency is not translated into a higher energy service efficiency because fishing vessels have to travel more to get the same fish. Thus, results provided by the EREA and NREA approaches are complementary and both are needed to provide a realistic picture of exergy efficiency. On the other hand, results obtained by the EEA approach are dominated by capital and environmental impacts, revealing the disproportionality between material and immaterial inputs in this methodology.

Keywords: Exergy, extended exergy accounting, efficiency, environmental assessment, agriculture

INTRODUCTION

Sustainable development is the desirable long term goal of bringing humankind to a compromise between its welfare and the protection of our planet. According to Pezzey [1] "A temptation when writing on 'defining sustainability' is to try to distill, from the myriad debates, a single definition which commands the widest possible academic consent. However, several years spent in fitful pursuit of this goal have finally persuaded me that it is an alchemist's dream, no more likely to be found than an elixir to prolong life indefinitely". Still, the most consensual definition of sustainable development is provided by the Brundtland Report [2]: "Sustainable development is development that meets the needs of the present without compromising the ability of future generations to meet their own needs".

The sustainability of society is critically dependent on the agricultural, forestry and fisheries (AFF) sector because all mass and energy flows that maintain our society come either directly from the environment or from the energy and mass surpluses obtained in this sector and the resource extraction and energy transformation sectors. The AFF sector provides all the food, feed and fiber and is also a major source of renewable energy resources. According to Committee on Twenty-First Century Systems Agriculture [3], the sustainability of the AFF sector should take into account the environmental, social and economic dimensions:

- Satisfy human food, feed, and fiber needs, and contribute to biofuels needs.
- Enhance environmental quality and the resource base.
- Sustain the economic viability of agriculture.
- Enhance the quality of life for farmers, farm workers, and society as a whole.

Several efforts have been made in measuring the sustainability of the AFF sector, but the focus is, in most cases, only on one of the pillars of sustainability. The complementary research of both social and economic sustainability dimensions with efficiency and environmental extraction is both scarce and complex since no common unit is easily found and accepted. Also, success in one dimension does not imply the same for the others. The economic viability of an agricultural sector is not directly correlated with the environmental efficiency of the same sector nor the wellbeing of farmers.

Exergy analysis is an assessment methodology that has been used to account for efficiency at several scales, which can be society as a whole, economic sectors or industries. The term efficiency usually reflects a narrow approach that considers only one type of output and input flows. However, an efficiency measure of the AFF sector that would take into account several types of flows, such as the extraction from the environment of renewable and non-renewable resources, undesired output flows such as waste and pollution and the capital and labour flows involved, could present useful insights about its sustainability. Further assessment of other aspects, such as the economic viability of agriculture or the quality of life of farmers, would still be needed for a complete evaluation of sustainability.

Exergy is the maximum amount of work that can be obtained from a system (a resource or a flow) when it is brought to equilibrium with the surrounding environment through a reversible process [4]. One of the key properties of exergy is that it allows the conversion of all inputs and outputs of a process into a common unit, allowing comparisons otherwise impossible.

The first methodology to quantify exergy flows was the energy resources' exergy accounting (EREA) methodology that started with energy carriers only [5]. Later, natural resources were included [6] in the natural resources' exergy accounting (NREA) methodology and more recently capital, labour and environmental externalities [7] in the extended exergy accounting (EEA) methodology.

EREA is an input assessment method that measures the exergy embedded (or intrinsic) in the energy carriers used by the transformation devices (inputs) and the useful exergy (outputs) (Figure 1). With EREA is possible to determine how well the system converts the final exergy to useful work from the available energy carriers and transformation devices; this knowledge can then be used to propose measurements to improve the efficiency of the system.

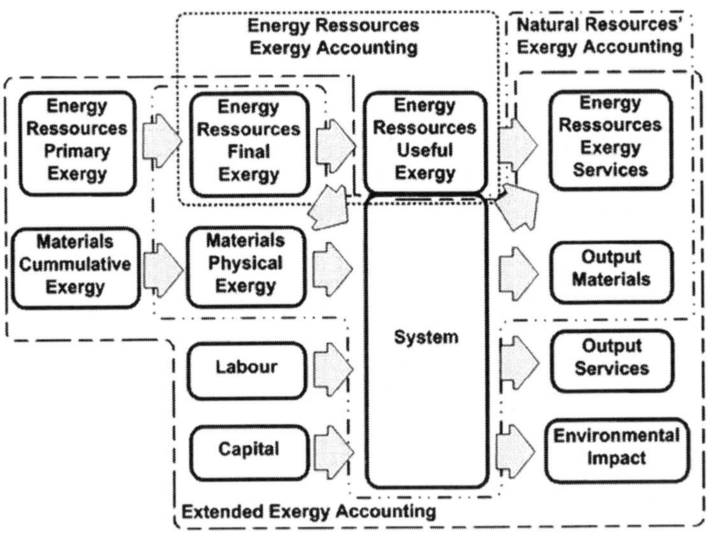

Figure 1: Exergy accounting methodologies.

EREA quantifies the useful exergy that is available for each end use, but does not inform on how well that exergy service is being used. To illustrate, consider two similar gasoline vehicles on a highway, the first with only one passenger and the second with five people on board. When assessing both systems by EREA the useful exergy is the same since the energy carrier-end use pairs are the same (gasoline-transportation). However, when measuring the system output, in this case the number of passengers times distance, the performances are widely different.

NREA is an input-output assessment where all energy and matter that enters and leaves the system is accounted for (Figure 1). Contrary to EREA, no information on how each energy carrier is used is needed but more data is required to account for all fluxes across system boundaries. By accounting both the input and output fluxes, the NREA approach measures the overall physical performance of the transformation process.

Finally, EEA is similar to NREA but also introduces capital and labour as inputs and environmental impact as a virtual input, integrating the social, economic and environmental pillars of sustainability. Translating capital, labour and environmental impact externalities into an exergetic cost is a key feature of EEA as it reflects the real thermodynamic system's efficiency and the usually hidden resource costs [8]. The production factors are no longer just energy resources (EREA) or energy and matter (NREA) but also immaterial flows which are essential to assess the system's behaviour (Figure 1). The exergy of inputs is no longer the physical and chemical exergies but the cumulative exergy introduced by [9] in the cumulative exergy consumption methodology (CExC). The latter takes into account the sum of all the exergetic content of resources that are consumed in a production chain. Consequently, the main focus of EEA is to find the real exergy embodied into the output flows under study, adding not only all exergy necessary to bring all inputs to the specific space and time domains but also all labour, capital and environmental impacts.

First exergetic studies focusing exclusively on agriculture adopted an EREA approach and included conversion in agriculture machinery, such as tractors, and electric pumps. Agricultural sectors of Saudi Arabia [10], Turkey [11], Jordan [12], Iran [13] and finally Malaysia [14] were studied longitudinally (except for Jordan) in order to verify the evolution of efficiency. The efficiencies of devices vary widely among studies.

First exclusive NREA approaches to agriculture also included virtual environmental impact as an input to integrate pollutant emissions [15]. The input-output ratio was approximately 10 (ten times more exergy in inputs than outputs) but solar radiation was responsible for 90 to 95% of the inputs with an estimated photosynthetic efficiency of 4.5%. A combination of EEA and cumulative exergy extraction from the natural environment was proposed by [16] to measure the agricultural performance of 29 countries from 1990 to 2003. As important findings, the study highlighted the three types of resources that agricultural production most extracts from the natural environment: organic content in top soil, feed and water. Finally, it concluded that the exergy efficiency in the livestock sector is much lower than in the crop sector. Some societal exergy studies that use the EREA approach, also included, in detail, the AFF sector [17,18]. These studies have allocated all energy carriers (e.g., diesel) used in the AFF sector to end-uses (e.g., mechanical drive). The AFF sectors of Sweden [19,20], Japan [21] and Italy [22] were assessed by the NREA methodology where the major energetic and material inputs were considered along with all the production from the sector. EEA studies also addressed the AFF sector. Efficiencies estimated by the EEA approach for the province of Siena, Italy [23]; the U.K. economy [24]; Chinese society [25,26,27,28,29], and the economy of Nova Scotia, Canada [30] were all below 100%. These EEA studies did not take into account solar radiation, but considered crops and wood as extractions from the environment and not as productions.

All three methodologies have been used to assess the performance of society [31] and of the AFF sector. However, analyses typically focus on results provided by only one methodology without justifying the use of that specific methodology in detriment of others. Also, there are major differences in the exergy inputs and outputs that are accounted for, across published NREA and EEA analyses of the agricultural sector. This is problematic if results provided by the three methodologies or even by the same methodology are inconsistent driving to different policy recommendations.

The present study evaluates the performance of the agricultural sector in Portugal, between 2000 and 2012, applying three exergy methodologies: the energy carrier's exergy accounting approach, the natural resource's exergy accounting approach and the extended exergy accounting.

METHODOLOGY AND DATA

Exergy Accounting of Energy Resources

Appendix A briefly revisits the exergy concept. More detail can be found in [32]. EREA measures the useful exergy services delivered to each end use. EREA estimates the system's useful exergy by multiplying the intrinsic exergy applied to the system by the specific real efficiency for each energy carrier in its dominant end use. The efficiencies reflect the actual performance of the machinery or device that uses the energetic carrier in each end use (Figure 2). The accounting methodology is done in four separate steps [17,32]. First, all final energy that enters the system is converted into final exergy, multiplying each carrier by its exergy factor. Second, end-uses are allocated to each energy carrier. Third, a real exergy (or second law) efficiency is attributed to each carrier-end use pair based on the actual final to useful transformation devices involved. Finally, all useful exergy obtained from each end use category are added to discover the overall system's useful exergy delivery and calculate an output-input ratio to assess the efficiency of the transformation process.

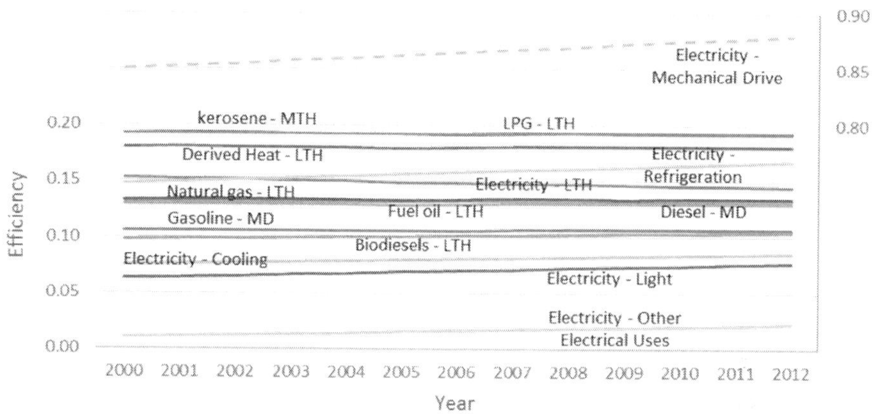

Figure 2: Energy carriers' conversion efficiency from 2000 to 2012 (full lines on the left axis and dashed line on the right axis).

End Use Categories

From all energy carriers presented in the statistical data, natural gas, biofuels, heat production and fuel oil were predominantly used in low temperature heat processes while liquefied petroleum gas and kerosene were mainly used to support medium temperature heat devices. Gasoline and diesel were prevalent in motors in fisheries and agricultural machinery, being used mainly for transportation in the fisheries subsector and mechanical drive in general (including transportation) in crops, husbandry and forestry [17]. Electrical energy was used for several end uses: 26.7% for low temperature heat (water and process heating), 51.7% for mechanical drive (irrigation, material handling and other process use), 5.0% for lighting, 4.2% for cooling (AC), 10.0% for refrigeration (industrial cooling) and 2.4% for other electrical equipment like phones, computers and network devices [33].

Conversion Efficiencies

Efficiencies vary by end use and by energy carrier since the technologies involved are different. Low temperature heat (LTH) efficiencies considered were 10% for solid biofuels and biodiesel, 13% for natural gas and fuel oil, 15% from electricity and 18% from derived heat. Medium temperature heat (MTH) achieved better efficiencies since higher temperature flows have higher exergetic content. Both liquefied petroleum gas and kerosene were considered to have a 19% transformation efficiency. Gasoline and diesel were used for transportation and mechanical drive (MD) with 11% and 13% efficiencies, respectively. Also electricity was used to drive mechanical appliances with an increasing efficiency from 85% in 2000 to 88% in 2012 (Figure 2). All efficiency values were obtained from [34,35].

Natural Resources' Exergy Accounting Approach

The NREA approach consists in quantifying all energy and matter that enters and leaves the system in the common unit of exergy. Contrary to EREA where the system is analysed in order to understand how the energy resources are used, in this methodology the system is evaluated only by quantifying the inputs and outputs, without any knowledge about the system's internal behaviour (Figure 3). The ratio between outputs and inputs will reflect how well energy and matter are being used and offers insights on the sustainability of the agricultural sector. Although the amount of data is usually a lot bigger than in EREA, the methodology is simpler to apply since we only have to address the exergetic content for each flux across the system's boundaries.

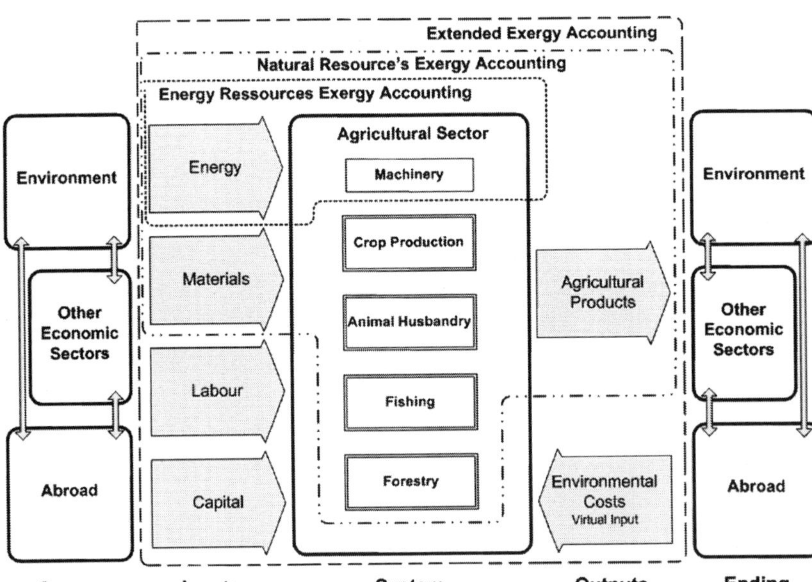

Figure 3: Exergy accounting methodologies for the Agricultural, Forestry and Fisheries Sector.

In this study, there is a clear distinction between the Agricultural, Forestry and Fisheries Sector and the environment (see Figure 3) focusing on the efficiency of anthropogenic activities. We consider crops, wood removals and fish from aquaculture as outputs because they are a product of an anthropogenic activity. In contrast with other studies [23,24,25,26,27,28,29,30] we do not count them as an extraction from the environment because they do not appear spontaneously. In contrast, we consider fish catch as a natural input from the environment.

Extended Exergy Accounting Approach

Extended exergy analysis adds, on top of natural NREA, the immaterial inputs that are crucial to run and maintain the system working and also adds the environmental impact as a virtual input (Figure 3). Labour and capital exergy estimations are based on two postulates elaborated by Sciubba [36], defending that the global influx of exergy resources are mainly used to sustain workers and, that the capital exergy flux is proportional to the labour exergy flux. The proposed equation to evaluate the exergy used to sustain all workers is (exergy of labour): $E_L=f\times e_{surv}\times N_h$, where f is an amplification factor that represents the society consumption over the survival mode, e_{surv} is the minimum exergy required to maintain healthy metabolic needs (2500 kcal/day per person or 1.05×10^7 J/day·person) and N_h is the total population. The amplification factor is estimated using the Human Development Index (HDI) [37] which, based on three social dimensions, reflects the development of a country, $f = \frac{HDI}{HDI_0}$, being HDI_0 the Human Development Index of a pre-industrial society ($HDI_0 \approx 0.055$) [36]. Labour equation implies that E_L is linearly dependent on the HDI and independent of the fraction of workers in the society and the type of work. The exergy equivalent of Labour (ee_L) for the society, or economic sector, can be calculated dividing the exergy of labour by the total hours worked (N_{wh}), $ee_L = \frac{E_L}{N_{wh}}$.

Seckin et al. [38] followed by Rocco et al. [8] slightly changed the equation proposed by Sciubba [36] for the exergy of the capital flux from $E_K = \frac{M_2}{S} \times E_L$ to $E_K = \frac{M_2 - S}{S} \times E_L$ where S is sum of all wages and M_2, defined by the European Central Bank [39], is an intermediate monetary aggregate which reflects the currency under circulation plus the liquid deposits (maturity up to 2 years and redeemable up to 3 months). The new formulation considers M_2-S because labour (reflected as the wages in the previous formula) was already internalized as an exergetic flux. The exergy equivalent of capital is obtained by dividing the total exergy of the capital flux by the monetary circulation, $ee_K = \frac{E_K}{M_2 - S}$. Capital equation assumes that the monetary aggregate M_2minus wages is eligible to the productive sector and represents the effective capital that drives all economic sectors. To translate money into exergy, the capital equation assumes the relationship developed for labour: $ee_K = \frac{E_K}{M_2 - S} = \frac{E_L}{S}$.

Any material stream that is not physically confined and represents a discharge of the system to the environment is considered a pollutant. The environmental impact in EEA consists in developing a state-of-the-art process to bring the pollutant wastes and emissions to equilibrium with the reference environment. The total amount of exergy used in the innovative process is accounted for as an environmental externality and a virtual input to the economic sector. The equivalent exergy of the environmental impact portrays all exergetic consumption to run such a process (real or virtual) namely energy, matter, capital and labour having no relation with the pollutant physical exergy.

Reference Environment

The reference environment should be carefully chosen since the resource flow has exergy due to the imbalance (lack of equilibrium) with this reference state. The flow will have zero exergy in complete equilibrium with the reference state. In this study, the pressure P_0 = 101325 Pa and temperature T_0 = 298.15 K were adopted as well as the standard chemical composition of the natural environment. This is the reference environment used by Kotas [40] to estimate chemical exergy values that were used in this study.

Time Window and Control Volume

Recent exergy studies of countries or large regions focus on the seven economic sectors of society plus the environment and abroad [24,30,41] (see Table 1).

Table 1: Description of Economic Sectors, Environment and Abroad

Extraction (Ex)	Extraction of minerals and ores, coal mining, oil and gas extraction, refining and fuel manufacturing.
Conversion (Co)	Power, cogeneration and heat plants, based on renewables and non-renewables.

Agriculture, Forestry and Fisheries (AFF)	Crop and animal production, forestry and logging, fisheries and aquaculture. Includes animal products like milk, honey and eggs.
Industry (In)	All manufacturing industry excluding refining.
Tertiary (Te)	Public and private sector services, includes schools, hospitals, construction, trade and commerce, tourism, real estate and finance.
Transportation (Tr)	Private and commercial transportation of people and goods.
Domestic (Do)	Includes households and population.
Environment (E)	Lithosphere, the hydrosphere and the atmosphere.
Abroad (A)	Other countries or regions outside de economic system boundaries which allow energy and mass transfers.

This study focuses on just one economic sector, the AFF, because of the relevance and preponderance in sustaining human life. Figure 4 exhibits the agricultural sector within the region's economy and specifies inputs and outputs considered for this study.

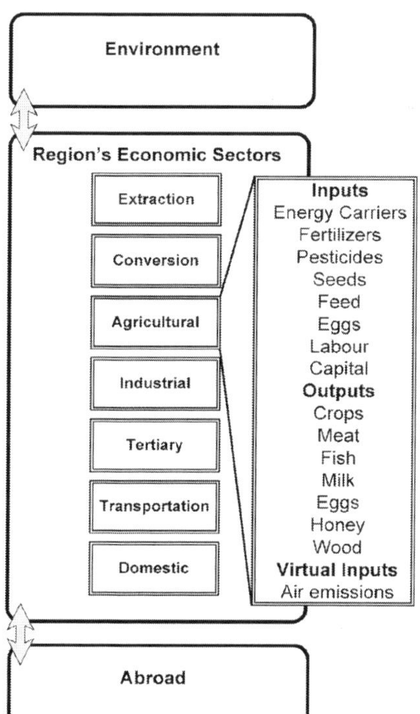

Figure 4: Inputs and outputs of the Agriculture, Forestry and Fishery (AFF) sector.

Time and regional boundaries were chosen in accordance with the available data, since few data were available posterior to 2012 and there was a lot of missing data prior to 2000.

Data

The statistical office of the European Union, Eurostat, provides in its database [42] the quantities of energy carriers, pesticides, fertilizers, harvested crops, meat slaughtered, fish catch, incubation eggs, milk collection, fish aquaculture, wood removals, atmospheric emissions, population, human labour and economic information. The energy of energy carriers was assumed to be the lower heating value (LHV) and the exergy factors for the energy carriers are the ratio of the standard chemical exergy of the organic fuels to the LHV, these factors being obtained from [40,43,44]. However, the consumption matrix of energy carriers is only available in two branches, being the first the agricultural plus the forestry subsectors and the second is the fisheries subsector. Also the data for pesticides and fertilizers is not disaggregated between agriculture and forestry subsectors. Due to this limitation and not wanting to apply some educated division between these carriers on both subsectors, all aggregated results will be presented in the conjunction of the agriculture and forestry subsectors.

Crop seeds, feed, produced eggs and honey carriers and some nutritional energy values for products not intended for human consumption were obtained from the statistics division of the Food and Agriculture Organization of the United Nations (FAOSTAT) [45]. Nutritional energy values assigned to each food element, in its raw state, are from USDA (United States Department of Agriculture) National Nutrient Database [46]. Moisture content of all food items was considered to be equal between USDA and Eurostat databases. All economic values such as compensation for employees, agricultural gross value added (GVA), monetary aggregate M2 and gross domestic product (GDP) were obtained at current prices and converted to constant euro GDP prices through a GDP price deflator obtained from the economic and financial affairs of the European commission (Ameco) [47]. The GDP price deflator is referenced to 2010 and measures the ratio between real GDP and the nominal GDP, providing a measure of inflation over the period.

Agricultural outputs were obtained in mass units, except wood which was available in volume units, and converted to exergy units using the energetic nutritional value presented in the USDA database [46] or in FAO (Food and Agriculture Organization of the United Nations) nutritive factors table [48], for food compounds. The food (and feed) energy content was considered equal to the metabolizable energy as defined by [46]. The energy content (combustible or gross energy) of the ingested food should be measured by a bomb calorimetry. However, foods are not fully digested and absorbed by the organism. From the ingested or gross energy, some is lost to faeces (faecal energy) and gases (combustible gas), the rest is the digestible energy. Subtracting from the digestible energy, the energy that is lost is urine and heat results in metabolizable energy [49]. Data available from the United States Department of Agriculture for food energy is based on the Atwater system which is equivalent to the metabolizable energy [46]. The choice to use the metabolizable energy was mainly due to the lack of a complete database of combustible or gross energy.

The aggregate of food energy values presented are not fully ingested by society. Food supply chains cover five steps: agricultural production, postharvest, handling and storage, processing and packaging and finally distribution and consumption. In each step food is lost or wasted and the total in Europe reaches 31%. For Europe, at the producer level, food losses and waste accounts 2% for cereals, 3.1% for meat, 3.5% for milk, 9.4% for fish and seafood, 10% for oilseeds and pulses and 20% for roots, fruits and vegetables [50]. Such losses are not edible for societal consumption.

To fill in the gaps for a given resource or product, in a year or more, data was interpolated (between known values) or extrapolated (in extremes of the data set) by a linear regression made from all the other known values.

Fluxes inside the sector were not considered and only fluxes between sectors that have economic relevance were accounted for. Although seeds, feed and incubations are produced by the sector, they were considered to leave the agricultural sector to other economic activities (ex. industry or tertiary) and returned as an input, essential to the production. However straw, green fodder and manure (used as a fertilizer) that are internal inputs and outputs of activities within the sector were not accounted for.

Solar radiation, vital for crop production, and water, essential for crops and livestock, were not included in the study. The sun radiation is not an anthropogenic controlled flux and it can be seen as a flux from the environment with no economic value. Water from rain follows the same thinking as it is not an anthropogenic activity but irrigation water should have been an input although the lack of trustworthy data prevented its use in the study. Animal skins production data (including wool) is only available for goats and sheep but the absence of a specific exergetic value for skins lead us to consider only the wool production flux that has a specific exergy of 5850 kJ/kg [51]. However, its exergetic value is very low ranging from 51 TJ in 2000 to 35 TJ in 2012. Other animal products such as lanolin or cow's leather were not accounted for, due to lack of data. For lanolin and cow's leather there are no specific exergetic values while for cow's leather there is also no production data.

Wood removals are considered as under bark and the exergy values were obtained from Dewulf et al. [52]. Dry matter exergy values are 22.11 MJ/kg for softwoods and 22.01 MJ/kg for hardwoods and densities of 450 kg/m^3 and 650 kg/m^3 respectively. Moisture content was assumed to be 20%.

Environmental remediation exergy costs were estimated by multiplying the pollutant air emission (obtained in mass units) by the specific extended cost of removing the pollutant from the atmosphere. Studying the transportation sector for Turkey, Seckin et al. [38] also created a virtual process to find the environmental extended exergy cost of three pollutants: carbon dioxide (57,600 kJ/kg), methane (322,400 kJ/kg) and nitrous oxide (10,600 kJ/kg). Dai et al. [53] studying the transportation sector in China determined the specific environmental remediation exergy cost of carbon monoxide (11,800 kJ/kg), nitrogen oxides (3610 kJ/kg) and sulphur oxides (5890 kJ/kg). For ammonia, the chemical exergy was used as the extended exergy cost of removing the pollutant (since no study is available in the literature) and valued by 19,841 kJ/kg.

RESULTS

Energy Resources' Exergy Accounting

Energy Resources

Energy resources are essential as they fuel all the machinery applied in the AFF sector. In the agriculture and forestry subsectors, energy use decreased drastically, more than 50%, mainly due to the decrease in the use of diesel to one third of the initial value (Figure 5).

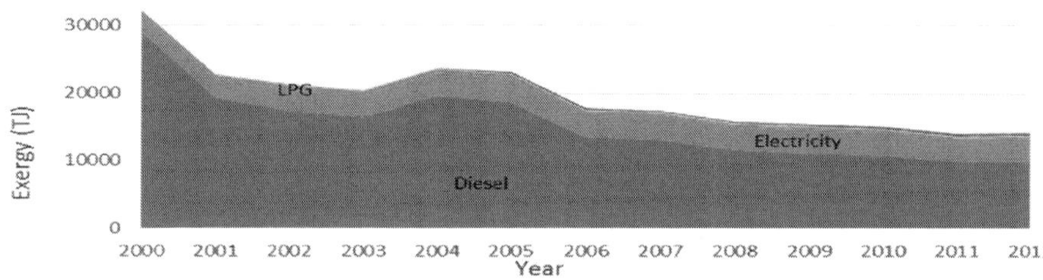

Figure 5: Exergy of energy resources' inputs to agriculture and forestry.

In contrast to agriculture and forestry, the consumption of energy resources in fishery has almost tripled due to the diesel carrier (Figure 6). In this subsector, LPG (Liquefied Petroleum Gas) and gasoline were also replaced by biodiesels and electricity. Notably, electricity grew from an 8% share in 2000 for agriculture and forestry to 23% in 2012 while in fishery and aquaculture (just fishery subsector from now on) it increased from 2.6% to 7.1% in 2012.

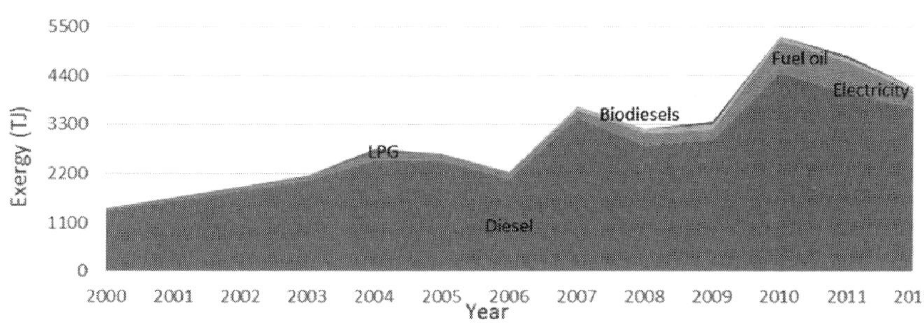

Figure 6: Exergy of energy resources' inputs to the fishery subsector.

End-Uses

Mechanical drive (e.g., tractors, irrigation pumps) and transportation (ex. boat engines) consume the biggest share of useful exergy in the sector. The decrease of 56% in the exergetic input of energy carriers in agriculture and forestry, from 2000 to 2012, was reflected in a 39% decrease in the amount of useful exergy applied to both subsectors (Figure 7). Fisheries increase energy carriers' consumption by 291% and obtained a 320% increase of useful exergy (Figure 8). Both improvements are associated to an increasing final-to-useful efficiency and are mostly due to the increased use of electricity, namely in irrigation.

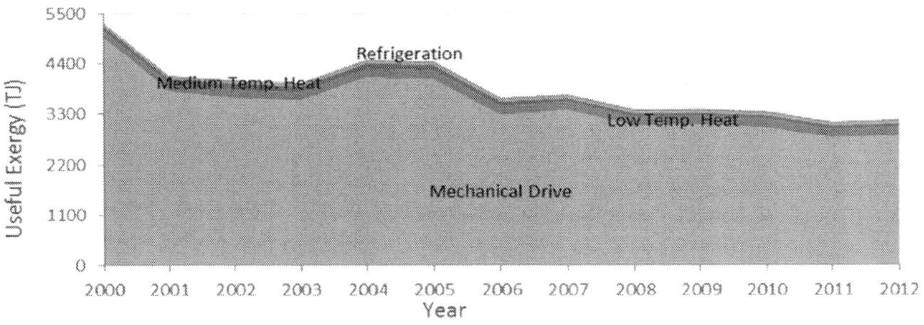

Figure 7: Useful Exergy by end use category in agriculture and forestry.

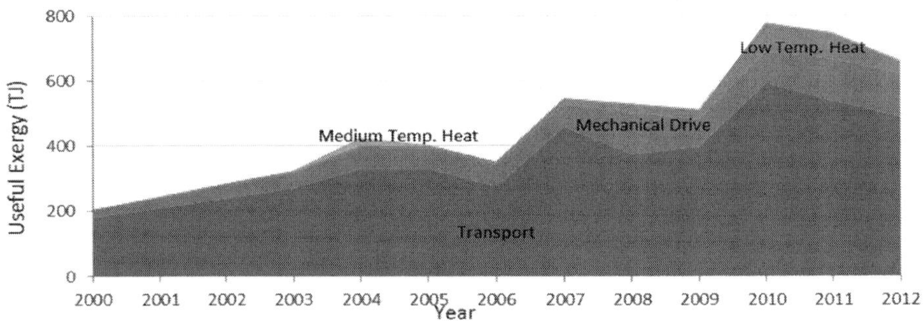

Figure 8: Useful Exergy by end use category in fisheries.

Sector and Subsector Efficiencies

All subsectors have increased their efficiency (Figure 9) with agriculture and forestry showing a better performance. The increase in the share of electricity and the reduction of the share of diesel (mostly associated to the transition from diesel-powered to electrically-powered irrigation pumps) allowed an increase in performance of 38% (from 16% to 22%) for agriculture and forestry and 30% (from 16% to 21%) overall. Fishery also increased their conversion efficiency from energy carriers to useful exergy from more than 14% to 16%.

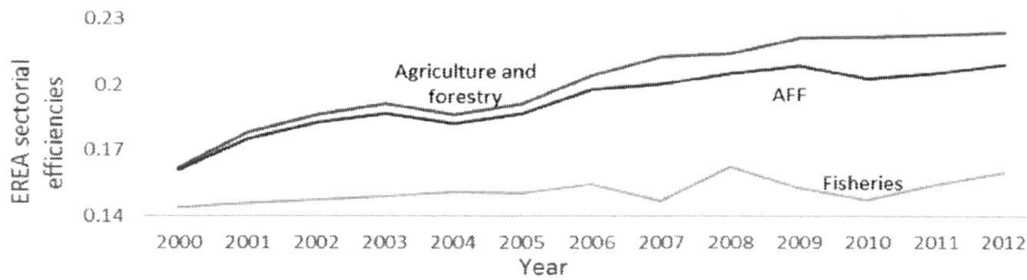

Figure 9: AFF sector and subsectors efficiencies by EREA approach.

To test whether the increase in efficiency was only due to technological improvements we estimated EREA subsectorial efficiencies assuming the carrier end-use pairs efficiencies were constant from the year 2000 onwards (Figure 10). In these conditions, the useful exergy per final exergy would have increased 35% for agriculture and forestry (from 16% to 22%), 8% for fishery (from 15% to 16%) and 27% overall (from 16% to 21%), showing that the overall increase in efficiency was due not only to better technologies but also reflect a wiser choice of energy carriers for the desired end-use.

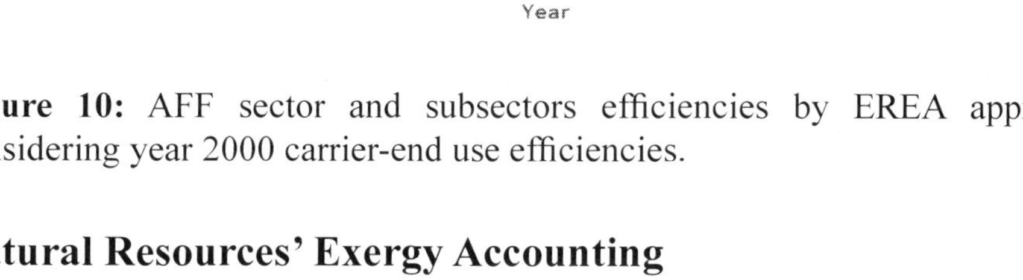

Figure 10: AFF sector and subsectors efficiencies by EREA approach considering year 2000 carrier-end use efficiencies.

Natural Resources' Exergy Accounting

Total Outputs

About two thirds of all agricultural, forestry and fisheries exergetic output is from wood removals. The changes in the overall annual production that ranges between 140 PJ to 160 PJ are controlled by changes in wood production (for a more detailed description see Appendix B). The aggregated production intended for food and feed decreased 14% mainly due to a 20% decrease in crops harvested despite a 20% increase in meat production (Figure 11).

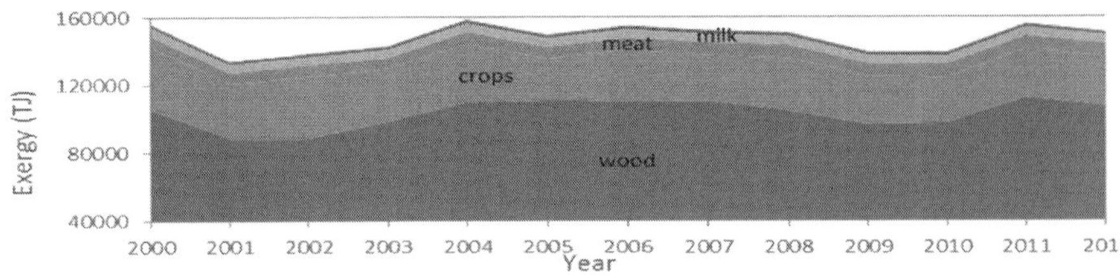

Figure 11: Natural resources exergy output from the AFF sector.

The production matrix for the Portuguese AFF sector has remained almost constant, being forestry the major producer and increasing its share. Fishery nutritional values are neglected while energetic values for animal husbandry are constant and the share of vegetables and fruits have decreased, mainly after 2004 (Figure 12).

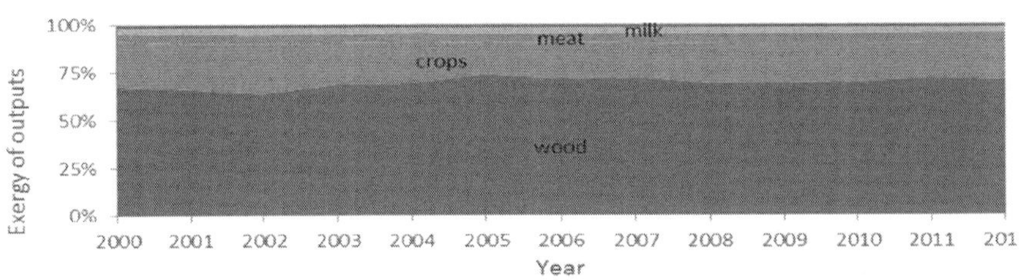

Figure 12: Natural resources exergy output from the AFF sector in percentage.

Total Inputs

Overall input to the agricultural sector is decreasing (24% over the 13 years) mainly due to fewer amounts of energy resources which alone represented 29% of all inputs in 2012 (Figure 13 and Figure 14). Seeds, fertilizers, pesticides and eggs for incubations were almost negligible in the sector's consumption matrix while feed is by far the biggest input (for a more detailed description see Appendix C).

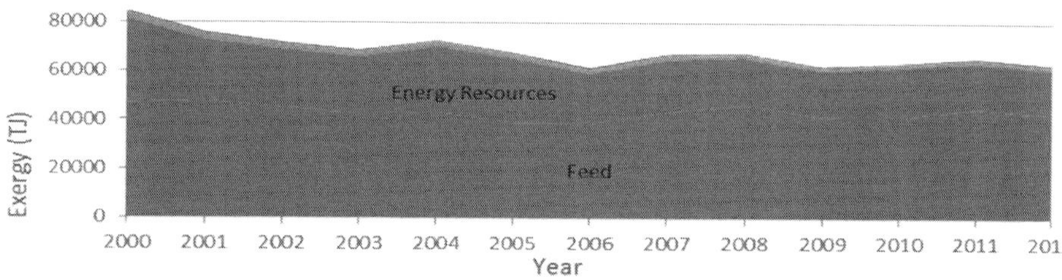

Figure 13: Natural resources exergy inputs to the AFF sector.

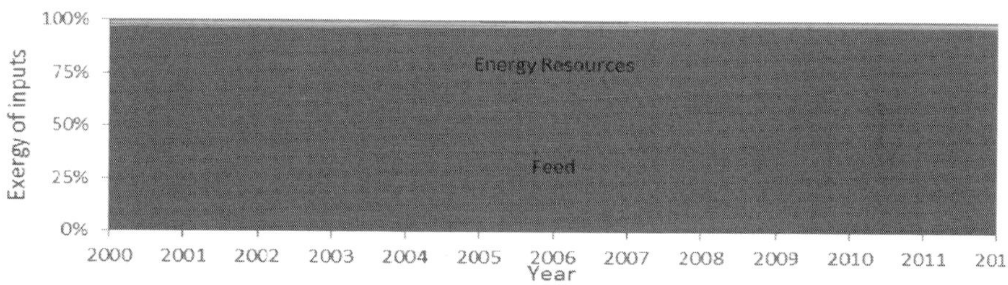

Figure 14: Natural resources exergy inputs to the AFF sector in percentage.

Sector and Subsector Efficiencies

The agricultural sector relies heavily on environmental conditions to sustain production. Water and radiation are offered freely by our natural environment and do not constitute a societal activity. Their contribution leads to efficiencies higher than 100% for agriculture and forestry. This subsector reached a 255% efficiency in 2012, a 36% increase since 2000, contrasting with fishery, which decreased its efficiency by 53% from a 43% efficiency in 2000 to 20% in 2012 (Figure 15).

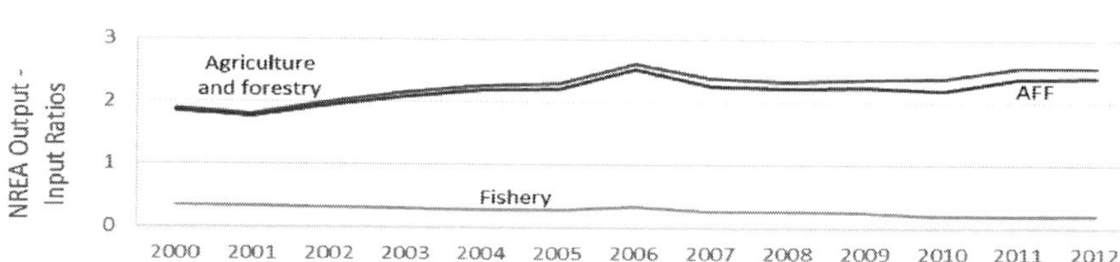

Figure 15: AFF sector and subsectors efficiencies by NREA approach.

Extended Exergy Accounting

The inclusion of three new production factors offers a new perspective of the AFF sector and allows comparisons between material and immaterial inputs. Exclusively for the three immaterial inputs of the EEA approach, the agriculture and forestry subsectors are quantified separately to allow comparisons between them.

Labour

Total exergy allocated to labour was calculated by the labour equation with HDI values from the United Nations Development Programme—Human Development Reports [37] and total population and number of workers data from the Eurostat database [42]. These values as well as other labour and capital statistics are available in Appendix D. To measure the associated labour exergy to the AFF sector and subsectors, the total societal labour exergy was multiplied by the fraction of workers in this sector and subsectors. This assumes that all workers have the same labour exergy independently of the sector or even the working hours, which is consistent with the equation used.

The number of workers and respective labour exergy is far superior for agriculture over forestry and fishery (Figure 16). Total labour exergy remained constant due to an increase in the HDI values and to a decrease in the number of workers in the AFF sector. Overall equivalent exergy of labour increased rapidly from 51.5 MJ/h in 2000 to 89.0 MJ/h in 2012 which means that the necessary societal exergy to produce one hour work increased 73% in 13 years.

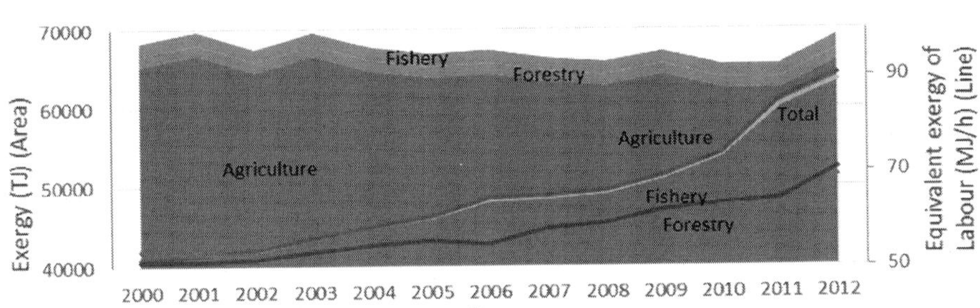

Figure 16: Exergy of labour into AFF subsectors and equivalent exergy of labour.

Although the cost of labour was similar, in the beginning of the century, for all three subsectors, in 2012 the necessary exergy to supply one hour work to agriculture was 30% higher than to forestry and fishery. This increase was due to the decrease in the number of hours worked per year by farmers from about 2000 to less than 1500 h in 2012.

Capital

In addition to labour, capital is a fundamental key for a productive system and should also be introduced as an input. To discover the monetary aggregate associated to the AFF sector, a direct relation was made between M2 and total gross value added (GVA) of the country (which in case of Portugal is almost identical). The monetary aggregate of the sector and subsectors under analysis was obtained assuming a linear relationship with the GVA of the AFF sector and subsectors allowing the calculation of different equivalent exergy of capital (ee_k) for each of the three subsectors. Capital exergy relates to labour exergy through the ratio $\frac{M_{2S}-S_S}{S_S}$ which is higher than 1, reaching 8 for forestry. The S subscript in the previous equation means that the considered values are sector or subsector related.

The exergy of capital has decreased 42% from 2000 to 2012, mainly due to a strong decrease in AFF's GVA (Figure 17). The equivalent exergy of capital has been constant for all subsectors with the agriculture subsector presenting the highest value. The explanation is that the wage per hour in agriculture subsector is more than five times lower than in fishery or forestry, although there was a 73% increase in the wage per hour in agriculture (in 13 years) compared to an 18% increase for forestry and 12% for fishery (at constant 2010 prices).

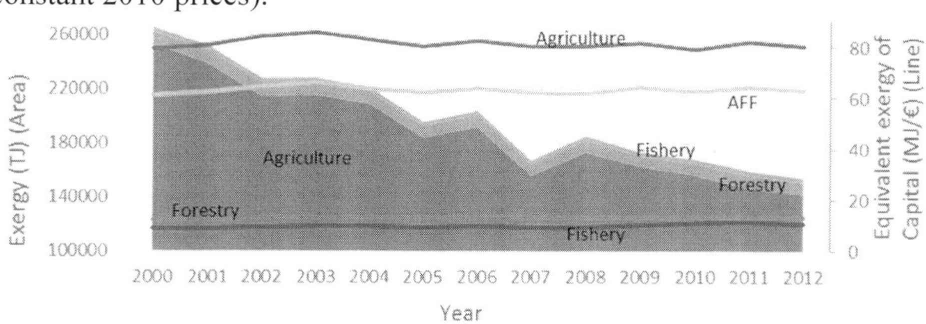

Figure 17: Capital exergy into AFF subsectors and equivalent exergy of capital.

Environmental Impact

The EEA considers that the exergy value of the waste or pollutant should not be his chemical exergy but the extended exergy cost to avoid the pollutant or bring the pollutant to the dead state. Such extended exergy costs are the sum of all physical exergy required to run the cleaning process plus the labour and capital exergy fluxes imperative to run the process [54]. Carbon dioxide is mainly emitted from the combustion of fossil and renewable fuels; emissions from biomass (not used as fuel) were not accounted for because in a life cycle perspective biomass is neutral. Emissions by animals comprise carbon dioxide and methane from enteric fermentation in the ruminants' digestive system. The methodology doesn't account carbon dioxide from animal husbandry but accounts methane emissions, being the total amount almost exclusively from livestock production (Figure 18). The total exergy needed for environmental clean-up from atmospheric emissions is the biggest input of the system showing a decrease of 10% over the years. Carbon dioxide represents almost two thirds of the equivalent exergy and methane the remaining part, being the exergy from the other pollutants residual. Methane emissions follow the trend line of animal husbandry production (almost constant) and carbon dioxide follows the decreasing trend line of energy resources.

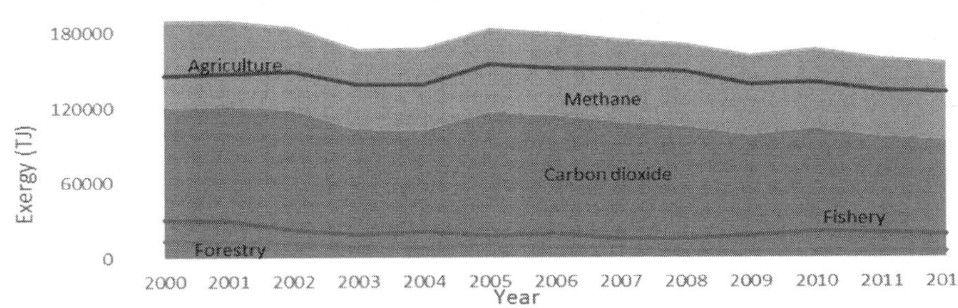

Figure 18: The virtual exergy input due to pollutant emissions into AFF subsectors.

Total Inputs

All three immaterial inputs addressed by EEA are quantitatively superior to NREA inputs. The biggest share of energetic and material inputs reached 14% in 2012, revealing the discrepancy between immaterial and material inputs and pointing out capital and labour as the main factors of production.

Environmental impact proved to be the strongest concern and quantitatively exceeded all other inputs (Figure 19). Overall values dropped 25% due to the decrease in energy resources, capital and air emissions.

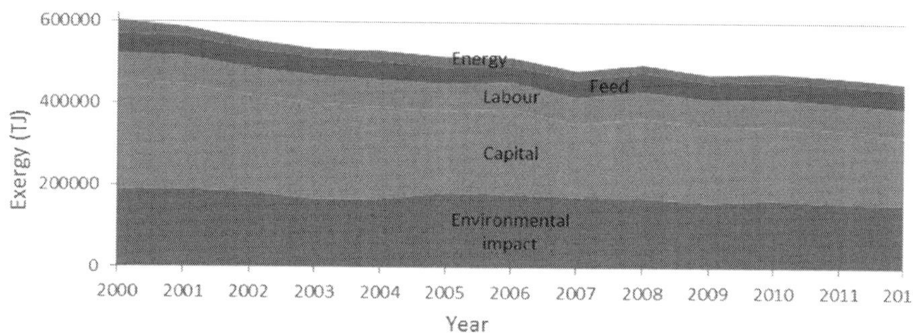

Figure 19: Extended exergy input to the AFF sector.

Sector and Subsector Efficiencies

The introduction of three highly valuable production factors to the input matrix dramatically lowered the efficiency of all AFF subsectors. The agriculture and forestry subsector increased their efficiency by 33% (from 27% in 2000 to 36% in 2012) while fishery increased its efficiency by 32% (from 2.7% in 2000 to 3.6% in 2012) (Figure 20). This improvement was mainly due to a decrease of the capital exergetic input.

Figure 20: AFF sector and subsectors efficiencies by EEA approach.

DISCUSSION AND CONCLUSIONS

Results obtained for the Agriculture, Forestry and Fisheries (AFF) sector and subsectors, regarding the ratio of exergy outputs to inputs, using the three exergy accounting methodologies are synthesized in Figure 21.

Results for the AFF sector are similar among methodologies. With the EREA approach there is a 30% increase in the overall efficiency due to a better match between the energy carrier and the end uses. NREA presented a similar growth in overall efficiency representing the sector's return over the applied energy and matter. The EEA methodology showed the same overall tendency regarding the evolution of efficiency, although the results are clearly dominated by the capital and environmental impact. For the fisheries subsector, the results are contradictory among methodologies.

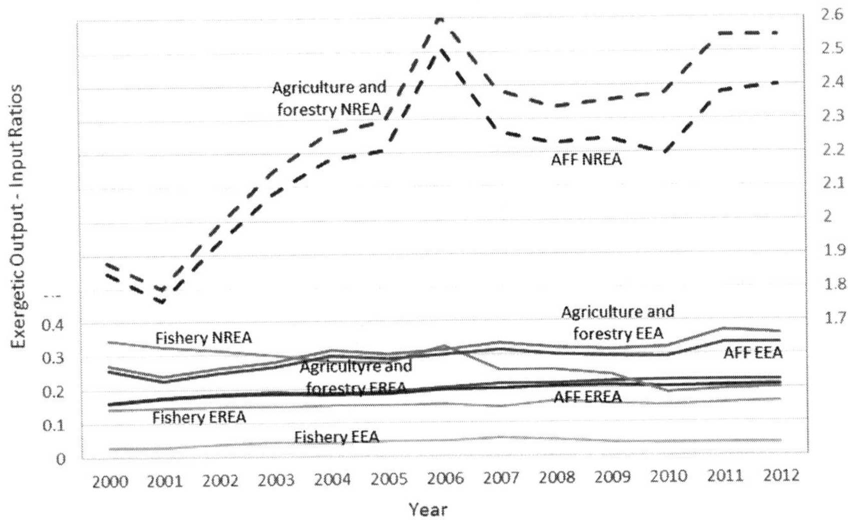

Figure 21: AFF sector and subsectors' efficiencies by methodology (full lines on the left axis and dashed lines on the right axis). EREA in shades of green, NREA in shades of blue and EEA in shades of orange.

Energy resources' exergy accounting (EREA) only addresses one type of inputs—energy carriers—to estimate the available useful exergy for its end uses and the overall efficiency. Although it estimates the actual amount of exergy that is delivered to a system function, no information regarding the output is introduced in the analysis. For Portugal, the overall efficiency reached 21% and is rapidly increasing over time due to a bigger share of the use of electricity whose efficiency is substantially higher, mainly for its predominant use in mechanical drives, mostly associated with irrigation systems. In fisheries, diesel fuel continues to be preponderant in fishing vessels, although with lower efficiencies. This explains the different performance levels of the two subsectors. The natural resources exergy accounting (NREA) approach uses all energetic and physical inputs and outputs to describe a system's efficiency without any knowledge about the system's processes and resources utilization. In the present analysis, all inputs were accounted for only by their intrinsic exergies values, not by their cumulative exergy values. The exergy that has been used to transform and bring the inputs to the boundaries of the AFF sector are integrated in the other economic sectors like industry for the fertilizers and pesticides, extraction and conversion for energy carriers and tertiary and transportation for almost all. The exergy input has decreased 24% since 2000 (Figure 13) with feed for animal husbandry representing the biggest share (68% in 2012), followed by the energy resources. The exergy of feed is on average 140% superior to the exergy of all crops harvested, which means that Portuguese agriculture does not produce enough to support its animal husbandry. Feed does not consider green fodder, pasture, meadows and straw because these are internally produced and consumed in the sector. The output-input ratio of animal husbandry (with feed and incubation eggs as inputs, only) evolved from 26%, in 2000, to 31% in 2012. This positive evolution indicates either a better use of feed in livestock production or a bigger share of non-accounted feed. The exergy output from the sector is mostly embedded in wood and harvested crops with wood exergy reaching a share of 71% in 2012 (Figure 11).

Although exergy is degraded, the output-input ratio of the agricultural and forestry subsectors can be higher than one (contrary to other economic sectors except extraction and conversion based on renewables) because solar radiation is not taken into account. Considering NREA, the sector's overall ratio increased more than 27% in this period being close to 2.5 in 2012, which means that for every exergy Joule that entered in the AFF sector, two and a half Joules were harvested. Agriculture and forestry performed much better than fishery (Figure 15) whose efficiency decreased 53% since 2000 (from 43% in 2000 to 20% in 2012). This decreasing efficiency verified for fisheries and aquaculture is exclusive for the NREA methodology. While in EREA we observe a better match between the energy carrier and the end use, in NREA the total amount of nutritional values from fish caught is decreasing for each exergy unit spent on the fishery activities.

The explanation for this puzzle is revealed by looking at the ratio of the output exergy (measured by the NREA methodology) to the useful exergy (measured by the EREA methodology) which is a proxy for the efficiency of the energy services in the AFF sector (see Figure 22). Contrary to agriculture and forestry which improved the efficiency of the exergy services by 59% (from 27 to 47), fishery reduced theirs to 31% of the initial value (from 5 to

1.5). While in the agriculture subsector each Joule of useful exergy produced 47 Joules of food in 2012, in fishery only 1.5 Joules were caught. The agriculture and forestry subsectors are less dependent on energy resources.

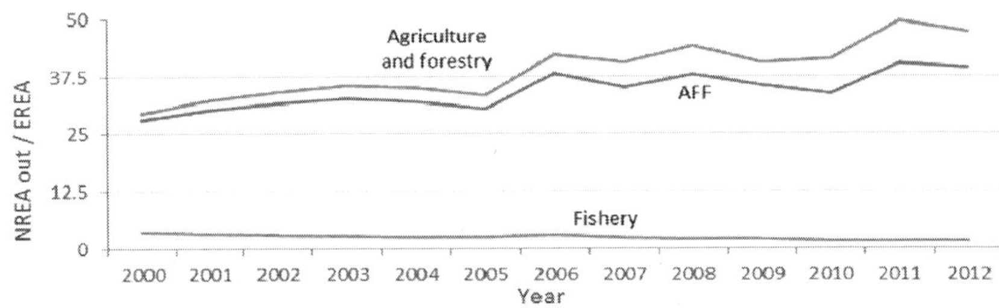

Figure 22: Ratio between output exergy from the sector and the useful exergy from energy resources.

The efficiency of the exergy services in fisheries is the amount of fish catch per unit of useful exergy. To explain this better, we will focus on the diesel carrier, which is responsible for almost 90% of the consumption matrix (in 2012) and used for transportation services. The fish catch per useful exergy ratio is equal to the fish catch per distance travelled times the thermodynamic efficiency of the boat engine (distance travelled per unit of useful exergy). In fisheries a higher thermodynamic efficiency is not translated into higher energy services efficiency presumably because fish population is decreasing leading fishing vessels to travel more (or more often) to get the same amount of fish.

Extended exergy accounting (EEA) allows us to compare the weight of different inputs. The exergy embodied in labour, which is less than an average daily work in other sectors (1500 h per year and per worker in 2012), has the same magnitude of all other material and energetic inputs taken together. Capital exergy is predominant over labour and although it is rapidly decreasing over time, it is larger than the overall exergy of outputs. Figure 18 also shows that actually, the bigger input in EEA is the environmental impact from the sector, being all three immaterial inputs seven times superior to the material ones. These results raise issues regarding the proportionality between material and immaterial inputs in the EEA methodology. By considering that the cost of labour and capital is a cumulative societal exergy and, by accounting all other inputs in its intrinsic form, the study reveals the disproportionality that results from this inconsistency. Also, measuring the capital input to the sector by all currency money (M2) minus wages creates an additional problem, because some of this money is used to buy all energetic and material inputs that entered the agricultural sector, and that were taken already into account for by their intrinsic exergies. Following the proposed methodology leads to double counting all natural resources inputs because both intrinsic exergy plus the currency money necessary to buy them are taken into account.

So, are the different exergy accounting methodologies providing consistent or contradictory results? Results provided by the EREA and NREA approaches can be contradictory but they are complementary because the EREA is focused on the thermodynamic efficiency while the NREA extends the boundary allowing the evaluation of the efficiency of the energy service. Both approaches are important in providing a realistic picture of the sectorial exergy efficiency. On the other hand, results obtained by the EEA approach are completely dominated by the exergies of capital and environmental impacts, revealing the disproportionality between material and immaterial inputs in this methodology. Improving the estimation of EEA externalities, labour and capital, for sectorial studies, remains a challenge for future studies.

We propose and apply to our case studies of the AFF sector, using the NREA and EEA approaches, methodological options to account for the renewable flows obtained from the environment, that are more adequate for sustainability assessments. While former NREA and EEA studies [19,21,22 ,23,24,25,26,27,28,29,30] account all crops, wood and fisheries as an input to the agricultural sector from the environment, this study acknowledges that all these flows are renewable and an output of human activities. Fish catch (excluding fish farming) is considered an extraction because no human activity helps to feed and raise the fish and excessive catches together with pollution have caused a decrease in fish population [55]. For a better assessment of sustainability, it would be important to include and correctly evaluate in NREA and EEA studies, the environmental positive impact of the carbon dioxide sequestered from atmosphere by the AFF sector because, although there is a substantial fraction of the carbon content in vegetables, fruits, crops and wood that is released after consumption, some carbon remains sequestered in living beings and wood used for veneer and saw logs.

ACKNOWLEDGMENTS

Work financially supported by FCT/MCTES (PIDDAC) through project UID/EEA/50009/2013.

AUTHOR CONTRIBUTIONS

Tiago Domingos provided significant comments throughout the study; Tania Sousa helped conceptually and technically with the application of the methodologies, contributed to the interpretation of the results and writing of the paper; and Ricardo Manso undertook the analysis, writing the code to manipulate datasets, and wrote most of the paper.

APPENDIX A. THE EXERGY CONCEPT

Exergy is measured in relation to a reference environment characterized by a temperature, T_0, a pressure p0, and a chemical composition. When the system is in equilibrium with the environment, the system is at the "dead state" and its exergy is null.

A system or stream can have several types of mechanical energy (or exergy), depending on the relation to the surrounding environment. If it has non-zero velocity then it possesses kinetic energy and both the specific energy and exergy are given by $ex_k = \frac{V^2 - V_0^2}{2}$ being V its velocity. It has a potential energy or exergy if it is bound to a conservative force (gravitational, elastic, electric, ...) being the gravitational potential energy the most common. Both the specific gravitational potential energy and exergy are the same and given by $ex_{pg}=g(z-z_0)$ being $g$gravitational acceleration and $z - z_0$ the height relative to the "dead state".

The physical exergy, which is the maximum amount of work that can be extracted by taking the resource from its initial pressure and temperature to the reference environment at T_0 and p_0, is $ex_{ph}=(u-u_0)+p_0(v-v_0)-T_0(s-s_0)$ [43], being u_0, v_0 and s_0 the specific internal energy, specific volume and specific entropy at the "dead state". All the change in internal energy of the system can be converted to work except: (1) the minimum energy that must be dissipated in the form of heat to decrease the entropy from s to s_0 and (2) the minimum mechanical energy that must be used to increase the volume from v to v_0.

For an open system, the physical exergy that accompanies the mass flow must take into account also the flow work and is given by $ex_{f,ph}=h-h_0-T_0(s-s_0)$ where h_0 is the specific enthalpy at the "dead state". As pointed out by Sciubba [23] this equation truly expresses the real attribute of exergy: the amount of useful work that can be extracted from a certain system is measured by its enthalpic content $(h-h_0)$ but taking into account that the system has to dissipate energy in the form of heat to get rid of entropy.

The exergy associated to heat transfer (Q) is $Ex_Q = Q\left(1 - \frac{T_0}{T}\right)$. This exergy is obtained by a direct application of the Carnot efficiency and represents the maximum amount of work that the heat flow can produce. Finally, the chemical exergy of a substance, or resource, is the maximum work that can be obtained by taking it from T_0 and p_0 to chemical equilibrium with the reference environment. The reference environment could be the lithosphere (Earth crust) if the substance is a solid compound, the hydrosphere (water on, under and above the surface) if it is a dissolved ionic compound or the atmosphere if it is a gaseous compound. The specific chemical exergy of a compound is computed by Romero and Linares [56] as: $ex_{ch} = \sum_i \Delta_f G^0 + \sum_{el} n_{el} b^0_{ch,el}$ were $\Delta_f G^0$ is the standard Gibbs energy of formation, n_{el}, the number of moles of element *el* in the compound and $b^0_{ch,el}$ the standard chemical exergy of each element.

The total specific exergy of a system or resource can now be written as the sum of all components: $ex_S = ex_k + ex_p + ex_{ph} + ex_{ch}$. In any real or irreversible process the entropy generated multiplied by T_0 is the exergy loss during the transformation process $ex_{loss} = T_0\sigma$.

APPENDIX B. OUTPUT FLOWS BY NREA

Appendix B.1. Crops Harvested

Harvested crops include cereals, root crops, industrial crops, fibre crops, vegetables, fruits, nuts, vineyards and olive trees. Overall nutritional energy values are obtained by multiplying each crop mass production by their specific nutritional energy. Table A1 presents all crop specific nutritional energies [46] of all products in the production database [42].

Table A1: Specific Nutritional Energy (SNE) in kJ/100 g for each harvested crop [46]

Crop	SNE	Crop	SNE	Crop	SNE	Crop	SNE
Almonds	2423	Mushrooms	130	Oranges	197	Radishes	66
Apples	218	Eggplants	104	Other berries	225	Raspberries	220
Apricots	201	Eggplants	104	Other brassicas	150	Red pepper	166

Bananas	371	Endives	71	Other citrus fruits	180	Rice	1548
Barley	1473	Figs	310	Other fresh vegetables	100	Rye and maslin	1414
Beans	368	Garlic	623	Other fruits	369	Sour cherries	209
Black currants	264	Grain maize	1527	Other leafy or stalked vegetables	117	Spinach	97
Broad and field beans	1377	Hazelnuts	2629	Other nuts	2170	Strawberries	136
Cabbage (white)	103	Kiwis	255	Other pulses	354	Sugar beet	293
Carrots	173	Leeks	255	Peaches	165	Sunflower seed	1289
Cauliflower and broccoli	140	Lemons and acid limes	126	Pears	239	Tomatoes	74
Celeriac	176	Lettuces	65	Peas	339	Triticale	1406
Cherries	263	Melons	141	Plums	192	Vineyards	288
Chestnuts	891	Oats and mixed grain	1628	Pomelos and grapefruit	134	Walnuts	2738
Chicory	96	Olive trees	481	Potatoes	321	Watermelons	127
Courgettes	80	Onions	166	Quinces	238	Wheat	1418
Cucumbers	65						

From the 65 accounted products, Figure A1 presents the 13 top crops that most contributed to the annual nutritional energy production. It is important to mention that grain maize represents nearly 40% of the total, that farmers sharply decreased wheat, oats and potatoes production and opted for olive trees and tomatoes. Overall crop production decreased 20%.

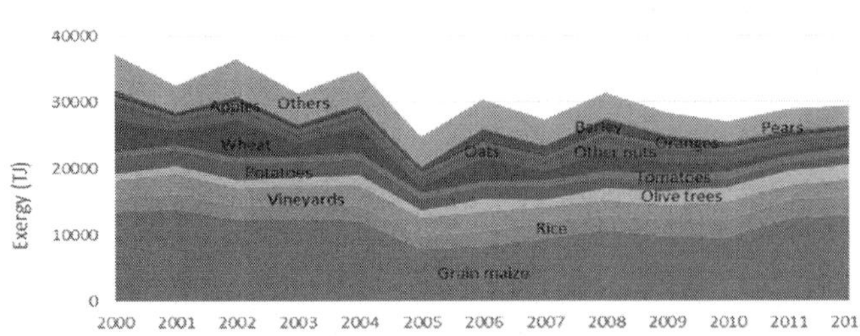

Figure A1: Nutritional energy output from crop production.

Appendix B.2. Meat Slaughtering

In the past, animal husbandry was a fundamental source of labour but in this 21st century they just represent the root of protein in our food. It's not just the high protein content and the retained nitrogen but also the specific animal amino acids and high biological value that favour their consumption. Meat production is available in mass units from the Eurostat database [42] and nutritional energy values [46] were chosen as a raw mix of meat for each animal (Table A2).

Table A2: Specific Nutritional Energy (SNE) in kJ/100 g for each slaughtered meat [46]

Meat	SNE	Meat	SNE	Meat	SNE
Meat of bovine animals	979	Meat of rabbits	569	Pigmeat	995
Meat of horses, asses, mules or hinnies	556	Meat of sheep and goats	1067	Poultry meat	979

Figure A2 presents the produced energy from animal slaughtering to human consumption. Livestock is slowly increasing with pigs and poultry becoming the major intervenient. Meat of rabbits and horses is residual and can hardly be seen on the graph.

Figure A2: Nutritional energy output from meat slaughtering.

Appendix B.3. Fish Catch

Portugal, being on the southwest corner of Europe and with its long border with the Atlantic Ocean always looked at the sea as a primary source of resource extraction. As animal husbandry, fish is a fundamental source of iron and protein, has a high biological value and with the advantage of having less and better quality fat content. Fish catch is available by catching zones, fish families and individually [42]. Table A3 presents the specific nutritional energy considered for each fish, the average of their nutritional values if a family or an assumed average if in a zone. Total energy from fish catch increased until 2007 and decreased since then, mainly due to the common fisheries policy who manages fish stock for the European Union as a whole. This policy sets quotas for each member state which is allowed to catch a certain amount of each type of fish. Figure A3 shows a partial catch matrix, with the most important contributors of the Portuguese fishery. The reducing quota of sardines, a Portuguese favourite, which represented, in the past, almost 50% of the incoming energy, nowadays is relegated to second place. Upper fish fluxes of the graph represent coastal fishes; squids, cuttlefishes and octopuses; clams, cockles and arkshells; flounders, halibuts and soles and twenty more fish species families.

Table A3: Specific Nutritional Energy (SNE) in kJ/100 g for fish [46]

Fish	SNE	Fish	SNE	Fish	SNE
Abalones, winkles, conchs	439	Lobsters, spiny-rock lobsters	469	Mussels	360
Aquatic mammals	462	Marine fishes not identified	500	Oysters	213
Carps, barbels and cyprinids	531	Miscellaneous aquatic animals	500	River eels	770
Clams, cockles, arkshells	360	Miscellaneous coastal fishes	500	Salmons, trouts, smelts	594
Cods, hakes, haddocks	343	Miscellaneous demersal fishes	500	Scallops, pectens	289
Crabs, sea-spiders	364	Miscellaneous diadromous fishes	650	Shads	824
Flounders, halibuts, soles	294	Miscellaneous freshwater fishes	465	Sharks, rays, chimaeras	544
Freshwater crustaceans	300	Miscellaneous marine crustaceans	300	Shrimps, prawns	297
Herrings, sardines, anchovies	661	Miscellaneous marine molluscs	330	Squids, cuttlefishes, octopuses	343
King crabs, squat-lobsters	377	Miscellaneous pelagic fishes	500	Tunas, bonitos, billfishes	602

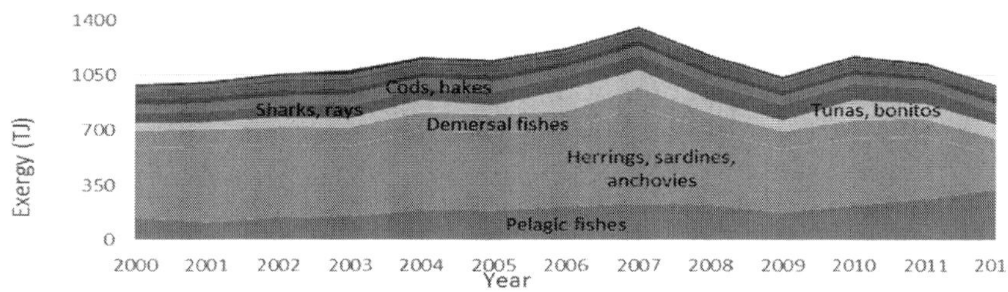

Figure A3: Nutritional energy output from fish catch.

Appendix B.4. Aquaculture

Aquaculture in Portugal is residual due to our borders with the Atlantic Ocean. Major energy supplied based on fish species is represented on Figure A4, being the total exergy not relevant to the fisheries and aquaculture subsector. Flounders, halibuts and soles are increasing its share while coastal fishes are reducing in a fluctuating production, with a small tendency to increase, over the years. Specific nutritional values are similar to the fisheries.

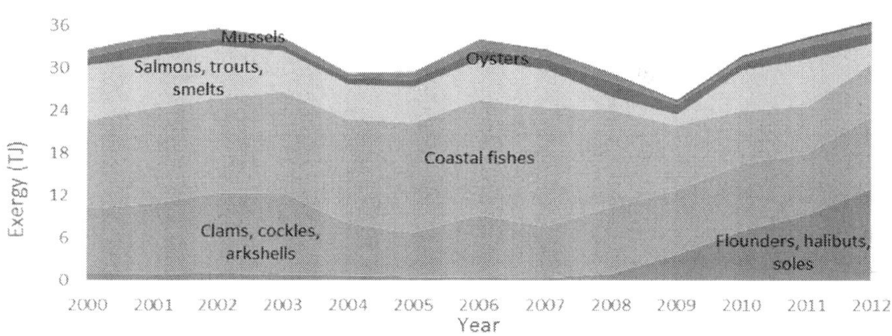

Figure A4: Nutritional energy output from aquaculture production.

Appendix B.5. Milk

Milk is a rich source of calcium and vitamin D with benefits for bones and teeth and processed into a diversity of dairy products such as cheese, butter, yogurt and cream. Specific nutritional values (Table A4) are for unprocessed milk at the producer level [46] and the production is in mass units [42].

Table A4: Specific Nutritional Energy (SNE) in kJ/100 g for milk [46]

Milk	SNE	Milk	SNE	Milk	SNE
Cows' milk	268	Ewes' milk	451	Goats' milk	288

Milk production remained solid over the years due to dairy produced quota, a measure by the European governments to control milk production. Cow's milk represents almost the totality of milk production (Figure A5) and, as an exergy input to the society, is five times bigger than total fish catch.

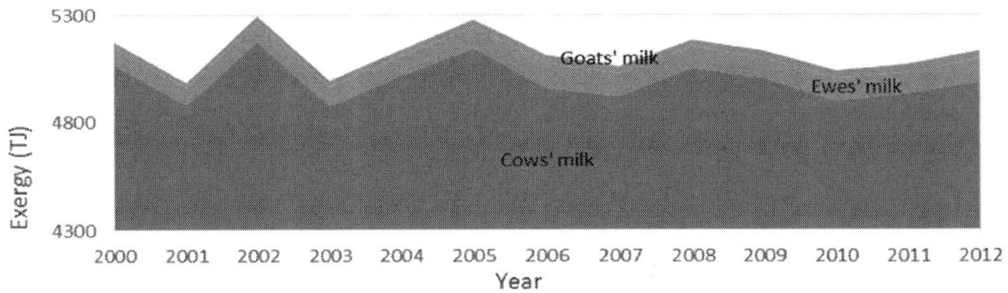

Figure A5: Nutritional energy output from collected milk.

Appendix B.6. Honey

Honey is produced by honey bees and from flowers' nectar. It is sweet due to high levels of fructose and glucose and possesses antiseptic and antibacterial properties. Overall honey production increased more than 50% through the 13 years (Figure A6). The specific nutritional energy of honey is 1272 kJ per 100 g of product [46].

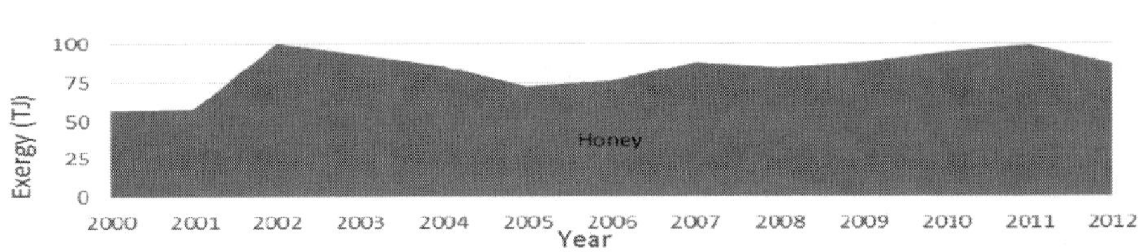

Figure A6: Nutritional energy output from produced honey.

Appendix B.7. Eggs

Eggs are one of the best sources of protein and represent the reference food in the biological value scale. Produced eggs are mainly for human food and for incubations to produce laying hens or poultry meat. It should be noted that incubation eggs are part of the produced eggs and an input to the system. Table A5 presents the specific nutritional energy of Hen eggs and an average for other bird's eggs.

Table A5: Specific Nutritional Energy (SNE) in kJ/100 g for eggs

Eggs	SNE	Eggs	SNE
Hen eggs	599	Other bird's eggs	776

Hens' eggs are predominant in production and incubation with a stable total exergy produced (Figure A7).

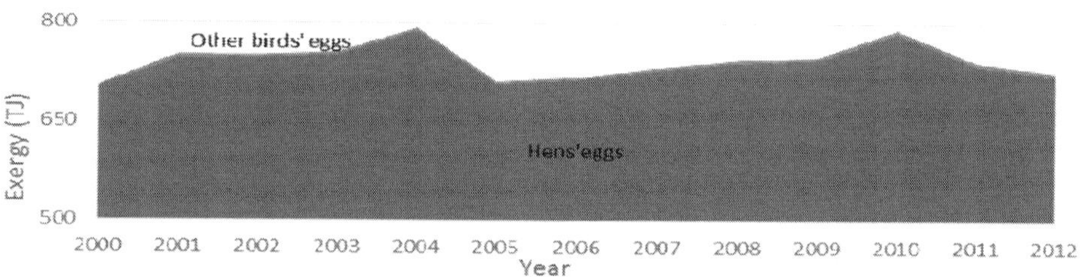

Figure A7: Nutritional energy output from produced eggs.

Appendix B.8. Wood

Forest area, in Portugal, accounts for 38% of the total area which is a high percentage compared with other European countries. Pulp and paper industry's sector represented, in 2012, 4.4% of GDP, 8% of total industrial production and 1.2% of economy's employment [57]. Wood removals are considered as under bark with a moisture content of 20%. Production is available in volume units and specific exergy values (Table A6) obtained from Dewulf et al. [52].

Table A6: Wood exergies per type of wood [52]

Wood	Exergy (MJ/kg)	Exergy (GJ/m³)	Wood	Exergy (MJ/kg)	Exergy (GJ/m³)
Coniferous	17.688	7.9596	Non-coniferous	17.608	11.4452

Total industrial round wood removals had minor fluctuations over the years but the non-coniferous (hardwoods) to coniferous (softwoods) ratio increased from 1.57 in 2000 to 4.12 in 2012 (Figure A8). Coniferous wood is mainly (70%) for saw logs and veneer logs and the remaining for pulpwood. Non-coniferous wood is more than 90% for pulpwood and the remaining for fuelwood. To highlight the importance of the forestry subsector, the total exergy collected from wood is roughly 3.5 times superior to the total exergy of harvested crops.

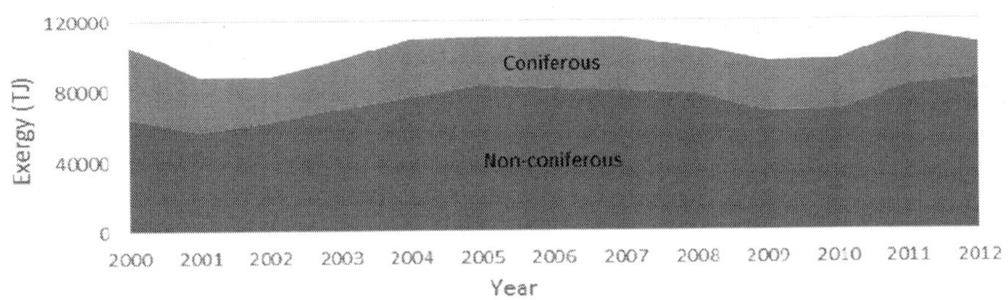

Figure A8: Exergy output from wood removals.

Appendix C. Input Flows by NREA

Appendix C.1. Fertilizers

Growing plants or trees extract nutrients from the soil that are taken away as the cultivated plants are harvested or tree fruits are removed. To allow soil productivity and enhance plant's growth, with increasing trees and plants densities, is necessary to artificially add nutrients by fertilising the soil. The fertilisers can be organic (e.g., compost, manure), bio-fertiliser (containing living microorganisms) or industrialized chemical compounds. The main nutrients are nitrogen (N), phosphorus (P) and potassium (K) and the compounds phosphate (P_2O_5) and potash (K_2O).

The chemical exergy values (Table A7) were estimated with the corresponding chemical formula, multiplying molecular mass by the standard chemical exergy (values from [44] or alternatively [58]).

Table A7: Fertilizers' chemical exergies

Fertilizer	Chemical Formula	Molecular Mass g/mol	Exergy kJ/mol	Exergy kJ/kg	Alternatively [58]	Exergy kJ/kg
Potassium/Potash	K_2O	94.203	413.1	4385		4400
Nitrogen	NH_4NO_3	80.04348	294.8	3682		3680
Phosphorus/Phosphate	P_2O_5	141.9446	319.5	2251		2700

Fertilizer usage has decreased, from 2000 to 2008, to 55% of its initial value and remained constant thereafter (Figure A9). Nitrogen is the most applied fertilizer accounting nowadays for almost half of the total exergy.

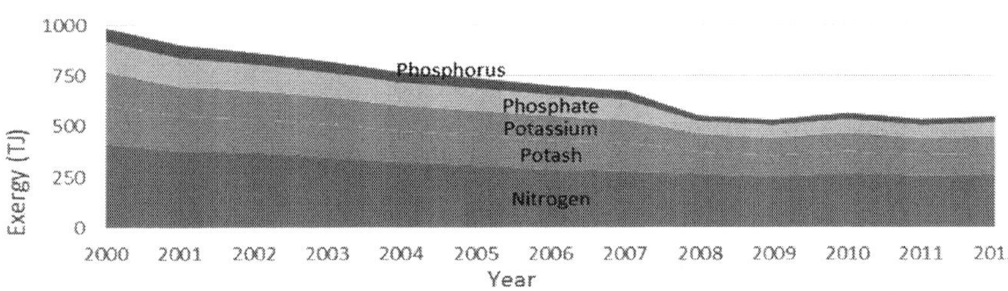

Figure A9: Exergy of fertilizers into agriculture.

Appendix C.2. Pesticides

Pesticides are biological or chemical substances that are mainly used in agriculture to protect plants from the harmful and damaging influences of microbes and plant diseases. They also kill and remove pests, insects and weeds that prevent the plants from a healthy grow. Plant grow regulators are used to protect the crop or fruit, after or before harvest, during storage and transport.

Eurostat presents data for six types of aggregated pesticides: fungicides and bactericides; herbicides, haulm destructors and moss killers; insecticides and acaricides; molluscicides; plant growth regulators and other plant protection products. Chemical exergy values were adopted from [58,59] being the herbicides value the average of six known herbicides and the insecticides value the average of three insecticides active substances (Table A8). Fungicides and bactericides represent almost two thirds of total usage which decreased 20% from 2002 onwards (Figure A10).

Table A8: Pesticides chemical exergies

Pesticide Type	Exergy kJ/kg	Reference
Fungicides and bactericides	27,900	[58]
Herbicides, haulm destructors and moss killers	24,100	[59]
Insecticides and acaricides	19,900	
Molluscicides, total	19,900	
Other plant protection products	24,100	
Plant growth regulators, total	24,100	

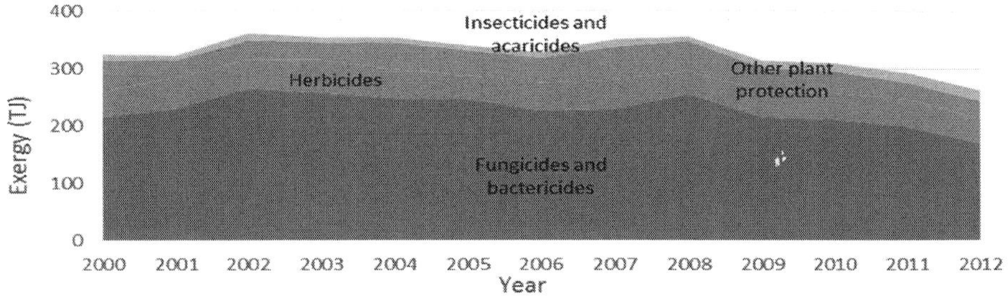

Figure A10: Exergy of pesticides into agriculture.

Appendix C.3. Seeds

As harvested cereals, roots or vegetables leave the soil new seeds must be planted. These seeds are part of the collected crops and will origin new plants to be harvested. Seeds used in agriculture are available in mass units for each plant type [45] and specific exergy values presented in Table A9. Figure A11 exhibits the top products that were used as seeds. Seeds follow the trend of harvest, as expected. The ratio of harvests to seed exergies exceeds 30.

Table A9: Specific Nutritional Energy (SNE) in kJ/100 g for seeds [46]

Seed	SNE	Seed	SNE	Seed	SNE	Seed	SNE
Barley	1473	Cow peas, dry	1432	Oats	1628	Sesame seed	2399
Beans, dry	1427	Grain, mixed	1455	Oilseeds	1800	Sorghum	1377
Broad and horse beans	1436	Groundnuts	1733	Peas, dry	1448	Soybeans	614
Buckwheat	1381	Hempseed	1800	Poppy seed	2231	Sunflower seed	1289
Cabbages and brassicas	150	Lentils	1448	Potatoes	321	Taro (cocoyam)	360
Canary seed	1624	Linseed	2076	Pulses	1423	Triticale	1406
Castor oil seed	1800	Lupins	1633	Rapeseed	2068	Vegetables, fresh	150
Cereals	1455	Maize	1527	Rice, paddy	1172	Vetches	1360
Chick peas	1499	Millet	1582	Rye	1414	Wheat	1418
Cottonseed	1059	Mustard seed	1963	Safflower seed	1314	Yams	422

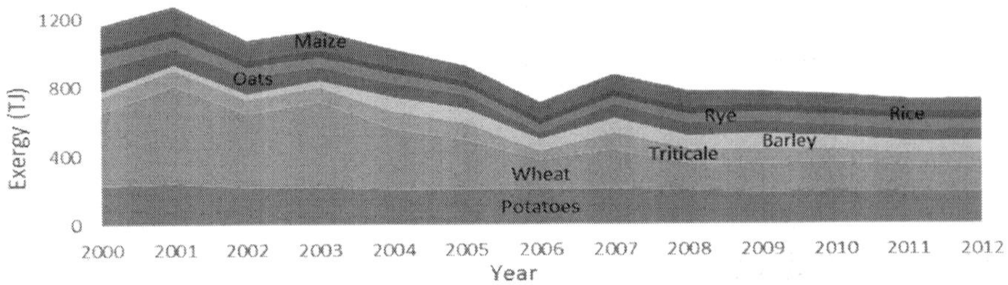

Figure A11: Nutritional energy of seeds into crop production

Appendix C.4. Feed

Animal feed represents food given to animals. Green plants that animals forage are not accounted as input to animal husbandry nor as an output because they never exit the sector. Feed used in agriculture is available in mass units for each of the constituents [45] and specific exergy values taken from [46,48] (Table A10).

It is important to notice that the exergetic contents of maize used to feed animals and total harvests are of the same magnitude (see Figure A12) which explains the need to import the majority of cereals used to produce feed. From the 54 accounted products used to produce feed, maize represented 57% of all inputs in 2000 and 69% in 2012. Total exergy over time has a slight tendency to decrease, reducing 10% from 2000 to 2012.

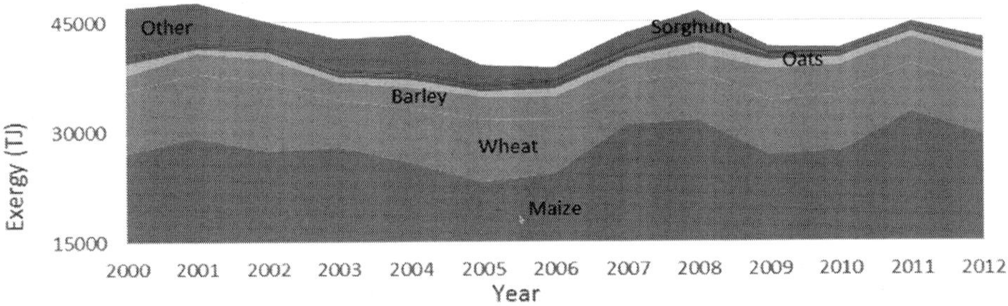

Figure A12: Nutritional energy of feed into animal husbandry.

Table A10: Specific Nutritional Energy (SNE) in kJ/100 g for feed components [46,48]

Feed Product	SNE	Feed Product	SNE	Feed Product	SNE	Feed Product	SNE
Apples	218	Fats, Animals, Raw	3257	Oilcrops Other	1639	Rye	1414
Aquatic Plants	180	Fish Body Oil	3776	Olive Oil	3700	Sesame seed	2399
Bananas	371	Fish, Liver Oil	3776	Onions	166	Sesameseed Oil	3700
Barley	1473	Freshwater Fish	400	Palm kernels	2152	Sorghum	1377
Beans	368	Fruits, Other	203	Peas	339	Soyabean Oil	3700
Butter, Ghee	3001	Groundnuts	1733	Pelagic Fish	400	Soyabeans	614
Cassava	667	Maize	1527	Potatoes	321	Sugar beet	293
Cephalopods	343	Marine Fish Other	400	Poultry Meat	979	Sugar cane	135
Cereals Other	1488	Meat, Other	858	Pulses	354	Sunflower seed	1289
Coconuts	770	Milk—Excluding Butter	353	Rape and Mustard Oil	3700	Sweet potatoes	359
Cottonseed	1059	Millet	1582	Rape and Mustardseed	2068	Tomatoes	74
Crustaceans	300	Oats	1628	Rice	1548	Vegetables Other	226
Demersal Fish	400	Offals, Edible	486	Roots Other	449	Wheat	1418
Eggs	682	Oilcrops Oil, Other	3700				

Appendix C.5. Incubation eggs

Poultry meat increased 40% over 13 years and consequently the incubated eggs. Part of the produced eggs is for incubations which ranged from 10% in 2000 to over 15% in 2012. Exergy on incubations increase 75% over the considered years (Figure A13). Data from the Eurostat database [42] is in number of egg units and, in order to measure the mass content, the egg's average mass [60] was multiplied by the respective number of eggs into incubation (Table A11).

Table A11: Specific Nutritional Energy (SNE) in kJ/100 g [46] and average mass for eggs [60]

Eggs	SNE	Mass (g)	Eggs	SNE	Mass (g)	Eggs	SNE	Mass (g)
Ducks' eggs	776	70	Guinea fowl eggs	663	9	Turkey eggs	716	79
Geese eggs	775	144	Hens'eggs	599	53			

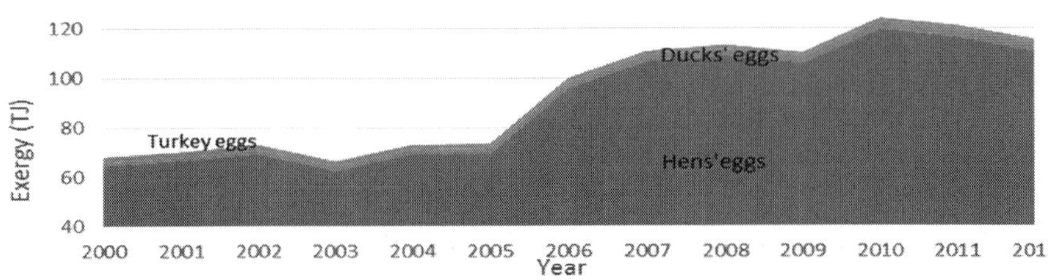

Figure A13: Nutritional energy of incubation eggs.

Appendix D. Additional Data

Missing data is relevant on HDI values (54%) and eggs for incubation which represented 40%, mainly on non-relevant fluxes of turkey and duck's eggs. There was also missing data on pesticides (about 15%), feed (7%), crops (5%), meat (4%), energy (3%), aquaculture (2%) and fish catch (1%). All other fluxes had no missing data.

Table A12 presents the HDI values for Portugal where the interpolated data is in italic. Table A13 presents the price deflator in a 2010 reference from which all monetary values were multiplied, since all were downloaded at current prices. Table A14 presents the population and workers allocated to each AFF subsector and Table A15 all the monetary values of GVA, M2 and wages.

Table A12: HDI values for Portugal (italic values interpolated)

2012	2011	2010	2009	2008	2007	2006	2005	2004	2003	2002	2001	2000
0.8220	0.8191	0.8159	*0.8103*	0.8053	*0.8029*	*0.7992*	0.7900	*0.7917*	*0.7880*	*0.7842*	*0.7805*	0.7800

Table A13: Price deflator values referenced to 2010

2012	2011	2010	2009	2008	2007	2006	2005	2004	2003	2002	2001	2000
99.333	99.730	100	99.360	98.283	96.604	93.809	90.913	87.984	85.925	83.068	79.714	76.859

Table A14: Population and workers allocated to each AFF subsector (thousands)

Workers	2012	2011	2010	2009	2008	2007	2006	2005	2004	2003	2002	2001	2000
Population	10,514	10,557	10,573	10,568	10,558	10,542	10,522	10,503	10,483	10,458	10,419	10,362	10,289
Agriculture	497.2	489.8	508.8	534.3	542.2	546.2	556.9	557.3	566.7	591	585.9	604.3	584.6
Forestry	10.9	10.9	10.6	10.7	12.0	11.8	11.8	11.7	11.7	11.8	11.7	11.9	11.6
Fishery	14.0	14.3	14.0	13.9	14.4	14.3	15.0	14.9	15.0	15.0	14.8	15.0	14.8

Table A15: Monetary values of all sectors and AFF sector at constant 2010 prices (Billion Euro for Total values and Million Euro for subsectorial values). Total—All sectors, GVA—gross value added, Wages—Compensation of employees

Monetary Values	**2012**	**2011**	**2010**	**2009**	**2008**	**2007**	**2006**	**2005**	**2004**	**2003**	**2002**	**2001**	**2000**
Total GVA	141	149	151	149	151	151	147	146	148	145	147	147	145
Total M2	163	174	173	170	179	163	159	157	150	147	147	153	151
Agriculture GVA	2224	2184	2413	2440	2497	2538	2887	2891	3272	3260	3337	3655	3869
Agriculture M2	2585	2537	2765	2776	2946	2735	3113	3097	3312	3294	3327	3784	4019
Agriculture Wages	817	756	787	784	782	787	778	795	780	775	768	825	823
Forestry GVA	647	689	657	613	662	682	708	716	785	814	893	893	841
Forestry M2	752	800	753	697	782	735	763	767	795	822	890	924	874
Forestry Wages	108	104	102	98	105	101	100	102	100	100	99	106	106
Fishery GVA	399	412	396	378	419	418	413	416	459	459	476	509	521
Fishery M2	464	478	454	430	494	450	445	446	464	464	474	527	541
Fishery Wages	170	160	156	160	175	178	176	179	176	175	173	186	186

REFERENCES

1. Pezzey, J.C.V. Sustainability constraints versus "optimality" versus intertemporal concern, and axioms versus data. *Land Econ.* 1997, *73*, 448.
2. Imperatives, S. *Report of the World Commission on Environment and Development: Our Common Future*; Oxford University Press: Oxford, UK, 1987.
3. National Research Council. *Toward Sustainable Agricultural Systems in the 21st Century*; National Academies Press: Washington, DC, USA, 2010.
4. Dewulf, J.; Van Langenhove, H.; Muys, B.; Bruers, S.; Bakshi, B.R.; Grubb, G.F.; Paulus, D.M.; Sciubba, E. Exergy: Its Potential and Limitations in Environmental Science and Technology. *Environ. Sci. Technol.* 2008, *42*, 2221–2232.
5. Reistad, G. Available Energy Conversion and Utilization in the United States. *ASME Trans. Ser. J. Eng. Power* 1975, *97*, 429–434.
6. Wall, G. *Exergy: A Useful Concept*; TH: Göteborg, Sweden, 1986.
7. Sciubba, E. Beyond thermoeconomics? The concept of Extended Exergy Accounting and its application to the analysis and design of thermal systems. *Exergy Int. J.* 2001, *1*, 68–84.
8. Rocco, M.V.; Colombo, E.; Sciubba, E. Advances in exergy analysis: A novel assessment of the Extended Exergy Accounting method. *Appl. Energy* 2014, *113*, 1405–1420.
9. Szargut, J. Analysis of cumulative exergy consumption. *Int. J. Energy Res.* 1987, *11*, 541–547.
10. Dincer, I.; Hussain, M.M.; Al-Zaharnah, I. Energy and exergy utilization in agricultural sector of Saudi Arabia. *Energy Policy* 2005, *33*, 1461–1467.
11. Utlu, Z.; Hepbasli, A. Assessment of the energy and exergy utilization efficiencies in the Turkish agricultural sector. *Int. J. Energy Res.* 2006, *30*, 659–670.
12. Ghandoor, A.A.; Jaber, J.O. Analysis of energy and exergy utilisation of Jordan's agricultural sector. *Int. J. Exergy* 2009, *6*, 491.
13. Avara, A.; Karami, M. Energy and exergy efficiencies in agricultural and utility sectors of Iran compared with other countries. In Proceedings of the 2010 2nd International Conference on Mechanical and Electrical

Technology (ICMET), Singapore, 10–12 September 2010; pp. 6–10.

14. Ahamed, J.U.; Saidur, R.; Masjuki, H.H.; Mekhilef, S.; Ali, M.B.; Furqon, M.H. An application of energy and exergy analysis in agricultural sector of Malaysia. *Energy Policy* 2011, *39*, 7922–7929.
15. Chen, G.Q.; Jiang, M.M.; Yang, Z.F.; Chen, B.; Ji, X.; Zhou, J.B. Exergetic assessment for ecological economic system: Chinese agriculture. *Ecol. Model.* 2009, *220*, 397–410.
16. Hoang, V.-N.; Alauddin, M. Analysis of agricultural sustainability: A review of exergy methodologies and their application in OECD countries. *Int. J. Energy Res.* 2011, *35*, 459–476.
17. Serrenho, A.C.; Warr, B.; Sousa, T.; Ayres, R.U.; Domingos, T. Structure and dynamics of useful work along the agriculture-industry-services transition: Portugal from 1856 to 2009. *Struct. Chang. Econ. Dyn.* 2016, *36*, 1–21.
18. Guevara, Z.; Sousa, T.; Domingos, T. Insights on Energy Transitions in Mexico from the Analysis of Useful Exergy 1971–2009. *Energies* 2016, *9*, 488.
19. Wall, G. Exergy conversion in the Swedish society. *Resour. Energy* 1987, *9*, 55–73.
20. Gong, M.; Wall, G. Exergy Analysis of the Supply of Energy and Material Resources in the Swedish Society. *Energies* 2016, *9*, 707.
21. Wall, G. Exergy conversion in the Japanese society. *Energy* 1990, *15*, 435–444.
22. Wall, G.; Sciubba, E.; Naso, V. Exergy use in the Italian society. *Energy* 1994, *19*, 1267–1274.
23. Sciubba, E.; Bastianoni, S.; Tiezzi, E. Exergy and extended exergy accounting of very large complex systems with an application to the province of Siena, Italy. *J. Environ. Manag.* 2008, *86*, 372–382.
24. Gasparatos, A.; El-Haram, M.; Horner, M. Assessing the sustainability of the UK society using thermodynamic concepts: Part 2. *Renew. Sustain. Energy Rev.* 2009, *13*, 956–970.
25. Chen, G.Q.; Chen, B. Extended-exergy analysis of the Chinese society. *Energy* 2009, *34*, 1127–1144.
26. Zhang, B.; Chen, G.Q. Physical sustainability assessment for the China society: Exergy-based systems account for resources use and environmental emissions. *Renew. Sustain. Energy Rev.* 2010, *14*, 1527–1545.

27. Dai, J.; Chen, B. Extended exergy-based ecological accounting of China during 2000–2007. *Procedia Environ. Sci.* 2011, *5*, 87–95.
28. Dai, J.; Fath, B.; Chen, B. Constructing a network of the social-economic consumption system of China using extended exergy analysis. *Renew. Sustain. Energy Rev.* 2012, *16*, 4796–4808.
29. Chen, B.; Dai, J.; Sciubba, E. Ecological accounting for China based on extended exergy. *Renew. Sustain. Energy Rev.* 2014, *37*, 334–347.
30. Bligh, D.C.; Ugursal, V.I. Extended exergy analysis of the economy of Nova Scotia, Canada. *Energy* 2012, *44*, 878–890.
31. Sousa, T.; Brockway, P.E.; Cullen, J.M.; Henriques, S.T.; Miller, J.; Serrenho, A.C.; Domingos, T. The need for robust, consistent methods in societal exergy accounting. *Ecol. Econ.* 2017, *141*, 11–21.
32. Sciubba, E.; Wall, G. A brief commented history of exergy from the beginnings to 2004. *Int. J. Thermodyn.*2007, *10*, 1–26.
33. Department of Energy. *Draft 2012 Integrated Energy Planning Report—Annexure A—Technical Report*; Department of Energy: Pretoria, South Africa, 2012.
34. Palma, M.; Sousa, T.; Guevara, Z. How Much Detail Should We Use to Compute Societal Aggregated Exergy Efficiencies? *Energies* 2016, *9*, 364.
35. Ayres, R.U.; Warr, B. Accounting for growth: The role of physical work. *Struct. Change Econ. Dyn.* 2005, *16*, 181–209.
36. Sciubba, E. A revised calculation of the econometric factors α- and β for the Extended Exergy Accounting method. *Ecol. Model.* 2011, *222*, 1060–1066.
37. Human Development Index (HDI). Human Development Reports. Available online: http://hdr.undp.org/en/content/human-development-index-hdi (accessed on 18 September 2014).
38. Seckin, C.; Sciubba, E.; Bayulken, A.R. Extended exergy analysis of Turkish transportation sector. *J. Clean. Prod.* 2013, *47*, 422–436.
39. European central Bank. Eurosystem statistics. Available online: https://www.ecb.europa.eu/stats/money_credit_banking/monetary_aggregates/html/hist_content.en.html (accessed on 13 June 2015).
40. Kotas, T.J. *The Exergy Method of Thermal Plant Analysis*; Butterworths: London, UK, 1985.
41. Milia, D.; Sciubba, E. Exergy-based lumped simulation of complex

systems: An interactive analysis tool. *Energy* 2006, *31*, 100–111.

42. Database—Eurostat. Available online: http://ec.europa.eu/eurostat/data/database (accessed on 10 January 2015).
43. Szargut, J.; Morris, D.R.; Steward, F.R. *Exergy Analysis of Thermal, Chemical, and Metallurgical Processes*; Hemisphere: New York, NY, USA, 1988.
44. Szargut, J. *Exergy Method: Technical and Ecological Applications*; WIT Press: Boston, MA, USA, 2005.
45. FAOSTAT. Available online: http://www.fao.org/faostat/en/#home (accessed on 23 January 2015).
46. US Department of Agriculture; Agricultural Research Service; Nutrient Data Laboratory. *USDA National Nutrient Database for Standard Reference, Release 27*; USDA: Washington, DC, USA, 2014.
47. EUROPA—Economic and Financial Affairs—Indicators -AMECO database. Available online: http://ec.europa.eu/economy_finance/ameco/user/serie/SelectSerie.cfm (accessed on 13 March 2015).
48. ESS: Nutritive Factors. Available online: http://www.fao.org/economic/the-statistics-division-ess/publications-studies/publications/nutritive-factors/en/ (accessed on 11 July 2015).
49. Food and Agriculture Organization of the United Nations. *Food Energy—Methods of Analysis and Conversion Factors*; Food and Agriculture Organization of the United Nations: Rome, Italy, 2003.
50. Koester, U.; Empen, J.; Holm, T. *Food Losses and Waste in Europe and Central Asia*; Food and Agriculture Organization of the United Nations: Rome, Italy, 2013.
51. Seckin, C.; Sciubba, E.; Bayulken, A.R. An application of the extended exergy accounting method to the Turkish society, year 2006. *Energy* 2012, *40*, 151–163.
52. Dewulf, J.; Bösch, M.E.; de Meester, B.; van der Vorst, G.; van Langenhove, H.; Hellweg, S.; Huijbregts, M.A.J. Cumulative Exergy Extraction from the Natural Environment (CEENE): A comprehensive Life Cycle Impact Assessment method for resource accounting. *Environ. Sci. Technol.* 2007, *41*, 8477–8483.
53. Dai, J.; Chen, B.; Sciubba, E. Extended exergy based ecological accounting for the transportation sector in China. *Renew. Sustain. Energy Rev.* 2014, *32*, 229–237.
54. Sciubba, E. A Thermodynamically Correct Treatment of Externalities

with an Exergy-Based Numeraire. *Sustainability* 2012, *4*, 933–957.

55. Food and Agriculture Organization of the United Nations. *The State of the World Fisheries and Aquaculture 2014: Opportunities and Challenges*; Food and Agriculture Organization of the United Nations: Rome, Italy, 2014.
56. Romero, J.C.; Linares, P. Exergy as a global energy sustainability indicator. A review of the state of the art. *Renew. Sustain. Energy Rev.* 2014, *33*, 427–442.
57. Celpa. *Boletim Estatístico*; Industria Papeleira Portuguesa: Lisboa, Portugal, 2014.
58. Özilgen, M.; Öner, E.S. *Biothermodynamics: Principles and Applications*; CRC Press: Boca Raton, FL, USA, 2017.
59. Dewulf, J.; van Langenhove, H.; van de Velde, B. Exergy-Based Efficiency and Renewability Assessment of Biofuel Production. *Environ. Sci. Technol.* 2005, *39*, 3878–3882.
60. Weight Equivalents: Eggs. Available online: Hannaone.com (accessed on 13 August 2018).

CHAPTER 10

BIOFUEL IMPACTS ON WORLD FOOD SUPPLY: USE OF FOSSIL FUEL, LAND AND WATER RESOURCES

David Pimentel , Alison Marklein, Megan A. Toth, Marissa Karpoff, Gillian S. Paul, Robert McCormack, Joanna Kyriazis , and Tim Krueger

5126 Comstock Hall, Cornell University, Ithaca, NY 14853, USA

ABSTRACT

The rapidly growing world population and rising consumption of biofuels are increasing demand for both food and biofuels. This exaggerates both food and fuel shortages. Using food crops such as corn grain to produce ethanol raises major nutritional and ethical concerns. Nearly 60% of humans in the world are currently malnourished, so the need for grains and other basic foods is critical. Growing crops for fuel squanders land, water and energy resources vital for the production of food for human consumption.

Citation: Pimentel, D.; Marklein, A.; Toth, M.A.; Karpoff, M.; Paul, G.S.; McCormack, R.; Kyriazis, J.; Krueger, T. Biofuel Impacts on World Food Supply: Use of Fossil Fuel, Land and Water Resources. *Energies* **2008**, *1*, 41-78.

Using corn for ethanol increases the price of U.S. beef, chicken, pork, eggs, breads, cereals, and milk more than 10% to 30%.

Keywords: Agriculture, biofuels, energy, food security, fossil fuels, natural resources, renewable energy

INTRODUCTION

With shortages of fossil energy, especially oil and natural gas, and heavy biomass energy consumption occurring in both developed and developing countries, a major focus has developed worldwide on biofuel production [1,2,3]. Emphasis on biofuels has developed globally, including those made from crops such as corn, sugarcane, and soybean, considered by some researchers as renewable energy sources. Wood and crop residues also are being used as fuel [4]. Though it may seem beneficial to use renewable plant materials for biofuel, the use of crop residues and other biomass for biofuels raises concerns about major environmental impacts [5]. Conflicts exist today in the use of land, water, energy, and other environmental resources required by both food and biofuel production. Although much of the land worldwide is occupied by grain and other crops, malnutrition is still the leading cause of death in the world today [6]. The World Health Organization [7] reports that more than 3.7 billion people (56% of the global population) are currently malnourished and that number is steadily increasing. Grains make up more than 80% of the world food supply, and unfortunately the Food and Agricultural Organization of the United Nations reports that per capita grain production has been declining for the past 23 years [8,9]. This suggests that the nutritional needs of the human population will require an increasing amount of agricultural resources as food. Food and biofuels are dependent on the same resources for production: land, water, and energy. In the U.S., about 19% of all fossil energy is utilized in the food system, including about 7% for agricultural production, 7% for processing and packaging foods, and about 5% for distribution and preparation of food [10]. In developing countries, about 50% of wood energy is used primarily for cooking in the food system [11]. Worldwide, the process of turning natural resources into food requires large amounts of energy, land, and water resources. The objective of this article is to analyze: (1) the reliance of both food and biofuel for the same land, water, and fossil energy resources in food versus biofuel production and (2) the characteristics of the environmental impacts caused by food and biofuel production.

FOOD SUPPLY AND MALNOURISHMENT IN THE WORLD

Serious problems face the world food supply today. The human population faces serious food shortages and malnutrition [7]. Considering this crisis, the Food and Agricultural Organization (FAO) of the United Nations confirms that worldwide food available per capita has been declining *continuously* based on availability of cereal grains during the past 23 years [8], which make up alarmingly about 80% of the world's food supply [9]. Although grain yields per hectare in both developed and developing countries are still gradually increasing, the rate of increase is slowing, while the world population and its food needs are rising rapidly [8,12]. For example, from 1950 to 1980, U.S. grain yields increased at about 3% per year, but since 1980, the annual rate of increase for corn and other other grains is only approximately 1% [13]. Therefore, the rate of increase in grain production is not keeping up with the rapid rate of world population growth of 1.2% [12].

In 2005, the World Health Organization reported the largest number and proportion of malnourished people, 3.7 billion [7]. This is the largest number and proportion of malnourished people ever reported. Their assessment of malnutrition includes deficiencies in calories, protein, iron, iodine, and vitamins A, B, C, and D [14,15]. Others, like FAO [16], report that 850 million people are malnourished (based only on protein/calorie malnourishment). However, this estimate is low because it does not account for the strenuous work and micronutrient deficiencies that affect billions of people worldwide [17]. Malnutrition is especially difficult for children because of their rapid growth and activity. Malnutrition is the underlying cause of more than half of the deaths of children under 5 years old worldwide [18]. Furthermore, the UN reports over 9 million people die from starvation each year [19]. Food shortages are also related to such problems as distribution, wars, droughts, and climate change.

In order to solve the problem of malnutrition in the world, the U.S. must address over-consumption of food, fuel, and other resources. The average American consumes more than 1,000 kg (2,200 lbs) of food per year [20]. This is the equivalent of an average of 3,800 kcal per day, whereas the average male only requires 2,400 kcal and female 2,000 kcal per day [20]. The U.S. over-consumption is compounded by the massive use of fats and sweets in the diet [21]. Growing crops for biofuel will only exacerbate the problem of resource allocation, for biofuel production depletes land, water, and energy resources for human food production.

WORLD CROPLAND RESOURCES

More than 99.7% of human food comes from the terrestrial environment, while less than 0.3% comes from the oceans and other aquatic ecosystems [22]. Most of the suitable land for biomass production is already in production. Worldwide, of the total 13 billion hectares of land area on earth the percentages in use are: cropland, 11%; pasture land, 27%; forest land, 32%; urban, 9%; and other 21%. The remaining land (21%) is unsuitable for crops, pasture, and/or forests because the soil is too infertile or shallow to support plant growth, or the climate and region are too cold, dry, steep, stony, or wet [23]. Thus, most suitable land for production is already is in use.

In 1999 the United Nations reported that only 0.22 ha of cropland per capita was available worldwide [24,25]; however, 0.5 ha is considered essential. In 1960, when the world population numbered only 3 billion, approximately 0.5 ha per capita was available for the production of a diverse, nutritious diet of plant and animal products [26].

As the human population continues to increase, there has been an expansion of diverse human activities, which has reduced available cropland and pasture land. For example, vital cropland and pasture land has been covered by transportation systems and urbanization. In the U.S., each person in the population uses 0.4 ha (1 acre) of land for urbanization and highways [27]. China's recent explosion in development provides an example of rapid declines in the availability of per capita cropland [28]. The current amount of available cropland in China is only 0.08 ha per capita (Table 1). This relatively small amount of cropland provides the people in China with a predominantly vegetarian diet, which requires less energy, land, and biomass than a typical American diet.

Table 1: Resources used and/or available per capita per year in the U.S., China, and the world to supply basic human needs [23]

Resources	U.S.	China	World
Land			
Cropland (ha)	0.48	0.08	0.22
Pasture (ha)	0.79	0.33	0.52
Forest (ha)	0.79	0.11	0.59
Total (ha)	2.78	0.46	1.97
Water (liters x 10^6)	2.0	0.46	0.60

Fossil fuel (BP, 2005)			
Oil equivalents (liters) 9,500		1,400	2,100

As a result of world population growth, the average available cropland per capita worldwide has now diminished to less than 0.22 ha (a total of 1.5 billion ha of cropland [25]. This is about half the amount currently available in the U.S.; the average cropland per capita in the U.S. has also declined to 0.48 ha, or less than the critical land area essential for diverse food production (Table 1).

The availability of cropland influences the kinds and types of foods produced. For example, at present, a total of 1,400 kg/yr per capita of agricultural product is produced to feed each American, while the Chinese food supply averages only 800 kg/yr per capita [28]. By all available measurements, the Chinese have reached or exceeded the limits of their agricultural system. Furthermore, the reliance of the Chinese on large inputs of fossil-fuel based fertilizers to compensate for shortages of arable land and severely eroded soils, and a limited fresh water supply, suggest severe problems looming in the near future [28]. Today, China imports large amounts of grain from the United States and other nations and is expected to further increase grain imports in the near future, because of land and water shortages.

WORLD WATER RESOURCES

Water is a vital resource for crop production and human survival [30]. The production of 9 t/ha of corn requires about 7 million liters of water (about 700,000 gallons of water per acre) [31]. However, world irrigation has declined about 10% during the past 10 years based on World Bank lending [32]. This is of critical importance because 17% of the crops that are irrigated provide 40% of the world food supply [22]. Thus, the world's capacity for food production is becoming limited by declining water resources.

ENERGY RESOURCES

Humans have relied on various sources of power for centuries, but solar energy is the ultimate source of global production. Biologically captured solar energy is vital to all natural ecosystems and to maintaining the diversity of all life systems. Other energy sources have ranged from human power, animal, wind, tidal, and water energy to wood, coal, gas, oil, and nuclear sources for fuel and power. Since about the year 1700, abundant fossil fuel energy supplies have made it possible to augment agricultural production to feed an increasing number of humans [9]. For example, energy availability has made possible the purification and transport of water, reduced human diseases, and a generally enhanced quality of human life.

Since the industrial revolution of the 1850's, the total rate of energy use from all sources has been growing even faster than world population growth. From 1970 to 1995, energy use increased at a rate of 2.5% per year (doubling every 30 years) compared with the worldwide population growth of 1.7% per year (doubling every 40 to 60 years) [9]. The use and availability of energy resources will start to decline slowly and continuously until little or no oil and gas resources remain [33,34,35].

About 50% of all the solar energy captured by worldwide photosynthesis is used by humans for food, forest products, and other systems. However, it is still inadequate to meet all current human needs – namely, food needs [36]. To make up for this shortfall, about 473 quads (1 quad = 1 x 10^{15} BTU) of fossil energy, mainly oil, gas, coal, and a small amount of nuclear, are utilized worldwide each year (Table 2). Of the 473 quads, about 100 quads (or about 22%) of the world's total energy are utilized in the United States [27].

Table 2: Fossil and solar energy use in the U.S. and world (quads = 10^{15} BTU) [27]

Fuel	U.S.	World
Petroleum	40.1	168
Natural Gas	23.0	103
Coal	22.3	115
Nuclear	8.2	28
Biomass	3.0	30
Hydroelectric power	3.4	27
Geothermal and windpower	0.4	0.8
Biofuels	0.5	0.9
Total	100.9	472.7

Solar Energy Captured by Green Plants

A large amount of vegetation biomass is required to supply the growing human population with usable biomass energy. Some estimate that approximately 0.5% of the solar energy reaching the earth is captured by green plants [37]. In the U.S., Pimentel and Patzek (2006) [38] estimate that 0.1% of total solar energy is captured by green plants (Table 3). The U.S. has a total land area of 917 million hectares and a current population of 302 million [12]. The U.S. population adds 3.3 million more humans per year with a rate of increase at 1.1%, which is more than double the rate of increase in China at 0.5% [12]. This growing population will result in more land used for urbanization, highways, and food production, all of which will further stress U.S. natural resources.

Table 3: Total amount of above ground biomass except for some crops that include underground biomass and solar energy captured each year in the United States. An estimated 32 x 10^{15} BTU of sunlight captured in the U.S. per year suggests that the green plants in the U.S. are collecting 0.1% of the solar energy [39,40,41,42]

Crops	901 x 10^6 tons	14.4 x 10^{15} BTU
Pasture	600 x 10^6 tons	9.6 x 10^{15} BTU
Forest	527 x 10^6 tons	8.4 x 10^{15} BTU
Total	2,028 x 10^6 tons	32.4 x 10^{15} BTU

Nearly one half (45%) of the land area in the United States is already used for crops and pastures [20]. The forest land area covers 30%. Urban areas and highways cover approximately 3%, with miscellaneous areas covering about 23% [20].

U.S. Energy Use

Each year, the U.S. population uses 3 times as much fossil energy as the total solar energy captured by harvested U.S. crops, forest products, and other vegetation (Table 1, Table 2 and Table 3). Industry, transportation, home heating and cooling, and food production account for most of the fossil energy consumed in the United States [27]. Per capita use of fossil energy in the United States per year amounts to about 9,500 liters of oil equivalents -- more than 7-times the per capita use in China (Table 1). In China, most fossil energy is used by industry, although a substantial amount, approximately 25%, is now used for agriculture and in the food production system [28].

Taken together, developed nations annually consume about 70% of the fossil energy worldwide, while the developing nations, which represent about 75% of the world population, use only 30% of world fossil energy [43]. The United States, with only 4.5% of the world's population, is a major consumer of fossil energy and annually consumes about 22% of the world's fossil energy output (Table 2).

Some developing nations that are experiencing especially high rates of population growth are increasing fossil fuel use to help augment their agricultural production of food and fiber. For example, since 1955 in China, there has been a 100-fold increase in fossil energy use in agriculture for fertilizers, pesticides, and irrigation [28]. However, as fossil energy resources decline, the use of fossil energy will become limited for food production, causing even more severe food shortages for the rapidly growing world population.

Worldwide fertilizer use per capita has declined by more than 22% since 1989, especially in the developing countries, due to fossil fuel shortages and high prices for fertilizers [44]. During the past 5 years, nitrogen fertilizer costs have more than doubled in the U.S. [45]. In addition, the overall projections of available fossil energy resources for manufacturing fertilizers and all other purposes are discouraging as the reserves of these finite fossil fuels rapidly diminish.

Fossil Fuel Supplies

World oil production has peaked and projections are that by 2040 oil will decline to about 62% below peak (W. Youngquist, petroleum geologist, Eugene, OR, 30 April, 2008). The world supply of oil is projected to last approximately 40-60 years, if use continues at current production rates [29]. Natural gas is considered adequate for about 30 years and coal for less than 100 years [29,46]. In the U.S., natural gas is already in short supply; it is projected that the U.S. will deplete its natural gas resources in about 20 years [33].

Initial drilling for oil and gas provided optimistic estimates of fossil fuel resources in the U.S. [46]. However, both the U.S. oil production rate and existing reserves have been declining for more than 30 years [27,47]. To date, approximately 90% of U.S. oil resources have already been exploited (W. Youngquist, personal communication, petroleum geologist, Eugene, Oregon, 2002) [48]. At present, the United States is importing more than 65% of its oil [27], which puts the U.S. economy at risk due to fluctuating oil prices and difficult international political situations.

BIOMASS RESOURCES

Worldwide, most biomass is burned for cooking and heating; however, it can also be converted into electricity. Assuming an optimal yield of 3 dry metric tons (t/ha) per year of woody biomass can be harvested sustainably, this would provide a gross energy yield of 13.5 million kcal/ha [49,50]. Harvesting this wood biomass requires an energy expenditure of approximately 30 liters of diesel fuel per ha, not including the embodied energy for cutting and collecting wood for transport to an electric power plant. Thus, the energy input per output ratio for such a system is calculated to be 1:25 [51].

Fuelwood provides an important source of biomass energy used for industry and homes in the United States, accounting for 3% of the total energy use in the U.S. [27]. About 60% of this biomass is wood-waste fuel from paper producing plants, lumber mills, and furniture producing plants [27]. Approximately 30% of the wood is utilized in homes and power plants as fuel [52].

The cost of producing 1 kWh of electricity from woody biomass in the U.S. is about $0.06, which is competitive with other electricity production systems that average $0.08 in the U.S. [27,53]. Approximately 3 kWh of thermal energy is expended to produce 1 kWh of electricity, an energy input/output ratio of 1:3. Per capita consumption of woody biomass for heat in the U.S. amounts to 625 kilograms (kg) per year. The diverse biomass resources (wood, crop residues, and dung) used in developing nations averages about 630 kg per capita per year [54]. Developing countries use only about 500 liters of oil equivalents of fossil energy per capita compared with nearly 9,500 liters of oil equivalents of fossil energy used per capita in the U.S. (Table 1) [55].

Woody biomass has the capacity to supply the U.S. with about 5 quads (1.5 x 10^{12} kWh thermal) of its total gross energy supply by the year 2050, provided that the amount of forest land stays constant [56]. A city of 100,000 people using the biomass from a sustainable forest (3 t/ha per year) for electricity requires approximately 200,000 ha of forest area, based on an average electrical demand of slightly more than 1 billion kWh (electrical energy [e]) (860 kcal = 1 kWh) [56].

Environmental impacts of burning biomass are less harmful than those associated with coal, but more harmful than those associated with natural gas [36]. Biomass combustion releases more than 200 different chemical pollutants, including 14 carcinogens and 4 co-carcinogens, into the atmosphere [57]. As a result of this, but especially in developing nations where people cook with fuelwood over open fires, approximately 4 billion people suffer from continuous exposure to smoke [58]. In the U.S., wood smoke kills 30,000 people each year [59], although many of the pollutants from electric plants that use wood and other biomass can be mitigated. These controls include the same scrubbers that are frequently installed on coal-fired plants.

Regional Benefits of Thermal Energy from Biofuels

Direct combustion is the most efficient use of wood biomass. Space heating in the U.S. is a major component of both residential and commercial buildings. In total, the amount of energy used in these sectors totals about 16 quads per year [27]. Currently, only an estimated 1% of U.S. homes are heated with wood biomass [60]. If 10% of the forest biomass in the Midwest, Northeast, and Southeast (that totals an estimated 100 million hectares) plus 10% of the grass biomass (that totals an estimated 26 million hectares) were harvested, this thermal energy would be a major asset to the nation [20]. The estimated sustainable forest and grass biomass produced per hectare per year is about 2 t/ha/yr without any fertilizer or other inputs (Table 3) [42]. Assuming half of the energy harvested from these biomass areas is utilized for harvesting, transport, and processing, about 0.032 quads of thermal energy can be utilized from the forest land, and 0.016 quads of thermal energy can be utilized from the grassland area. This would total about 0.05% of total energy use in the U.S. (Table 2). Areas in the Northeast that use wood for space heating, have become reforested as many farmlands were abandoned, and are now major sources of wood biomass.

Increasing the use of wood biomass for heating homes will increase the number of illnesses, like asthma and death [57]. Some communities, like Aspen, Colorado, already have banned the use of wood fuels for heating, due to many adverse health effects [61].

WORLD FOREST BIOMASS

The total sustainable world biomass energy potential has been estimated to be about 92 quads (10^{15} BTU) per year [62], which represents 19% of total global energy use (Table 2). The total forest biomass produced worldwide is 38 quads per year [62], which represents 8% of total energy use (Table 2). In the U.S., only 1% to 2% home heating is achieved with wood [63]. Of the 106 million U.S. homes that heat their homes, 2.0 million use wood as the main heating fuel (3%) [64].

Global forest area removed each year totals 15 million ha [65]. Global forest biomass harvested is just over 1,431 billion kg per year, of which 60% is industrial roundwood and 40% is fuelwood [66]. About 90% of the fuelwood is utilized in developing countries [62]. A significant portion (26%) of all forest wood is converted into charcoal [67]. The production of charcoal causes a 30% to 50% loss of energy [68], and produces large quantities of smoke. The resulting charcoal is cleaner burning and causes less smoke than burning wood fuel directly in the home [67]. Charcoal is dirty to handle but light in weight.

In developing countries, about 2 kcal of wood are utilized in cooking 1 kcal of food [69]. Thus, more biomass and more land and water are needed to produce the biofuel for cooking than the resources needed to produce the food.

In the U.S., about 30% of the land area is forested and produces 239 billion kg of wood biomass per year [20]. The total forest products harvested per year per person in the U.S. is 914 kg, but only 71 kg is used for firewood [70]. The rest is used for the production of lumber and pulp for paper production.

The average yield of forest biomass in U.S. timberland is 1.93 dry metric tons/ha/yr [42]. Under sustainable, moist, forest conditions in both temperate and tropical ecosystems, approximately 3 dry metric tons (t/ha) per year of woody biomass can be harvested sustainably [50,71,72,73].

CROPLAND AND PASTURE BIOMASS

Of the total world land area in cropland, pasture, and forest, about 38% is cropland and pasture and about 30% is forests [23]. Devoting a portion of this cropland and forest land to biofuels will stress both managed ecosystems and will not be sufficient to solve the fuel problem [1].

Corn Ethanol and Energy Inputs

In the United States, ethanol constitutes 99% of all biofuels [74]. The average, energy and monetary costs, for a large modern ethanol plant are enormous, about $65 million. For capital expenditures, new plant construction costs from $1.05 to $3.00 per gallon of ethanol [75]. Fermenting and distilling corn ethanol requires large amounts of water. The corn is finely ground and approximately 15 liters of water are added per 2.69 kg of ground corn. After fermentation, to obtain a liter of 95% pure ethanol from the 10% ethanol and 90% water mixture, 1 liter of ethanol must be extracted from the approximately 10 liters of the ethanol/water mixture. To be mixed with gasoline, the 95% ethanol must be further processed and more water must be removed, requiring additional fossil energy inputs to achieve 99.5% pure ethanol (Table 4 and Table 5). Thus, a total of about 12 liters of wastewater must be removed per liter of ethanol produced, and this relatively large amount of sewage effluent has to be disposed of at an energy, economic, and environmental cost.

Table 4: Energy inputs and costs of corn production per hectare in the United States

Inputs	Quantity	kcal x 1000	Costs $
Labor	11.4 hrs[a]	426[b]	300.00[c]
Machinery	55 kg[d]	1,018[e]	310.00[f]
Diesel	88 L[g]	1,003[h]	500.00
Nitrogen	155 kg[k]	2,480[l]	255.00[m]
Phosphorus	79 kg[n]	328[o]	150.00[p]
Potassium	84 kg[q]	274[r]	78.00[s]
Lime	1,120 kg[t]	315[u]	60.00
Seeds	21 kg[v]	520[w]	230.00[x]
Irrigation	8.1 cm[y]	320[z]	350.00[aa]
Herbicides	6.2 kg[bb]	620[ee]	372.00
Insecticides	2.8 kg[cc]	280[ee]	180.00
Electricity	13.2 Wh[dd]	34[ff]	27.00
Transport	204 kg[gg]	169[hh]	180.00
TOTAL		8,228	$2,992.00
Corn yield	9,400 kg/ha[ii]	33,840 kcal input:output 1:4.11	

a)NASS, 2003 [76].

b)It is assumed that a person works 2,000 hrs per year and utilizes an average of 8,000 liters of oil equivalents per year.

c)It is assumed that labor is paid $26.32 an hour.

d)Pimentel and Pimentel, 2008 [4].

e)Prorated per hectare and 10 year life of the machinery. Tractors weigh from 6 to 7 tons and harvesters 8 to 10 tons, plus plows, sprayers, and other equipment.

f)Estimated.

g)Estimated.

h)Input 11, 400 kcal per liter.

i)Estimated

j)Input 10,125 kcal per liter.

k)NASS, 2003 [76].

l)Patzek, 2004 [77].

m)Cost $1.65 per kg.

n)NASS, 2003 [76].

o)Input 4,154 kcal per kg.

p)Cost $1.90 per kg.

q)NASS, 2003 [76].

r)Input 3,260 kcal per kg.

s)Cost $0.93 per kg.

t)Brees, 2004 [78]

u)Input 281 kcal per kg.

v)Pimentel and Pimentel, 2008 [4].

w)Pimentel and Pimentel, 2008 [4].

x)USDA, 1997 [79].

y)USDA, 1997 [80].

z)Batty and Keller, 1980 [81].

aa) Irrigation for 100 cm of water per hectare costs $1,000 (Larsen et al., 2002) [82].

bb) Larson and Cardwell, 1999 [83].

cc)USDA, 2002 [84].

dd)USDA, 1991 [85].

ee) Input 100,000 kcal per kg of herbicide and insecticide.

ff) Input 860 kcal per kWh and requires 3 kWh thermal energy to produce 1 kWh electricity.

gg) Goods transported include machinery, fuels, and seeds that were shipped an estimated 1,000 km.

hh) Input 0.83 kcal per kg per km transported.

ii) Average. [20,47].

Table 5: Inputs per 1000 liters of 99.5% ethanol produced from corn[a]

Inputs	Quantity	Kcal x 1000	Dollar $
Corn grain	2,690 kg[b]	2,355[b]	856.22
Corn transport	2,690 kg[b]	322[c]	21.40[d]
Water	15,000L[e]	90f	21.16[g]
Stainless steel	3 kg[i]	165[p]	10.60[d]
Steel	4 kg[i]	92[p]	10.60[d]
Cement	8 kg[i]	384[p]	10.60[d]
Steam	2,646,000 kcal[j]	2,646[j]	21.16[k]
Electricity	392 kWh[j]	1,011[j]	27.44[l]
95% ethanol to 99.5%	9 kcal/L[m]	9[m]	40.00
Sewage effluent	20kg BOD[n]	69[h]	6.00
Distribution	331 kcal/L[q]	331	20.00[q]
TOTAL		7,474	$1,045.18

a)Output: 1 liter of ethanol = 5,130 kcal (Low heating value). The mean yield of 2.5 gal pure EtOH per bushel has been obtained from the industry-reported ethanol sales minus ethanol imports from Brazil, both multiplied by 0.95 to account for 5% by volume of the #14 gasoline denaturant, and the result was divided by the industry-reported bushels of corn inputs to ethanol plants [86].

b)Data from table 4.

c)Calculated for 144 km roundtrip.

d)Pimentel, 2003 [87].

e)15 liters of water mixed with each kg of grain.

f)Pimentel et al., 2004 [31].

g)Pimentel et al., 2004 [31].

h)4 kWh of energy required to process 1 kg of BOD [88].

i)Estimated from the industry reported costs of $85 millions per 65 million gallons/yr dry grain plant amortized over 30 years. The total amortized cost is $43.6/1000L EtOH, of which an estimated $32 go to steel and cement.

j)Illinois Corn, 2004 [89]. The current estimate is below the average of 40,000 Btu/gal of denatured ethanol paid to the Public Utilities Commission in South Dakota by ethanol plants in 2005.

k)Calculated based on coal fuel. Below the 1.95 kWh/gal of denatured EtOH in South Dakota, see j).

l)$.07 per kWh [47].

m)95% ethanol converted to 99.5% ethanol for addition to gasoline (T. Patzek, personal communication, University of California, Berkeley, 2004).

n)20 kg of BOD per 1000 liters of ethanol produced [90].

p)Newton, 2001 [91].

q)DOE, 2002 [92].

Manufacture of a liter of 99.5% ethanol uses 46% more fossil energy than it produces and costs $1.05 per liter ($3.97 per gallon) (Table 5). The corn feedstock alone requires more than 33% of the total energy input.

The largest energy inputs in corn-ethanol production are for producing the corn feedstock plus the steam energy and electricity used in the fermentation/distillation process. The total energy input to produce a liter of ethanol is 7,474 kcal (Table 5). However, a liter of ethanol has an energy value of only 5,130 kcal. Based on a net energy loss of 2,344 kcal of ethanol produced, 46% more fossil energy is expended than is produced as ethanol. The total cost, including the energy inputs for the fermentation/distillation process and the apportioned energy costs of the stainless steel tanks and other industrial materials, is $1,045 per 1000 liters of ethanol produced (Table 5).

The subsidies for corn ethanol total more than $6 billion per year (Koplow, 2006). This means that the subsidies per liter of ethanol are 60 times greater than the subsidies per liter of gasoline. The actual crux of state behavior regarding ethanol rests on federal subsidies. A calculation of the environmental effects resulting from subsidizing ethanol production is difficult and the subsidy process itself is complex and fluid. The basic methodology of such calculations, however, should be to first identify:

a)the amount and type of ethanol that would be produced in the U.S. in the absence of subsidies and tariffs

b)the amount and type of ethanol that would be consumed in the U.S. in the absence of subsidies and tariffs

c)the composition of energy that would most likely fill this void

d)the environmental costs and benefits of the current ethanol used in comparison to the environmental costs and benefits of the most likely alternative energy scenario.

These calculations assume that the behavior of Americans and their overall energy is an inelastic quantity, i.e. that energy consumption will neither increase nor decrease as a function of energy source. This assumption likely amounts to a conservative estimate of ethanol's environmental impact, as long as consumption behavior can be expected to constrict in the instance of a domestic shortage of energy, in other words, when a greater percentage of our energy comes from oil. The calculations also assume that Americans will follow an ideal-rational behavior model and, if given the option between purchasing ethanol and gasoline, will generally opt for whichever is cheaper. While there could be a non-economic, normative value to ethanol, this would be nearly impossible to calculate, and could just as easily exist as a normative value against ethanol.

In 2006, nearly 19 billion liters of ethanol were produced on 20% of U.S. corn acreage [27]. These 19 billion liters represents only 1% of total U.S. petroleum use.

However, even if we completely ignore corn ethanol's negative energy balance and high economic cost, we still find that it is absolutely not feasible to use ethanol as a replacement for U.S. oil consumption. If all 341 billion kg of corn produced in the U.S. [20] were converted into ethanol at a rate of 2.69 kg per liters of ethanol, then 129 billion liters of ethanol could be produced. This would provide only 7% of total oil consumption in the U.S. Of course, in this situation there would be no corn available for livestock or human consumption.

The environmental impacts of corn ethanol are enormous:

1)Corn production causes more soil erosion than any other crop grown [94].

2)Corn production uses more nitrogen fertilizer than any other crop grown and is the prime cause of the dead zone in the Gulf of Mexico [94]. In 2006, approximately 4.7 million tons of nitrogen was used in U.S. corn production [45]. In addition, about 1.7 million tons of phosphorus was used in the U.S. for corn production in 2006.

3)Corn production uses more insecticides than any other crop grown [95]. (Total: 7,530 kg used in corn in 2005 [96], or about 0.45 kg/ha.)

4)Corn production uses more herbicides than any other crop grown [77]. (Total: 186, 876 kg applied in 2005 [96], or 6.4 kg/ha.)

5)More than 1,700 gallons of water are required to produce 1 gallon of corn ethanol [97].

6)Enormous quantities of carbon dioxide are produced during corn ethanol production by the large quantity of fossil energy used in production, during fermentation, and when the soil is tilled, leaving soil organic matter exposed and oxidized. In addition, the conversion of cropland for biofuel production contributes to the release of greenhouse gases [98]. All this speeds global warming [99].

7)Related to the total operation, including the burning of the ethanol, the air pollution problem is significant [97,100,101,102]. Burning ethanol emits pollutants into air such as peroxyacetyl nitrate (PAN), acetaldhyde, alkylates, and nitrous oxide. These can have significant human health effects, as well as impacts on other organisms and ecosystems [103].

Grass and Cellulosic Ethanol

Tilman et al. (2006) [104] suggest that all 235 million hectares of grassland, plus crop residues, can be converted into cellulosic ethanol. This is a suggestion that causes concern among scientists. Tilman et al. suggest that crop residues, like corn stover, can be harvested and utilized as a fuel source, but this would be a disaster for the agricultural ecosystem because crop residues are vital for protecting topsoil. Leaving the soil unprotected would intensify soil erosion by 10-fold or more [105], and may increase soil loss as much as 100-fold [106]. Furthermore, even a partial removal of the stover can result in increased CO_2 emissions and intensify acidification and eutrophication due to increased runoff [107,108]. Already, the U.S. crop system is losing soil 10 times faster than the sustainable rate [94]. Soil formation rates, at less than 1 t/ha/yr, are extremely slow [94,109]. Increased soil erosion caused by the removal of crop residues for use as biofuels will facilitate the soil-carbon oxidation and contribute to the greenhouse problem [110].

Tilman et al. [104] assume about 1,032 liters of ethanol can be produced through the conversion of the 4 t/ha/yr of grasses harvested. However, Pimentel and Patzek (2008) [97] report a negative 68% return in ethanol produced compared with the fossil energy inputs in switchgrass conversion (Table 6 and Table 7). The cost of producing a liter of ethanol using switchgrass was 93¢ (Table 7). The two major energy inputs for switchgrass conversion into ethanol were steam and electricity production (Table 7).

Table 6: Average inputs and energy inputs per hectare per year for switchgrass production

Input	Quantity	10^3 kcal	Dollars
Labor	5 hr[a]	200[b]	$65[c]
Machinery	30 kg[d]	555	50[a]
Diesel	150 L[e]	1,500	75
Nitrogen	80 kg[e]	1,280	45[e]
Seeds	1.6 kg[f]	100[a]	3[f]
Herbicides	3 kg[g]	300[h]	30[a]
TOTAL	10,000 kg yield[i]	3,935	$ 268[j]
	40 million kcal yield input/output ratio 1:02[k]		

a)Estimated.

b)Average person works 2,000 hours per year and uses about 8,000 liters of oil equivalents. Prorated this works out to be 200,000 kcal.

c)The agricultural labor is paid $13 per hour.

d)The machinery estimate also includes 25% more for repairs.

e)Calculated based on data from Brummer et al., 2000 [111].

f)Data from Samson, 1991 [112].

g)Calculated based on data from Henning, 1993 [113].

h)100,000 kcal per kg of herbicide.

i)Samson et al., 2000 [114].

j)Brummer et al. 2000 [111] estimated a cost of about $400/ha for switchgrass production. Thus, the $268 total cost is about 49% lower than what Brummer et al. estimates and this includes several inputs not included in Brummer et al.

k)Samson et al. (2000) [114] estimated an input per output return of 1:14.9, but we have added several inputs not included in Samson et al. Still the input/output return of 1:11 would be excellent if the sustained yield of 10 t/ha/yr were possible.

Table 7: Inputs per 1000 liters of 99.5% ethanol produced from U.S. switchgrass[a]

Inputs	Quantity	kcal x 1000	Dollars $
Switchgrass	5,000 kg[b]	1,968[c]	500
S. Grass transport	5,000 kg[b]	600[c]	30[d]
Water	250,000L[e]	140[f]	40[m]
Stainless steel	3 kg[g]	165[g]	11[g]
Steel	4 kg[g]	92[g]	11[g]
Cement	8 kg[g]	384[g]	11[g]
Grind switchgrass	5,000 kg	200[h]	16[h]
Sulfuric acid	240 kg[i] 0 168[n]		
Steam	8.1 tons[i]	4,404	36
Lignin	1,250 kg[j]	minus 1,500	minus 12
Electricity	666 kWh[i]	1,703	46
95% ethanol to 99.5%	9 kcal/L[k]	9	40
Sewage effluent	40 kg BOD[l]	138[o]	12
Distribution	331 kcal/L[p]	331	20
TOTAL		8,634	$929

a)Output: 1 liter of ethanol = 5,130 kcal. The ethanol yield here is 200 L/t dry biomass (dbm). Iogen suggests 320 L/t dbm of straw that contains 25% of lignin. This yield is equal to the average yield of ethanol from corn, 317 L/t dbm (2.5 gal/bu). In view of the difficulties with breaking up cellulose fibers and digesting them quickly enough, the Iogen yield seems to be exaggerated, unless significantly more grinding, cell exploding with steam, and hot sulfuric acid are used.

b)Data from table 6.
c)Calculated for 144 km roundtrip.
d)Pimentel, 2003 [87].
e)15 liters of water mixed with each kg of biomass.
f)Pimentel et al., 2004 [31].
g)Newton, 2001 [91].
h)Calculated based on grinder information [115].
i)Estimated based on cellulose conversion [116].
j)Wood is about 25% lignin and removing most of the water from the lignin by filtering, the moisture level can be reduced to 200% [117].

k)95% ethanol converted to 99.5% ethanol for addition to gasoline (T. Patzek, personal communication, University of California, Berkeley, 2004).

l)20 kg of BOD per 1000 liters of ethanol produced [90].

m)Pimentel, 2003 [87].

n)Sulfuric acid sells for $7 per kg.

o)4 kWh of energy required to process 1 kg of BOD (Blais et al., 1995) [88].

p)DOE, 2002 [92].

Converting all 235 million ha of U.S. grassland into ethanol at the optimistic rate by Tilman et al. would provide only 12% of annual consumption of U.S. oil [20,27]. Sound data, however, confirm that the output in ethanol would require 1.5 liters of oil equivalents to produce 1 liter of ethanol (Table 6 and Table 7). In addition, to achieve the production of this much ethanol, we would have to displace the 100 million cattle, 7 million sheep, and 4 million horses that are now grazing on 324 million ha of U.S. grassland and rangeland [20]. Already, overgrazing is a serious problem on U.S. grassland and a similar problem exists worldwide [118]. Thus, the assessment of the quantity of ethanol that can be produced on U.S. and world grasslands by Tilman et al. (2006) [104] appears to be unduly optimistic.

Several problems exist the conversion of cellulosic biomass into ethanol. First, it takes from 2 to 5 times more cellulosic biomass to achieve the same quantity of starches and sugars as are found in the same quantity of corn grain. Thus, 2 to 5 times more cellulosic material must be produced and handled compared with corn grain. In addition, the starches and sugars are tightly held in lignin in the cellulosic biomass. The starches and sugars can be released using a strong acid to dissolve the lignin. Once the lignin is dissolved the acid action is stopped with an alkali. Now the solution of lignin, starches, and sugars can be fermented.

Some claim that the lignin can be used as a fuel. Clearly, this would not be when dissolved in water. The lignin in the water mixture can be extracted using various energy intensive technologies. Usually less than 25% of the lignin can be extracted from the water mixture [97].

Methanol

Methanol can be produced from a gasifier-pyrolysis reactor using biomass as a feedstock [119,120]. The yield from 1 ton of dry wood is about 370 liters of methanol [121,122]. For a plant with economies of scale to operate efficiently, more than 1.5 million ha of sustainable forest would be required to supply this woody biomass annually [36]. However, biomass is not available in such enormous quantities, even from extensive forests, at acceptable prices for methanol to compete as a viable energy source. Most methanol today is produced from natural gas, not biomass [123]. About 10,402 kcal are required to produce 1 liter of methanol that contains 7,430 kcal.

Soybean Biodiesel

Processed vegetable oils from soybean, sunflower, rapeseed, oil palm, and other oil plants can be used as fuel in diesel engines. Unfortunately, producing vegetable oils for use in diesel engines is costly in terms of economics and energy (Table 8 and Table 9) [97,124]. A slight net return on energy from soybean oil is possible only if the soybeans are grown without commercial nitrogen fertilizer. The soybean, since it is a legume, will under favorable conditions produce its own nitrogen. Soy, however, has a 63% net fossil energy loss (Table 9).

Table 8: Energy inputs and costs in soybean production per hectare in the U.S

Inputs	**Quantity**	**kcal x 1000**	**Costs $**
Labor	7.1 hrs[a]	284[b]	112.00[c]
Machinery	20 kg[d]	360[e]	181.00[f]
Diesel	38.8 L[a]	442[g]	25.00
Gasoline	35.7 L[a]	270[h]	16.00
LP gas	3.3 L[a]	25[i]	1..00
Nitrogen	3.7 kg[j]	59[k]	28.00[l]
Phosphorus	37.8 kg[j]	156[m]	29.00[n]
Potassium	14.8 kg[j]	48[o]	6.00[p]
Limestone	2000 kg[v]	562[d]	56.00[v]
Seeds	69.3 kg[a]	554[q]	59.00[r]
Herbicides	1.3 kg[j]	130[e]	32.00
Electricity	10 kWh[d]	29[s]	1.00
Transport	154 kg[t]	40[u]	56.00
TOTAL		2,959	$602.00
Soybean yield	2,890 kg/ha[w]	10,404 kcal input:output 1:3.52	

a)Ali and McBride, 1990 [125].

b)It is assumed that a person works 2,000 hrs per year and utilizes an average of 8,000 liters of oil equivalents per year.

c)It is assumed that labor is paid $13 an hour.

d)Pimentel and Pimentel, 2008 [4].

e)Machinery is prorated per hectare and a 10 year life of the machinery. Tractors weigh from 6 to 7 t and harvesters from 8 to 10 tons, plus plows, sprayers, and other equipment.

f)College of Agri., Consumer & Environ. Sciences, 1997 [126].

g)Input 11,400 kcal per liter.

h)Input 10,125 kcal per liter.

i)Input 7,575 kcal per liter.

j)Economic Research Statistics, 1997 [127].

k)Patzek, 2004 [77].

l)Hinman et al., 1992 [128].

m)Input 4,154 kcal per kg.

n)Cost 77¢ per kg.

o)Input 3,260 kcal per kg.

p)Costs 41¢ per kg.

q)Pimentel et al., 2002 [129].

r)Costs about 85¢ per kg.

s)Input 860 kcal per kWh and requires 3 kWh thermal energy to produce 1 kWh electricity.

t)Goods transported include machinery, fuels, and seeds that were shipped an estimated 1,000 km.

u)Input 0.83 kcal per kg per km transported.

v)Mississippi State University Extension Service, 1999 [130].

w)USDA, 2004 [131].

Table 9: Inputs per 1,000 kg of biodiesel oil from soybeans

Inputs	Quantity	kcal x 1000	Costs $
Soybeans	5,556 kg[a]	5,689[a]	1,157.00[a]
Electricity	270 kWh[b]	697[c]	18.90[d]

Steam	1,350,000 kcal[b]	1,350[b]	11.06[e]
Cleanup water	160,000 kcal[b]	160[b]	1.31[e]
Space heat	152,000 kcal[b]	152[b]	1.24[e]
Direct heat	440,000 kcal[b]	440[b]	3.61[e]
Losses	300,000 kcal[b]	300[b]	2.46[e]
Stainless steel	11 kg[f]	605[g]	18.72[h]
Steel	21 kg[f]	483[g]	18.72 [h]
Cement	56 kg[f]	2,688[g]	18.72 [h]
TOTAL		12,564	$1,251.74
The 1,000 kg of biodiesel produced has an energy value of 9 million kcal. In addition, 200 ml (2,080 kcal) of methanol must be added to the soy oil for transesterification. With an energy input requirement of 14.7 million kcal, there is a net loss of energy of 63%. If a credit of 7.4 million kcal is given for the soy meal produced, then the net loss is less.			
The cost per kg of biodiesel is $1.25.			

a)Data from table 8.

b)Data from Singh, 1986 [132].

c)An estimated 3 kWh thermal is needed to produce a kWh of electricity.

d)Cost per kWh is 7¢.

e)Calculated cost of producing heat energy using coal.

f)Calculated inputs.

g)Calculated from Newton, 2001 [91].

h)Calculated.

The U.S. provides $500 million in subsidies for biodiesel for the production of 850 million liters of biodiesel [93], which is 74 times greater than the subsidies per liter of diesel fuel. As mentioned, the subsidies per liter of ethanol are 60 times greater than the subsidies per liter of gasoline.

The environmental impacts of producing soybean biodiesel are second only to that of corn ethanol:

1)Soybean production causes significant soil erosion, is second only to corn production [94].

2)Soybean production uses large quantities of herbicides and is second only to corn production [133]. These herbicides cause major pollution problems with natural biota in the soybean production areas [5,134].

Another important consideration in the use of soybeans as a potential biofuel source is cropland competition with food production. The USDA (2005) [96] reports a soybean yield worldwide of 2.2 tons per hectare. With an average oil extraction efficiency of 18% [135,136], the average oil yield per year would be approximately 0.4 tons per hectare. This converts into 454 liters of oil per hectare. Based on current U.S. diesel consumption of 227 billion liters/year [137], this would require more than 500 million hectares of land in soybeans or more than half the total U.S. planted just for soybeans! In other words, all 71 billion tons of soybeans produced in the U.S. [20] could only supply 2.6% of total U.S. oil consumption.

Rapeseed and Canola Biodiesel

The European Biodiesel Board estimates a total biodiesel production of 4.89 million tons for the year 2006 [138]. Well suited to colder climates, rapeseed is the dominant crop used in European biodiesel production. Often confused with canola, rapeseed is an inedible crop of the Brassica family, yielding oil seeds high in erucic acid. Canola is in the same family, but is a hybrid created to lower saturated fat content and erucic acid content for human consumption in cooking oil and margarine [137].

Frondel and Peters (2007) [139] describe rapeseed-based biodiesel yields in Europe averaging 1390 liters per hectare in 2005. Using the density of biodiesel defined as 0.88 kg/l [139], it can be estimated that the average annual production of rapeseed biodiesel in Europe is 1.1 million tons total. Because of its high oil content (30%), rapeseed is preferred as a biodiesel feedstock source [137]. While Europe currently dominates the rapeseed production in the world, as the market for high-yield oilseed feedstock for biodiesel grows, interest in canola and rapeseed oil is likely to increase in many northern states [137].

Rapeseed and canola require the application of fertilizers and pesticides in production. This disturbs the acid-base equilibrium of the soil and also can lead to algal blooms, as well as cause toxic pollution. Additionally, the energy required to make these pesticides and fertilizers detracts from the overall net energy produced [139]. Although soybeans contain less oil than canola, about 18% soy oil compared with 30% oil for rapeseed/canola, soybeans can be produced with nearly zero nitrogen inputs (Table 9). This makes soybeans more advantageous for the production of biodiesel, as nitrogen fertilizer is one of the most energy costly inputs in crop production [140].

The biomass yield of rapeseed/canola per hectare is also lower than that of soybeans - about 1,600 kg/ha for canola compared with 2,890 kg/ha for soybeans (Table 10 and Table 11) [131]. The production of 1,568 kg/ha rapeseed/canola requires an input of about 4.4 million kcal per hectare and costs about $573/ha (Table 10). About 3,333 kg of rapeseed/canola oil are required to produce 1,000 kg of biodiesel (Table 10). Therefore, all 333 million tons of rapeseed/canola produced in the U.S. in 2006 [141] could be used to make 100 million liters of biodiesel, or 0.005% of the total oil used in the U.S. The total energy input to produce the 1,000 liters of rapeseed/canola oil is 14 million kcal. This suggests a net loss of 58% of energy inputs (Table 11). The cost per kg of biodiesel is also high, at $1.63. The subsidies for biodiesel are $500 million for the production of 850 million liters of biodiesel [93]. Thus, the subsidies per liter of biodiesel are 74 times greater than the subsidies per liter of diesel fuel.

Table 10: Energy inputs and costs in canola production per hectare in the North America

Inputs	**Quantity**	**kcal x 1000**	**Costs $**
Labor	7 hrs[a]	280[b]	91.00[c]
Machinery	20 kg[d]	360[e]	148.00[f]
Diesel	65 L[a]	740[g]	35.00
Nitrogen	120 kg[a]	1,920[h]	75.00[i]
Phosphorus	101 kg[a]	417[j]	71.00[k]
Potassium	14.8 kg[l]	48[m]	4.59[n]
Sulfur	22 kg[a]	10[l]	10.00
Limestone	1000 kg[a]	281[d]	23.00
Seeds	5 kg[o]	40[p]	35.00
Herbicides	1.5 kg[q]	150[p]	30.00
Insecticides	1 kg[q]	100	20.00
Electricity	10 kWh[a]	29[r]	0.70
Transport	100 kg[s]	26[t]	30.00
TOTAL		4,401	$573.29
Canola yield	1,568 kg/ha[u]	5,645 kcal input:output 1:1.06	

a)Smathers, 2005 [142].

b)It is assumed that a person works 2,000 hrs per year and utilizes an average of 8,000 liters of oil equivalents per year.

c)It is assumed that labor is paid $13 an hour.

d)Pimentel and Pimentel, 1996 [143].

e)Machinery is prorated per hectare with a 10 year life of the machinery. Tractors weigh from 6 to 7 t and harvesters from 8 to 10 tons, plus plows, sprayers, and other equipment.

f)College of Agri., Consumer & Environ. Sciences, 1997 [126].

g)Input 11,400 kcal per liter.

h)Patzek, 2004 [77].

i)Hinman et al., 1992 [128].

j)Input 4,154 kcal per kg.

k)Cost 70¢ per kg.

l)Pimentel and Pimentel, 2007 [9].

m)Input 3,260 kcal per kg.

n)Costs 31¢ per kg.

o)Molenhuis, 2004 [144].

p)Pimentel et al., 2002 [129].

q)Estimated.

r)Input 860 kcal per kWh and requires 3 kWh thermal energy to produce 1 kWh electricity.

s)Goods transported include machinery, fuels, and seeds that were shipped an estimated 1,000 km.

t)Input 0.83 kcal per kg per km transported.

u)USDA, 2004 [131].

Table 11: Inputs per 1,000 kg of biodiesel oil from canola

Inputs	Quantity	kcal x 1000	Costs $
Canola	3,333 kg[a]	9,355[a]	$1,419.00[a]
Electricity	270 kWh[b]	697[c]	18.90[d]
Methanol	120L[i]	1,248[i]	111.60
Steam	1,350,000 kcal[b]	1,350[b]	11.06[e]
Cleanup water	160,000 kcal[b]	160[b]	1.31[e]
Space heat	152,000 kcal[b]	152[b]	1.24[e]

Direct heat	440,000 kcal[b]	440[b]	3.61[e]
Losses	300,000 kcal[b]	300[b]	2.46[e]
Stainless steel	11 kg[f]	158[g]	18.72[h]
Steel	21 kg[f]	246[g]	18.72 [h]
Cement	56 kg[f]	106[g]	18.72 [h]
TOTAL		14,212	$1,625.34
The 1,000 kg of biodiesel produced has an energy value of 9 million kcal. The methanol input is a required addition to the canola oil for transesterification. With a total energy input requirement of 14.2 million kcal, there is a net loss of energy of 58%. If a credit of 4.6 million kcal is given for the canola meal produced, then the net loss is less.			
The cost per kg of biodiesel is $1.63.			

a)Data from Table 10.

b)Data from Singh, 1986 [132].

c)An estimated 3 kWh thermal is needed to produce a kWh of electricity.

d)Cost per kWh is 7¢.

e)Calculated cost of producing heat energy using coal.

f)Calculated inputs.

g)Calculated from Newton, 2001 [91].

h)Calculated.

i) Hekkert et al., 2005 [123].

Oil Palm

There is a major effort to plant and harvest oil palms for biofuels in some tropical developing countries, especially Indonesia, Malaysia, Thailand, Colombia, and some in West Africa [145]. In the last 20 years, the production of vegetable oil has more than doubled.

Palm oil makes up 30%, over 35% when including palm kernel oil, of biological oils and fats produced worldwide [146]. Global oil palm production for the U.S. Department of Agricultures' 2007-2008 growing season totaled more than 45.3 million tons; of which about 4.7 million metric tons was palm kernel oil [146]. Indonesia and Malaysia are the world's leading producers. Together these countries total over 84% of global palm oil production [147]. Relative to the world total agricultural oil, over 84% is palm oil production [147]. Relative to world total agricultural oil export, both Indonesia and Malaysia lead country exports with 15% and

14% respectively, the dominant majority of which is palm oil [146]. The oil palm once established (after 4 years) will produce about 4,000 kg of oil per hectare per year [148]. The energy inputs for maintaining the hectare of oil palm are indicated in Table 12. The data suggest that about 7.4 million kcal are required to produce 26,000 kg of palm oil bunches. This 26,000 kg is a sufficient quantity of palm nuts to produce 4,000 kg of palm oil. A total of 6.9 million kcal are required to process 6,500 kg of palm nuts to produce 1 ton of palm oil (Table 13). Thus, the net return on fossil energy invested in production and processing totals 30%, which is clearly a better return than corn ethanol and soybean biodiesel. However, an estimated 200 ml (2,080 kcal) of methanol is a required addition to the 1,000 kg of palm oil, for transesterification. This results in a negative 8% net energy output for palm oil (Table 13).

Table 12: Energy inputs for palm oil production in Thailand and Indonesia

Inputs	Quantity	kcal x 1000
Labor	1,000 hrs[a]	700[b]
Machinery	55 kg[c]	1,018[c]
Diesel	130 L[d]	1,430[e]
Nitrogen	150 kg[d]	2,400[f]
Phosphorus	81 kg[d]	336[g]
Potassium	206 kg[d]	672[h]
Magnesium	56 kg[d]	400[a]
Irrigation	8 cm[d]	320[a]
Herbicides	4.0 kg[d]	400[i]
Insecticides	1.0 kg[d]	100[j]
Transport	200 kg[a]	166[k]
Total		7,445
Palm Oil Bunches 26,000 kg[d]		

a)Estimated.

b)BP World Energy, 2005 [29].

c)Estimated.

d)Pleanjai et al., 2004 [149].

e)11,430 kcal

f)16,000 kcal/kg

g)4,154 kcal/kg

h)3,260 kcal/kg

i)100,000 kcal/kg

j)100,000 kcal/kg

k)Calculated.

Table 13: Inputs per 6.5 tons of palm nuts that produces 1 ton of palm oil [149]

Inputs	Quantity	kcal x 1000
Palm nuts	6,500 kg[a]	1,861[a]
Palm nuts transport	6,500 kg[b]	332[c]
Water	30,000 L[d]	68[e]
Stainless steel	3 kg[f]	165[e]
Steel	4 kg[f]	92[e]
Cement	8 kg[f]	384[e]
Steam	3,000,000 kcal[d]	3,000[d]
Electricity	380 kWh[d]	980[d]
Sewage effluent	9 kg BOD[g]	31[h]
Total		6,913[i]
Energy content of 1 ton of palm oil		8,300

a)Data from table 12.

b)Factory located 144 km away.

c)Calculated.

d)Plaenjai et al., 2004 [149].

e)Estimated.

f)From Pimentel and Patzek, (2007) [38].

g)Plaenjai et al., 2004 [149].

h)4 kWh of energy required to process 1 kg of BOD.

i)An estimated 200 ml (2,080 kcal) was added to the 1,000 kg of palm oil for transesterification. This results in a negative 8% net energy output for palm oil.

There are several environmental and social issues associated with oil palm plantations. In major palm oil producing countries the creation of new plantations represents the greatest loss of natural forests and mixed agro-forestry systems [150]. Additionally, the removal of tropical rainforests to plant the oil palm results in an increase in CO_2. The amount of CO_2 released increases tremendously when burning is used to clear forest [145]. Furthermore, the removal of natural and mixed agro-forestry systems and the planting of oil palms reduces the biodiversity of the local ecosystem; species composition of vertebrate taxa changed up to 60% in some instances [150]. The degree of loss is compounded with larger plantation sizes; individual monoculture oil palm plantations can reach sizes up to 20,000 ha [150]. Moreover, pollution in processing oil palm is quite substantial. Mill waste entering water bodies is equivalent to that generated by an estimated 1.5 million people [150]. Finally, the increased use of oil palm for fuel reduces the availability of the oil and increases the price for human use [145]. Oil Palm production has a better net fossil energy return than many competing biofuel crops. However, production is accompanied by various social, health, and environmental consequences.

Jatropha

Jatropha is receiving a great deal of attention now as new source of biodiesel. The shrub, endemic to Mexico [151], produces seeds that are 30% oil, which is excellent for biofuel use. The shrub will grow in heavy rainfall regions as well as in arid regions. In arid regions, the yield is reduced to only 0.5 tons/ha per year [152]. A potential benefit of Jatropha is that it cannot be used for a food crop. This allows the plant to be grown only for fuel. The most serious drawback to the shrub as a biofuel is that it is highly toxic to humans and livestock, plus it uses land and water resources that are critical for food crop production [153]. Therefore, further research is necessary to evaluate the long-term safety of producing, handling, and processing this feedstock.

ALGAE FOR OIL PRODUCTION

Some cultures of algae consist of 30% to 50% oil [154]. Thus, there is growing interest using algae to increase U.S. oil supply based on the theoretical claims that 47,000 to 308,000 liters/hectare/year (5,000 to 33,000 gallons/acre) of oil could be produced using algae [155,156]. The calculated cost per barrel would be $15 [157]. Currently, oil in the U.S. market is selling for over $100 per barrel. If the production and price of oil produced from algae were true, U.S. annual oil needs could theoretically be met if 100% of all U.S. land were in algal culture!

Despite all the algae-related research and claims dating back to 1970's, none of the projected algae and oil yields has been achieved [154]. To the contrary, one calculated estimate based on all the included costs using algae would be $800 per barrel, not $15 per barrel, as mentioned. Algae, like all plants, require large quantities of nitrogen fertilizer and water. In addition, significant fossil energy inputs would be needed for the functioning production system [158].

One difficulty in culturing algae is that the algae shade one another's cells. There are, therefore, different levels of light saturation in the cultures, even under optimal conditions such as in Florida [159]. These differences influence the rate of growth of the algae. In addition, wild strains of algae sometimes invade and dominate the algae culture strains, causing a decrease in oil production by the algae [159].

Another major problem with the culture of algae in ponds or tanks, is the harvesting of the algae. Since algae consist mostly of water, harvesting algae from the cultural tanks and separating the oil from the algae are difficult and energy intensive processes. This problem was observed at the University of Florida (Gainesville) when algae were being cultured in managed ponds for the production of nutrients for hogs. After two years, the unsuccessful algal-nutrient culture was abandoned. The best algal biomass yields under tropical conditions are about 50 t/ha/yr [159]. However, the highest yield of alga biomass produced per hectare, based on *theoretical* calculation, is 681 tons/ha/yr [156]. For comparison, the total yield for rice (including rice and straw) grown in the tropics is nearly 50 t/ha/yr of continuous culture [160]. Rice production in the tropics can produce 3 crops on the same hectare of land per year requiring about 400 kg/ha of nitrogen fertilizer and 240 million liters of water [160]. Obviously, a great deal of laboratory and field research is needed for the algae and oil theoretical system.

IMPACTS OF CULTURING CROPS AND GRASSES FOR BIOFUELS

Managing crops and grasses for biofuels results in many and varied environmental impacts to the land, water, and biodiversity. Some impacts of biofuel crop production on soil erosion and air pollution are discussed below.

Soil Erosion Associated with Biofuels

Soil erosion and land degradation are of particular concern to agriculturalists and foresters because of increasing biofuel production. Serious soil erosion is taking place in U.S. agricultural systems, with an estimated soil loss that is 11 times faster than a sustainable rate [161]. The prime cause of high erosion rates in the U.S. is due to the depletion of biomass cover that protects the soil from rainfall and wind energy. In particular, row crops such as corn and soybeans are particularly susceptible to erosion [5]. Tilling the soil for planting of row crops leaves the soil unprotected from wind and rainfall. After harvesting, soybeans have little crop residue which covers only 20% of the cropland. Corn stover covers about 60% of the land after harvest [162].

The intensive application of nitrogen fertilizers in corn production was perceived to sequester soil organic carbon in the soil. However, after a 40 to 50 year application of synthetic nitrogen fertilizer in Illinois, a net decline occurred in soil carbon despite the massive residue incorporation into the soil [163].

Some investigators, like Tilman et al. (2006) [104] and Perlack et al. (2005) [164], suggest that crop residues can be harvested for biofuel production. If this is done, soil erosion, with the removal of crop residues, will increase from 10-fold to 100-fold [165]. Removing crop residues would therefore devastate U.S. agriculture.

Close grown crops, like wheat with an average yield of 7 t/ha/yr, protect the soil from erosion with an erosion rate of about 5 t/ha/yr. This is better than the soil erosion rate of row crops, like corn, with an average soil erosion rate of 15 t/ha/yr [109]. After germinating, spring wheat has an early, fast growing development of a relatively dense stand of vegetation cover of 150 to 200 plants/m^2 that is capable of protecting the soil from rain and wind energy. Unfortunately, wheat is not a good biofuel crop because of its low yield [20].

Some crops, such as grass, provide nearly complete cover of the soil once established. These crops are usually grown as perennials and cover the soil all year for about 5-year periods. The soil erosion rate from continuous grass is reported to lose soil at only 0.1 to 1 t/ha/yr [110]. Yet, grass and other such crops are unfortunately not generally productive as biofuel crops.

Some biofuel crops, like sugarcane, have high soil erosion rates, which are reported to be about 148 t/ha/yr in Australia [166].

No-till and ridge-till planting of corn and similar crops will reduce the soil erosion rates from 18 t/ha/yr to 2 t/ha/yr [167]. Herbicides and other pesticides are needed in no-till corn production, but with ridge-till herbicides may not be necessary.

In addition, the water-holding capacity and nutrient levels of the soil decline when erosion occurs. With conventional corn production, erosion reduced the volume of moisture in the soil about 50% compared with organic corn production [168].

When conservation technologies, like organic agriculture, are employed, increased yields may result because water, nutrients, and soil organic matter are retained. For example, in Pennsylvania, yields of corn and soybeans were 33% to 50% higher in the organic systems when soil organic matter increased in the organic systems over time even when drought conditions occurred [168].

In the U.S., annual estimates of soil loss were estimated to be 1 to 2 billion tons per year, and 10 years later the erosion rate had increased to 3 billion tons annually [169].

Undisturbed forests often have a dense cover over the soil consisting of leaves, twigs, and other organic matter; these forest ecosystems have soil erosion rates that typically range from less than 0.1 t/ha/yr to 0.2 t/ha/yr [5]. The combination of organic mulch, tree cover, and tree roots makes most natural forest soils, even on steep slopes of 70%, resistant to erosion and rapid water runoff. Forests lose significant quantities of water, soil, and nutrients when cut and harvested [170]. In addition, erosion rates increase. Therefore, the use of forests for producing biofuels will increase rates of soil erosion. However, short-rotation woody-crops have been shown to improve groundwater quality and soil-water runoff in comparison to row crops [171].

Air Pollution

Smoke produced when fuelwood and crop biomass are burned is a serious pollution hazard because of the nitrogen, particulates, and other chemicals in the smoke. Although only 3% of U.S. heating energy comes from wood, about 15% of the air pollutants in the U.S. are produced from the burning of wood [172]. Emissions from burning wood and other biomass are a threat to public health because of the highly respirable nature of the more than 200 chemicals that the emissions contain [173].

Of special concern are the relatively high concentrations of potentially carcinogenic poly cyclic organic compounds (POMs, e.g. benzo{a}pyrene) and particulates. Sulfur and nitrogen oxides, carbon monoxide, and aldehydes are also released, but usually in smaller quantities. According to Naeher et al. (2007) [173], wood smoke contains an estimated 14 carcinogens and 4 co-carcinogens. Great concern is expressed for the 4 billion people in developing countries who cook and heat with wood and crop residues [6].

The burning of ethanol in automobiles also causes major pollution problems. Ethanol has only 66% of the energy that gasoline has and, therefore, significantly more fuel has to be burned to travel the same distance as gasoline [97]. In addition, ethanol produces more nitrous oxides and ozone than gasoline as measured from the exhausts of automobiles [100,101]. These are serious air pollutants affecting human health, and these pollutants and their impact on human health were recently confirmed in new studies by Jacobson [102].

In addition, ethanol and other biofuels exacerbate the current global warming situation worse. Nobel chemist Paul J. Crutzen [174] reported that biofuels made from corn and rapeseed release about twice as many nitrous oxides as previously reported. The authors of the report from the University of Edinburg, conclude that growing biofuels is probably of no benefit and is actually making the climate situation worse.

CONCLUSION

The rapidly growing world population and rising consumption of fossil fuels is increasing demand for both food and biofuels. This will exaggerate both food and fuel shortages. Producing biofuels requires huge amounts of both fossil energy and food resources, which will intensify conflicts among these resources.

Using food crops such as corn grain to produce ethanol raises major nutritional and ethical concerns. Nearly 60% of humans in the world are currently malnourished, so the need for grains and other basic foods is critical [7]. Growing crops for fuel squanders land, water, and energy resources vital for the production of food for people. Using corn for ethanol increases the price of U.S. beef, chicken, pork, eggs, breads, cereals, and milk from 10% to 30% [175]. In addition, Jacques Diouf, Director General of the U.N. Food and Agriculture Organization, reports that using food grains to produce biofuels is already causing food shortages for the poor of the world [176,177]. Growing crops for biofuel not only ignores the need to reduce natural resource consumption, but exacerbates the problem of malnourishment worldwide by turning precious grain into biofuel.

Recent policy decisions have mandated increased production of biofuels in the United States and worldwide. For instance, in the Energy Independence and Security Act of 2007, President Bush set "a mandatory Renewable Fuel Standard (RFS) requiring fuel producers to use at least 36 billion gallons of biofuel in 2022." This would require 1.6 billion tons of biomass harvested per year and would require harvesting 80% of all biomass in the U.S., including all agricultural crops, grasses, and forests (Table 3). With nearly total biomass harvested, biodiversity and food supplies in the U.S. would be decimated!

The release of large quantities of carbon dioxide associated with the planting and processing of plant materials for biofuels is reported to reduce the nutritional quality of major world foods, including corn, wheat, rice, barley, potatoes, and soybeans. When crops are grown under high levels of carbon dioxide, protein levels in the crops may be reduced as much as 15% [178].

Many problems associated with biofuels have been ignored by scientists and policy makers. For one, the biofuels that are being created in order to diminish the dependence on fossil fuels, actually depend on fossil fuels. In most cases, more fossil energy is required to produce a unit of biofuel compared with the energy that is produced (Table 4, Table 5, Table 6, Table 7, Table 8, Table 9, Table 10 and Table 11). Furthermore, the U.S. is importing oil and natural gas to produce biofuels, which is not making the U.S. oil independent. Secondly, some publications use incomplete or insufficient data to support the claims of the pro-biofuel publications. For instance, claims that cellulosic ethanol provides net energy [104] have not been experimentally verified because most of the calculations are theoretical. Finally, environmental problems such as water pollution from fertilizers and pesticides, global warming, soil erosion and air pollution are intensifying with biofuel production. There is simply not enough land, water, and energy to produce biofuels.

Based on careful up-to-date analysis of all fossil energy inputs, most conversions of biomass into ethanol and biodiesel result in a negative energy return. Four of the negative energy returns are: corn ethanol (minus 48%); switchgrass (minus 68%) soybean biodiesel (minus 63%); and rapeseed (minus 58%). Conversely, palm oil production in Thailand suggests a positive 30% energy return.

Some balance and improvement in the food and biofuel situation is possible. There is desperate need for petroleum and natural gas conservation as well as sound soil, water, and biological resource conservation. There are many opportunities for implementing known sustainable agricultural and forestry technologies. These include:

l)Soil conservation by leaving crop residues on the surface of the soil. Planting cover crops after the main crop, like corn has been harvested. Other soil conservation technologies that should be employed when possible are: planting on the contour; planting on terraces; using grass strips; crop rotations; mulches; and similar technologies and combinations.

2)Water conservation is vital because of the enormous amounts of water crops utilize. In addition to soil conservation technologies, increasing the organic matter in the soil will increase the water holding capacity significantly and will increase the number and abundance of soil biota.

3)Planting trees as hedge-rows and wind breaks can significantly reduce soil loss, water runoff, and evaporation.

4)Maintaining forest stands will conserve soil, water, biodiversity in the ecosystem.

Because the green plants in the U.S. collect a total of only 32 quads of solar energy per year, they cannot provide the nation with a replacement of the more than 100 quads of fossil energy use per year. This means that even if all available crops, forests, and grasses were harvested for biofuel, we would still be dependent on oil and natural gas supplies, which are projected to disappear in about 40 years. U.S. citizens should reduce their per capita energy use by one-half because this change will become a reality when oil and natural gas supplies disappear in about 40 years. Instead of subsidizing corn ethanol and soybean biodiesel by more than $10 billion per year, these funds should be invested in research in renewable energy technologies.

ACKNOWLEDGEMENTS

The authors wish to thank Holli Edgley and Elizabeth Keokosky at Cornell University for their contributions to the development of this paper. We are also grateful to the following people for their comments and suggestions on earlier versions of the paper: Harry de Gorter, Department of Applied Economics and Management, Cornell University, Ithaca, NY; Marcelo Dias De Oliveira, Rice University, Houston, TX; Andrew Ferguson, Optimum Population Trust, Manchester, UK; Mario Giampietro, Universitat

Autonoma de Barcelona, Spain; David Hammer, Electrical and Computer Engineering, Cornell University, Ithaca, NY; Roel Hammerschlag, Climate and Energy Program, Stockholm Environment Institute (U.S.), Somerville, MA; Conly Hansen, Director Center for Profitable Uses of Agricultural, Utah State University, Logan, UT; Phillip McMichael, Development Sociology, Cornell University, Ithaca, NY; Rattan Lal, Ohio State University, Columbus, OH; Michele Whitecraft, Cornell University, Ithaca, NY; Walter Youngquist, Petroleum Consultant, Eugene, OR. In addition, we offer our sincere gratitude to the Cornell Association of Professors Emeriti for the partial support of our research through the Albert Podell Grant Program.

REFERENCES

1. Barbara, J.S. The False Promise of Biofuels. Special Report from the International Forum on Globalization and the Institute for Policy Studies. 2007. 30 pages. http://www.ifg.org/pdf/biofuels.pdf.
2. Cleveland, C.J.; Costanza, R.; Hall, C.A.S.; R. Kaufmann, R. Energy and the United States Economy: A Biophysical Perspective. *Science* 1983, *225*, 890–897.
3. Hall, C.A.S.; Cleveland, C.J.; Kaufmann, R. Energy and Resource Quality: The Ecology of the Economic Process. Wiley Interscience: New York, 1986; 577p.
4. Pimentel, D.; Pimentel, M. *Food, Energy and Society*; CRC Press; Taylor and Francis Group: Boca Raton, Florida, 2008.
5. Pimentel, D. Soil Erosion: A Food and Environmental Threat. *Environment, Development and Sustainability*2006, *8*, 119–137.
6. Pimentel, D.; Cooperstein, S.; Randell, H.; Filiberto, D.; Sorrentino, S.; Kaye, B.; Nicklin, C.; Yagi, J.; Brian, J.; O'Hern, J.; Habas, A.; Weinstein, C. Ecology of Increasing Diseases: Population Growth and Environmental Degradation. *Human Ecology* 2007, *35*(6), 653–668.
7. WHO. Malnutrition Worldwide. World Health Organization. 2005. http://www.mikeschoice.com/reports/malnutrition_worldwide.htm. (12/7/07).
8. FAO. Food and Agricultural Organization of the United Nations. 1961–2006; FAO: Rome.
9. Pimentel, D.; Pimentel, M. Ecological Engineering: Growth of Human Population. Encyclopedia of Ecology. Elsevier, 2007; In Press.
10. Pimentel, D.; Gardner, J.B.; Bonnifield, A.J.; Garcia, X.; Grufferman, J.B.; Horan, C. M.; Rochon, E.T.; Schlenker, J.L.; Walling, E.E. Energy Efficiency and Conservation for Individual Americans. In *Environment, Development and Sustainability*; 2008.
11. Nonhebel, S. Renewable Energy and Food Supply: Will There Be Enough Land? *Renewable and Sustainable Energy Reviews* 2005, *9*(2), 191–201.
12. PRB. World Population Data Sheet. 2007; Population Reference Bureau: Washington, DC.
13. USDA. 1980–2006; *Agricultural Statistics*; U.S. Department of

Agriculture: Washington, DC.

14. Sommer, A.; West, K.P. *Vitamin A Deficiency: Health, Survival and Vision*; Oxford University Press: New York, 1996.
15. Tomashek, K.M.; Woodruff, B.A.; Gotway, C.A.; Bloland, P.; Mbaraku, G. Randomized Intervention Study Comparing Several Regimens for the Treatment of Moderate Anemia among Refugee Children in Kigoma Region, Tanzania. *American Journal of Tropical Medicine and Hygiene* 2001, *64*, 164–171.
16. Environment News Service. 850 Million Still Hungry on World Food Day. 2002. http://www.ens-newswire.com/ens/oct2002/2002-10-16-03.asp. (11/4/07).
17. Svedberg, P. *Poverty and Undernutrition: Theory, Measurement, and Policy*; Oxford University Press: New York, 2000.
18. Bryce, J.; Boschi-Pinto, C.; Shibuya, K.; Black, R.E. WHO Estimates of the Causes of Death in Children. *The Lancet* 2005, *365*(9465), 1147–1152.
19. Kirby, A. UN Warns of Future Water Crisis. *BBC News Online (March 5, 2003).m.* http://news.bbc.co.uk/2/hi/science/nature/2818615.stm/ (4/5/08).
20. USDA. *Agricultural Statistics, 2006*; U.S. Department of Agriculture; U.S. Government Printing Office: Washington, DC, 2006.
21. Blair, D.; Sobal, J. Luxus Consumption: Wasting Food Resources Through Overeating. *Agriculture and Human Values* 2006, *23*, 63–74.
22. FAO. The Salt of the Earth: Hazardous for Food Production. Food and Agricultural Organization of the United Nations. World Food Summit, 2002. http://www.fao.org/WorldFoodSummit/english/newsroom/focus/focus1.htm. (3/09/08).
23. FAOSTAT. Food and Agricultural Organization. 2001; United Nations: Rome.
24. UN. Population Division. Department of Economics and Social Affairs. United Nations Secretariat. The World at Six Billion. Part 1. 1999. http://www.un.org/esa/population/publications/sixbillion/sixbilpart1.pdf. (11/4/07).
25. IWMI. Global Irrigated Area. International Water Management Institute. 2007. http://www.iwmigiam.org/stats. (3/09/08).
26. Giampietro, M.; Pimentel, D. Energy Utilization. In *Encyclopedia of Agricultural Science*; Arntzen, C.J., Ritter, E.M., Eds.; Academic

Press: San Diego, CA, 1994; Vol. 2, pp. 63–76.

27. USCB. *Statistical Abstract of the United States, 2007*; U.S. Census Bureau; U.S. Government Printing Office: Washington, DC.
28. Pimentel, D.; Wen, D. China and the World: Population, Food and Resource scarcity. In *Dare to Dream: Vision of 2050 Agriculture in China*; Tso, T.C., Kang, H., Eds.; China Agricultural University Press: Beijing, 2004; Chapter 8; pp. 103–16.
29. BP. *BP Statistical Review of World Energy*; British Petroleum: London, 2005; 41 pages.
30. Gleick, P.H. Basic Water Requirements for Human Activities: Meeting Basic Needs. *Water International* 1996, *21*, 83–92.
31. Pimentel, D.; Berger, B.; Filberto, D.; Newton, M.; Wolfe, B.; Karabinakis, E.; Clark, S.; Poon, E.; Abbett, E.; Nandagopal, S. Water Resources: Current and Future Issues. *BioScience* 2004, *54*(10), 909–918.
32. FAO. Agriculture, Food and Water. Chapter 2: How the World is Fed. FAO Corporate Document Repository. 2001. Food and Agricultural Organization of the United Nations. http://www.fao.org/docrep/006/y4683e/y4683e06.htm (8/28/08).
33. Youngquist, W.; Duncan, R.C. North American Gas: Data Show Supply Problems. *Natural Resources Research* 2003, *12*(4), 229–240.
34. Campbell, C. *Excel Spread Sheets of Oil and Gas of Production and Discovery*; Oil Depletion Analysis Centre: London, England, 2006.
35. Heinberg, R. Oil Apocalypse: What happens when demand for oil outstrips Supply? 2007. http://www.richardheinberg.com/ (1/29/08).
36. Pimentel, D. 2001; The Limitations of Biomass Energy. In *Encyclopedia of Physical Science and Technology*, 3rd ed.; Meyers, R., Ed.; Academic Press: San Diego, 2001; Vol. 2, pp. 159–171.
37. Flow of Energy. Primary Production. 2005. http://www.globalchange.umich.edu/globalchange1/current/lectures/kling/energyflow.html. (12/17/06).
38. Pimentel, D.; Patzek, T. Ethanol Production: Energy and Economic Issues Related to U.S. and Brazilian Sugarcane. *Natural Resources Research* 2007, *16*(3), 235–42.
39. Jölli, D.; Giljum, S. 2005; Unused Biomass Extraction in Agriculture, Forestry and Fishery. SERI Studies No. 3; Sustainable Europe Research Institute: Vienna.

40. Crop Production. In *National Agricultural Statistical Services*; September 7 2007; United Stated Department of Agriculture. September 8 2007. < http://www.usda.gov/nass/PUBS/TODAYRPT/crop0907.txt>. (1/29/08).
41. Crop Harvest. *Biological System Engineering: Crop Systems*; Washington State University, 19 September 2007. < http://www.bsyse.wsu.edu/cropsyst/manual/parameters/crop/harvest.htm> (1/29/08).
42. Forest Service. *The Forest Inventory and Analysis RPA Assessment Tables*; U.S. Forest Service; U.S. Department of Agriculture: Washington, DC, 2007.
43. IEA. International Energy Administration. International Energy Annual. USDOE: Washington, DC, 2006.
44. IFIA. International Fertilizer Industry Association. Statistics. 2008. http://www.fertilizer.org/ifa/statistics/IFADATA/summary.asp (6/27/08).
45. USDA. 2007; Major Land Uses. Economic Research Services. June 9 2006; United States Department of Agriculture, September 9 2007. http://www.ers.usda.gov/data/majorlanduses/. (1/29/08).
46. Youngquist, W. *GeoDestinies. The inevitable Control of Earth Resources over Nations and Individuals*; National Book Company: Portland, Oregon, 1997.
47. USCB. *Statistical Abstract of the United States, 2004-2005*; U.S. Census Bureau; U.S. Government Printing Office: Washington, DC.
48. Pimentel, D.; Pleasant, A.; Barron, J.; Gaudioso, J.; Pollock, N.; Chae, E.; Kim, Y.; Lassiter, A.; Schiavoni, C.; Jackson, A.; Lee, M.; Eaton, A. U.S. Energy Conservation and Efficiency: Benefits and Costs. *Environment, Development and Sustainability* 2004, *6*, 279–305.
49. Ferguson, A.R.B. Biomass and Energy. Manchester (United Kingdom). *Optimum Population Trust* 2001, *4*(1), 14–18.
50. Ferguson, A.R.B. Implications of the USDA 2002 Update on Ethanol From Corn. Manchester (United Kingdom). *Optimum Population Trust* 2003, *3*(1), 11–15.
51. Hendrickson, O.Q.; Gulland, J.F. Residential Wood Heating: Forest, the Atmosphere and Public Consciousness. Air and Waste Management Association Conference: The Emission Inventory: Living in a Global Environment. New Orleans, Louisiana, 1993.

52. EIA. Energy Information Administration. Annual Energy Review. U.S. Department of Energy: Washington, DC, 2007. http://www.eia.doe.gov/emeu/aer/pdf/aer.pdf (6/4/08).
53. Schill, S.R. Harnessing the Power of Biomass. Biomass Magazine. August 2007. http://www.biomassmagazine.com/article-print.jasp?article_id=1229. (2/7/08).
54. Kitani, O. Biomass Resources. In *CIGAR Handbook of Agricultural Engineering*; Kitani, O., Jungbluth, T., Pearth, R.M., Ramdani, A., Eds.; American Society of Agricultural Engineering: St. Joseph, MI, 1999; pp. 6–11.
55. Pimentel, D. (Ed.) *Biofuels, Solar and Wind as Renewable Energy Systems: Benefits and Risks*; Springer: Dordrecht, The Netherlands, 2008; p. 504.
56. Pimentel, D. Ecological Footprint, World population, and the Environment. 2007. http://www.brass.cf.ac.uk/uploads/fullpapers/Pimentel.pdf. (1/22/08).
57. Burning Issues. What Are the Medical Effects of Exposure to Smoke Particles? 2006. http://burningissues.org/health-effects.html. (12/07/07).
58. Smith, K.R. Health impacts of household fuelwood use in developing countries. 2006. http://ehs.sph.berkeley.edu/krsmith/publications/2006%20pubs/Uasylva.pdf. (1/29/08).
59. EPA. Wood smoke. U.S. Environmental Protection Agency. 2002. http://www.webcom.com/-bi/brochure.pdf+wood+smoke+pollution-hl=ene=UTF-8. (10/20/02).
60. USCB. U.S. Census Bureau. Gas Most Popular Home Heating Fuel, Census Bureau Shows. In *U.S. Census Bureau News*; U.S. Department of Commerce: Washington, DC, Tuesday; November 16 2004.
61. Health Effects of Wood Smoke. Department of Ecology. Air Quality Program. Washington State Energy Office: Olympia, Washington, 1997. http://www.michigan.gov/documents/Health_Effects_of_Woodsmoke_WA_2004_142269_7.pdf (6/30/08).
62. Parikka, M. Global Biomass Fuel Reserve. *Biomass Bioenergy* 2004, *27*, 613–620.
63. USCB. U.S. Census Bureau. Home Heating Fuels. 1990. http://www.census.gov/apsd/cqc/cqc.27.pdf. (12/07/07).
64. EIA. Energy Information Administration "2001". Residential Energy

Consumption Survey. Housing Characteristic Tables. 2002. http://www.eia.doe.gov/emeu/recs/recs2001/detail_tables.html (2/15/08).

65. Forest Degradation Data. Forest Degradation Data. 2007. http://home.alltel.net/bsundquist1/df0.html(1/21/08).

66. FAOSTAT. Food and Agricultural Organization. 2005; United Nations: Rome.

67. Arnold, J.E.M.; Jongma, J. Fuelwood and Charcoal in Developing Countries: An Economic Survey. 2007. http://www/fap\o.org/docrep/12015e01.htm. (1/29/08).

68. Demirba, A. Biomass Resource Facilities and Biomass Conversion Processing for Fuels and Chemicals. *Energy Convers Mgmt.* 2001, *42*, 1357–1378.

69. Fujino, J.; Yamaji, K.; Yamamoto, H. Biomass-balance table for evaluating bioenergy resources. *Applied Energy* 1999, *62*(3), 75–89.

70. Howard, J.L. U.S. Timber Production, Trade, Consumption, and Price Statistics 1965 to 2002. Research Paper FPL-RP-615; Madison, Wisconsin; U.S. Department of Agriculture, U.S. Forest Service, Forest Products, Laboratory, 2003; 90 pages.

71. Birdsey, R.A. *Carbon Storage and Accumulation in the United States Forest Ecosystems*; USDA Forest Service: Washington, DC, 1992.

72. Repetto, R. Accounting for Environmental Assets. *Scientific American* 1992, *263*, 94–100.

73. Trainer, F.E. Can Renewable Energy Sources Sustain Affluent Society? *Energy Policy* 1995, *23*, 1009–1026.

74. Farrell, A.E.; Plevin, R.J.; Turner, B.T.; Jones, A.D.; O'Hare, M.O.; Kammen, D.M. Ethanol Can Contribute to Energy and Environmental Goals. *Science* 2006, *311*, 506–508.

75. Shapouri, H.; Gallagher, P. USDA's 2002 Ethanol Cost-of-Product Survey: Agricultural Economic Report Number 841. U.S. Department of Agriculture: Washington, DC, 2005.

76. NASS. National Agricultural Statistics Service. 2003. http://usda.mannlib.cornell.edu (11/05/04).

77. Patzek, T.W. Thermodynamics of the Corn-Ethanol Biofuel Cycle. *Critical Review in Plant Sciences* 2004, *23*(6), 519–567. http://petroleum.berkeley.edu/papers/patzek/CRPS416-Patzek-Web.pdf.

78. Brees, M. Corn Silage Budgets for Northern, Central and Southwest

Missouri. 2004. http://www.agebb.missouri.edu/mgt/budget/fbm-0201.pdf (9/1/04).

79. USDA. *1997 Census of Agriculture*; U.S. Department of Agriculture, 1997. http://www.ncfap.org. (8/28/2002).
80. USDA. 1997; *Farm and Ranch Irrigation Survey (1998)*; 1997 Census of Agriculture; Volume 3, Special Studies, Part 1; 280p.
81. Batty, J.C.; Keller, J. Energy Requirements for Irrigation. In *Handbook Of Energy Utilization in Agriculture*; Pimentel, D., Ed.; CRC Press: Boca Raton, FL, 1980; pp. 35–44.
82. Larsen, K.; Thompson, D.; Harn, A. Limited and Full Irrigation Comparison for Corn and Grain Sorghum. 2002. http://www.colostate.edu/Depts/SoilCrop/extension/Newsletters/2003/Drought/sorghum.html(9/2/2002).
83. Larson, W.E.; Cardwell, V.B. *History of U.S. Corn Production*; University of Minnesota: Minneapolis, 1999.
84. USDA. *Agricultural Statistics*; U.S. Department of Agriculture; U.S. Government Printing Office: Washington, DC, 2002.
85. USDA. *Corn-State. Costs of Production*; U.S. Department of Agriculture, Economic Research Service, Economics and Statistics System: Washington, DC, Stock #94018; 1991.
86. Patzek, T. The Real Corn-Ethanol Transportation System. 2006. http://petroleum.berkeley.edu/patzek/BiofuelQA/Materials/TrueCostofEtOH.pdf.
87. Pimentel, D. Ethanol Fuels: Energy Balance, Economics, and Environmental Impacts are Negative. *Natural Resources Research* 2003, *12*(2), 127–134.
88. Blais, J.F.; Mamouny, K.; Nlombi, K.; Sasseville, J.L.; Letourneau, M. Les Mesures Deficacite Energetique dans le Secteur de Leau. J.L. In *Les Mesures deficacite Energetique pour Lepuration des eaux Usees Municipales*; Sassville, J.L., Balis, J.F., Eds.; Scientific Report 405; Vol. 3. 1995; INRS-Eau: Quebec.
89. Illinois Corn. Ethanol's Energy Balance. 2004. http://www.ilcorn.org/Ethanol/Ethan_Studies/Ethan_Energy_Bal/ethan_energy_bal.html (8/10/04).
90. Kuby, W.R.; Markoja, R.; Nackford, S. Testing and Evaluation of On-Farm Alcohol Production Facilities. Acures Corporation. Industrial Environmental Research Laboratory. Office of Research

and Development. U.S. Environmental Protection Agency: Cincinnati, OH, 1984; 100p.

91. Newton, PW. Human Settlements Theme Report. Australian State of the Environment Report. 2001. http://www.environment.gov.au/soe/2001/publications/theme-reports/settlements/index.html (10/6/05).
92. DOE. 2002; Review of Transport Issues and Comparison of Infrastructure Costs for a Renewable Fuels Standard. U.S. Department of Energy: Washington, D.C., 2002. http://tonto.eia.doe.gov/FTPROOT/service/question3.pdf (10/8/2002).
93. Koplow, D. Biofuels – At What Cost? Government Support for Ethanol and Biodiesel in the United States. The Global Studies Initiative (GSI) of the International Institute for Sustainable development (IISD). 2006. http://www.globalsubsidies.org/IMG/pdf/biofuels_subsidies_us.pdf (2/16/07).
94. NAS. *Frontiers in Agricultural Research: Food, Health, Environment, and Communities*; National Academy of Sciences: Washington, DC, 2003; Also online: http://books.nap.edu/openbook.php?isbn=0309084946(7/14/08).
95. McLaughlin, S.B.; Walsh, M.E. Evaluating Environmental Consequences of Producing Herbaceous Crops for Bioenergy. *Biomass and Bioenergy* 1998, *14*(4), 317–324.
96. USDA. *Statistics of Oilseeds, Fats, and Oils*; USDA-NASS Agricultural Statistics; 2005; Chapter III. http://www.usda.gov/nass/pubs/agr05/acro05.htm. (10/20/2007).
97. Pimentel, D.; Patzek, T. Ethanol Production Using Corn, Switchgrass and Wood; Biodiesel Production Using Soybean. In *Biofuels, Solar and Wind as Renewable Energy Systems: Benefits and Risks*; Pimentel, D., Ed.; Springer: Dordrecht, The Netherlands, 2008; Chapter 15; pp. 375–396.
98. Searchinger, T.; Heimlich, R.; Houghton, R.A.; Dong, F.; Elobeid, A.; Fabiosa, J.; Tokgoz, S.; Hayes, D.; You, T. Use of U.S. Croplands for Biofuels Increases Greenhouse Gases Through Emissions From Land-use Change. *Science* 2008, *319*, 1238–1240.
99. Socolow, R.; Hotinski, R.; Greenblatt, J.B.; Pacala, S. Solving the Climate Problem. *Environment* 2004, *46*(10), 8–19.
100. Hodge, C. Ethanol Use in US Gasoline Should be Banned, Not Expanded. *Oil and Gas Journal* 2002, 20–30.

101. Hodge, C. More evidence mounts for banning, not expanding, use of ethanol in gasoline. *Oil and Gas Journal* 2003, 20–25.

102. Jacobson, M.Z. Effects of Ethanol (E85) versus Gasoline Vehicles on Cancer and Mortality in the United States. *Environmental Science and Technology* 2007, *41*(11), 4150–4157.

103. Davis, J.M.; Thomas, V.M. Systematic Approach to Evaluating Trade-offs Among Fuel Options: The lessons of MTBE. *Annals of the New York Academy of Sciences* 2006, 498–515.

104. Tilman, D.; Hill, J.; Lehman, C. Carbon-negative Biofuels from Low-input High-diversity Grassland Biomass. *Science* 2006, *314*, 1598–1600.

105. Rasnake, M. Tillage and Crop Residue Management. 1999. www.ca.uky.edu/agc/pubs/agr/agr99/agr99.htm(1/31/08).

106. Fryrear, D.W.; Bilbro, J.D. Wind Erosion Control With Residues and Related Practices. In *Managing agricultural residues*; Unger, P.W., Ed.; Lewis Publ.: Boca Raton, FL, 1994; pp. 7–17.

107. Lal, R. World Crop Residues Production and Implications of its Use as a Biofuel. *Environment International*2005, *31*(4), 575–584.

108. Kim, S.; Dale, B.E. Life Cycle Assessment of Various Cropping Systems Utilized for Producing Biofuels: Bioethanol and Biodiesel. *Biomass Engineering* 2005, *29*, 426–439.

109. Troeh, F.R.; Hobbs, A.H.; Donahue, R.L. *Soil and Water Conservation*; Prentice Hall: New Jersey, 2004.

110. Lal, R. Soil Carbon Sequestration Impacts on Global Climate Change and Food Security. *Science* 2004, *34*, 1623–1627.

111. Brummer, E.C.; Burras, C.L.; Duffy, M.D.; Moore, K.J. *Switchgrass Production in Iowa: Economic Analysis, Soil Suitability, and Varietal Performance*; Iowa State University: Ames, Iowa, 2000.

112. Samson, R. Switchgrass: A Living Solar Battery for the Prairies. Ecological Agriculture Projects, Mcgill University (Macdonald Campus, Ste-Anne-de-Bellevue, QC, H9X 3V9 Canada. Copyright @ 1991 REAP Canada.

113. Henning, J.C. Big Bluestem, Indiangrass and Switchgrass. Department of Agronomy, Campus Extension, University of Missouri: Columbia, MO, 1993.

114. Samson, R.; Duxbury, P.; Drisdale, M.; Lapointe, C. Assessment of Pelletized Biofuels. PERD Program, Natural Resources Canada;

Contract 23348-8-3145/001/SQ. 2000.

115. Wood Tub Grinders. Wood Tub Grinders. 2004. http://p2library.nfesc.navy.mil/P2_Opportunity_Handbook/7_III_13.html (8/3/04).

116. Arkenol. Our Technology: Concentrated Acid Hydrolysis. 2004. www.arkenol.com/Arkenol%20Inc/tech01.html(8/2/04).

117. Crisp, A. Wood residue as an energy source for the forest products industry. Australian National University, 1999. http://sres.anu.edu.au/associated/fpt/nwfp/woodres/woodres.html (7/10/06).

118. Brown, L.R. World's Rangelands Deteriorating Under Mounting Pressure. Earth Policy Institute, 2002. http://www.earthpolicy.org/Updates/Updates6.htm. (1/24/08).

119. Hos, J.J.; Groenveld, M.J. Biomass Gasification. In *Biomass*; Hall, D.O., Overend, R.P., Eds.; John Wiley & Sons: Chichester, UK, 1987; pp. 237–255.

120. Jenkins, B.M. Pyrolysis Gas. In *CIGAR Handbook of Agricultural Engineering*; Kitani, O., Jungbluth, T., Pearth, R.M., Ramdani, A., Eds.; American Society of Agricultural Engineering: St. Joseph, MI, 1999; pp. 222–248.

121. Ellington, R.T.; Meo, M.; El-Sayed, D.A. The Net Greenhouse Warming Forcing of Methanol Produced from Biomass. *Biomass and Bioenergy* 1993, *4*(6), 405–418.

122. Osburn, L.; Osburn, J. Biomass Resources for Energy and Industry. 2001. http://www.ratical.org/renewables/biomass.html (10/20/02).

123. Hekkert, M.P.; van Giessel, J.F.; Ros, M.; Wietschel, M.; Meeus, M.T.H. The Evolution of Hydrogen Research: Is Germany Heading for an Early Lock-in? *Journal of Hydrogen Energy* 2005, *30*, 1045–1052.

124. Ozaktas, T. Compression Ignition Engine Fuel Properties of a Used Sunflower Oil-diesel Fuel Blend. *Energy Sources* 2000, *22*(4), 377–382.

125. Ali, M.B.; McBride, W.D. Soybeans: State Level Production Costs, Characteristics, and Input Use, 1990. Economic Research Service; Stock no. ERS SB873; 1990; 48p.

126. College of Agricultural, Consumer and Environmental Sciences. Machinery Cost Estimates: Summary of Operations. University of Illinois at Urbana-Champaign, 1997. www.aces.uiuc.edu/~vo-ag/custom.htm(11/8/01).

127. Economic Research Statistics. Soybeans: Fertilizer Use by State.

Agricultural Chemical Usage: 1996 Field Crops Summary. 1997. http://usda.mannlib.cornell.edu/data-sets/inputs/9X171/97171/agch0997.txt(11/11/01).

128. Hinman, H.; Pelter, G.; Kulp, E.; Sorensen, E.; Ford, W. Enterprise Budgets for Fall Potatoes, Winter Wheat, Dry Beans and Seed Peas under Rill Irrigation. Farm Business Management Reports, Columbia; Washington State University: Pullman, WA, 1992.

129. Pimentel, D.; Doughty, R.; Carothers, C.; Lamberson, S.; Bora, N.; Lee, K. Energy Inputs in Crop Production: Comparison of Developed and Developing Countries. In *Food Security & Environmental Quality in the Developing World*; Lal, R., Hansen, D., Uphoff, N., Slack, S., Eds.; CRC Press: Boca Raton, FL, 2002; pp. 129–151.

130. Mississippi State University Extension Service. Agronomy Notes. 1999. http://msucares.com/newsletters/agronomy/1999/199910.html (7/10/06).

131. USDA. *Agricultural Statistics, 2004*; (CD-ROM); U.S. Department of Agriculture/ National Agriculture Statistics Service: Washington, DC, A1.47/2:2004.

132. Singh, R.P. Energy Accounting of Food Processing. In *Energy in Food Processing*; Singh, R.P., Ed.; Elsevier: Amsterdam, 1986; pp. 19–68.

133. Duffy, M. Does planting GMO seed boost farmer's profits? Iowa State University, Leopold Center for Sustainable Agriculture; Leopold Letter; 1999; Volume 11, No. 3.

134. Artuzi, J.P.; Contiero, R.L. Herbicides Applied in Soybean and the Productivity of Corn in Succession. *Pesquisa Agronpecuaria Brasilera* 2006, *41*(7), 1119–1123.

135. USDA. *Nutritive Value of American Foods*; Agriculture Handbook No. 456; Agricultural Research Service; U.S. Department of Agriculture: Washington, DC, 1975.

136. Fyksen, J. Soybean Issues Tackled: 2007 Season, Varieties for Biofuel, Foliar Fungicides. 2008. http://agriview.com/articles/2008/01/24/crop_news/crops04.txt. (2/15/08).

137. Tickell, J. *Biodiesel America: How to Achieve Energy Security, Free America from Middle-east Oil Dependence, and Make Money Growing Fuel*; Murphy, M., Graziano, C., Eds.; Yorkshire Press, 2006; pp. 121–140.

138. EBB. European Biodiesel Board. 2007. http://www/greencarcongress.

com/2007/07european-biodie.html. (1/29/08).

139. Frondel, M.; Peters, J. Biodiesel: A New Oildorado? *Energy Policy* 2007, *35*, 1675–1684.
140. Chamsing, A.; Salokhe, V.M.; Singh, G. Energy Consumption Analysis for Selected Crops in Different Regions of Thailand. (2004); In *Agricultural Engineering International: the CIGAR Ejournal*; Manuscript EE 06 013; Vol. VIII, November 2006.
141. USDA. 2007; The Price of Urea, the Primary Solid Nitrogen Fertilizer Used in the U.S. , Rose from $200 to $453 Per Ton. http://www/ers.usda.gov/AmberWaves/November07/Findings/TightSupply.htm. (3/15/08).
142. Smathers, R.L. Winter Rapeseed after Summer Fallow. 2005 Northern Idaho Crop Costs and Returns Estimate. College of Agriculture and Life Sciences. University of Idaho; EBB1-WR-05.
143. Pimentel, D.; Pimentel, M. *Food, Energy and Society*; Colorado University Press: Boulder, CO, 1996.
144. Molenhuis, J. Business Analysis and Cost of Production Program. In *Ontario Ministry of Agriculture, Food and Rural Affairs.*; 2004. http://www.omafra.gov.on.ca/english/busdev/bear2000/Budgets/Crops/Oilseeds/wcanolahybrid_static.htm (7/10/06).
145. Thoenes, P. Biofuels and Commodity Markets – Palm Oil Focus. FAO Commodities and Trade Division. Food And Agriculture Organization of the United Nations: Rome, 2007.
146. USDA. 2008; Foreign Agricultural Service. Oilseeds: World Markets and Trade Monthly Circular. Major Vegetable Oils: World Supply and Distribution (Commodity View). February 8 2008; U.S.Department of Agriculture: Washington, DC. http://www.fas.usda.gov/oilseeds/circular/2008/February/Oilseeds0208.pdf(3/15/08).
147. MPOA. Malaysian Palm Oil Association and Sustainable Palm Oil: RSPO Public Forum on Sustainable Palm Oil. 6 January 2005; Mont Kiara Business Centre: Kuala Lumpur.
148. Carter, C.; Finley, W.; Fry, J.; Jackson, D.; Willis, L. Palm Oil Markets and Future Supply. *European Journal of Lipid Science and Technology* 2007, *109*(4), 307–314.
149. Pleanjai, S.; Gheewala, S.H.; Garivait, S. Environmental Evaluation of Biodiesel Production from Palm Oil in a Life Cycle Perspective. In Presented at the Joint International Conference on "Sustainable

Energy and Environment" in Hua Hin, Thailand, 1-3 December, 2004. 19 pages.

150. Donald, P.F. Biodiversity Impacts of Some Agricultural Commodity Production Systems. *Conservation Biology* 2004, *18*(1), 17–38.

151. World Agroforestry Centre. 2007. http://www/worldagroforestrycentre.org/sites?TreeDB/aft/speciesPrinterFriendly. asp?ID=1013 (11/4/07).

152. Haverkort, A.; Bindraban, P.; Bos, H. Food, Fuel or Forest? In Proceedings of the seminary held at Wageningen, the Netherlands, March 2, 2007. 66 pages.

153. Pacific Island Ecosystems. Jatropha curcas: Risk Assessment Results. 2007. http://www.hear.org/pier/wra/pacific/jatropha_curcas_htmlwra.htm. (11/4/07).

154. Dimitrov, K. GreenFuel Technologies: A Case Study for Industrial Photosythetic Energy Capture. Brisbane, Australia. 2007. http://moritz.botany.ut.ee/~oilli/b/Dimitrov.pdf. (3/11/08).

155. Briggs, M. Widescale Biodiesel Production from Algae. 2004. http://unh.edu/p2/biodielsel/article_algae.html. (1/7/08).

156. Vincent Inc. Valcent Products: Intial Data From the Vertigro Field Test Bed Plant Reports Average Production of 276 Tons of Algae Bio Mass on a Per Acre/per Year Basis. 2007. http://money.cnn.com/news.newsfeeds/articles/marketwire/0339181.htm. (1/7/08).

157. Green Car. Green Car Congress. 2006. http://www.greencarcongress.com/2006/07/malaysia_and__in.html. (1/7/08).

158. Goldman, J.C.; Ryther, J.H. Mass Production of Algae: Engineering Aspects. In *Biological Solar Energy Conversion*; Mitsui, A., *et al.*, Eds.; Academic Press: New York, 1977; pp. 367–378.

159. Biopact. An In-Depth Look at Biofuels from Algae. 2007. http://biopact.com/2007/01/in-depth-look-at-biofuels-from-algae-html. (1/7/08).

160. CIIFAD. More Rice with Less Water. Cornell International Institute for Food, Agriculture and Development, 2007. http://ciifad.cornell.edu/SRI/estrmats/philmanual.pdf. (1/7/08).

161. Peak Soil. Peak Soil: Why biofuels are not sustainable and a Threat to America's National Security. 2008. http://www.ea2020.org/drupal/node/39. (1/31/08).

162. Leiting, K.B. Agronomy – Corn and Soybeans Crop Residue Management. Natural Resource Conservation Service: Albuerque, New Mexico, 2007. http://www/nm.nrcs.usda.gov/technical/tech-

notes/agro/ag67.pdf. (10/9/07).

163. Khan, S.A.; Mulvaney, R.L.; Ellsworth, T.R.; Boast, C.W. The Myth of Nitrogen Fertilization for Soil Carbon Sequestration. *Journal of Environmental Quality* 2007, *36*, 1821–1832.
164. Perlack, R.D.; Wright, L.L.; Turhollow, A.F.; Graham, R.L.; Stokes, B.J.; and Erbach, D.C. 2005; Biomass as Feedstocks for a Bioenergy and Bioproducts Industry: The Technical Feasibility of A Billion-Ton Annual Supply. (Report No. ORNL/TM-2005/66). Oak Ridge National Laboratory.
165. Lal, R.; Pimentel, D. Biofuels from Crop Residues. *Soil and Tillage Research* 2007, *93*(2), 237–238.
166. Prove, B.G.; Truong, P.N.; Evans, D.S. Strategies for Controlling Caneland Erosion in the Wet Tropical Coast of Queensland. In Proceedings. Australian Society Sugar Cane Technology,1986 Conference; pp. 77–84.
167. Wortmann, C.S.; Jasa, P.J. 2003; *Choosing the Right Tillage System for Row Crop Production*; Institute of Agriculture and Natural Resources at the University of Nebraska: Lincoln, Nebraska.
168. Pimentel, D.; Hepperly, P.; Hanson, J.; Douds, D.; Seidel, R. Environmental, Energetic, and Economic Comparisons of Organic and Conventional Farming Systems. *Bioscience* 2005, *55*(7), 573–582.
169. Liebman,M.NewCroppingSystemsforCellulosicFeedstockProduction and Environmental Stewardship. 2008. http://129.186.41.143/20th/home.files/liebman.pdf. (1/31/08).
170. Mongabay. Forest Erosion. 2004. http://www.mongabay.co/0903.htm. (12/07/07).
171. Thornton, F. C.; Green, T. H.; Joslin, J. D.; Houston, A.; Bock, B. R.; Schoenholtz, S.; Tyler, D. D.; Pettry, D. Environmental Impacts of Converting Cropland to Short-rotation Woody Crop Production: First Year Results. In Proceedings of Bioenergy 1996 - The Seventh National Bioenergy Conference, Nashville, TN, September 15-20, 1996; pp. 210–216.
172. IEA. International Energy Administration. International Energy Annual. USDOE: Washington, DC, 2001.
173. Naeher, L.P.; Brauer, M.; Lipsett, M.; Zelikoff, J.T.; Simpson, C.D.; Hoenig, J.Q.; Smith, K.R. Woodsmoke Health Effects: A Review. *Inhalation Toxicology* 2007, *19*, 67–106.

174. Crutzen, P.J.; Mosier, A.R.; Smith, K.A.; Winiwerter, W. N_2O Release from Agro-biofuel Production Negates Global Warming Reduction by Replacing Fossil fuels: Discussion. *Atmospheric Chemical Physics* 2007, 7, 11191–11205.
175. Brown, L.R. Why ethanol production will drive world food prices even higher in *Earth Policy Institute*. 2008. http://www.earthpolicy.org/Updates/2008/Update69.htm. (1/24/08).
176. Diouf, J. Biofuels a Disaster for World Food. In EU Coherence: EU Coherence Policy for Development. 2007. http://eucoherence.org/renderer.do/clearState/false/menuld/227351/return. (10/31/07).
177. United Nations Radio. Climate Change and Bio-Fuel Threaten Food Security. FAO Chief. 26 November 2007. www.un.org/radio/news/html/16436.html (1/22/2008).
178. Southwestern University. Biology Research Finds Rising CO_2 Levels Could Decrease the Nutritional Value of Major Food Crops. 2008. http://www.blackwell-synergy.com/doi/full10.1111/j.1365-2486.2007.01511.x. (2/15/08).

CHAPTER 11

AGRICULTURAL ENERGY DEMAND MODELING IN IRAN: APPROACHING TO A MORE SUSTAINABLE SITUATION

LeilaFarajian[a], RezaMoghaddasi[a], and SafdarHosseini[b]

[a]Department of Agricultural Economics, College of Agriculture and Natural Resources, Science and Research Branch, Islamic Azad University, Tehran, Iran

[b]Department of Agricultural Economics, University of Tehran, Iran

ABSTRACT

Energy has been heavily subsidized in Iranian economy for decades. Due to low oil prices, and, thus, budget limit, fossil fuel prices are under gradual liberalization. This will result in higher cost of production in all fuel consuming activities including agriculture. So it seems that farmers

Citation: Farajian L, Moghaddasi R, Hosseini S. Agricultural energy demand modeling in Iran: Approaching to a more sustainable situation. Energy Reports 2018; 4: 260-265.

will plan to substitute fossil fuels by non-fossil alternatives. This study aims at modeling and prediction of energy demand (as a key factor of production) in Iranian agriculture. Box–Jenkins methodology is applied to model agricultural consumption of four major sources of energy including: gasoline, kerosene, gasoil and electricity for the period spanning from 1988 to 2014. Also same data are predicted for 2015–2026 in order to check model adequacy and provide information regarding status of energy demand in Iranian agriculture in the future. Main results indicate a downward trend in consumption of four energy types except electricity which is predicted to go up. This implies that, as far as energy consumption is concerned, Iranian agriculture is approaching to a more sustainable situation. Rearrangement of fuel price support policy from nonrenewable to renewable sources and provision of more governmental support for switching to lower-carbon and environment-friendly energy sources is recommended.

Keywords: Energy consumption, BJ methodology, Modeling, Agriculture, Iran

INTRODUCTION

Energy demand forecasting is crucial for energy planning, formulating strategies and recommending energy policies (Bhattacharyya and Timilsina, 2009). Considering the fact that world population is increasing and predicted to reach 8.5 billion by 2030 (United Nations, 2015), energy demand will also be increasing. As the world becomes more dependent on technology and economies getting more developed, the energy consumption and thus dependency on energy will continue to rise. In conventional terms, demanding more energy implies a more tight dependence on fossil fuels like natural gas and oil which are, however, left with limited reserves and release greenhouse gases which are harmful to the environment and the global climate change. All these reasons force countries to adopt new sustainable energy policies, and to find alternative ways to balance the energy demand in most efficient, cheapest and strategic way (Oguz, 2013).

The agriculture sector has at its core the production process for foodstuff (e.g., grains, fruits and vegetables, meat, fish, poultry, and milk), and non-food vegetable products of economic value (e.g., tobacco, jute, hemp). However, the sector also comprises or has close links with processes that take place before and after this core production process, such as fertilizer production, post-harvest processing, and transport of foodstuff. Defined broadly, the agriculture sector has as its primary goal the delivery of food on the table for the population or for export. Thus, any measure that will reduce the fossil energy consumption while delivering the food service is in principle a potential candidate for analysis as a way of greenhouse gas (GHG) mitigation option (Sathaye and Meyers, 1995).

Agricultural activities require energy as an important input to production. Agriculture uses energy directly as fuel or electricity to operate machinery and equipment, to heat or cool buildings, and for lighting on the farm, and indirectly in the fertilizers and chemicals produced off the farm. Modern agriculture needs modern energy — the two are closely linked. For many developing countries, agriculture is the dominant sector in developing the economy. Increasing productivity and the modernization of agricultural production systems are the primary drivers of global poverty reduction and energy plays a key role in achieving this. Energy input to modern and sustainable agricultural production and processing systems is a key factor in moving beyond subsistence farming towards food security, added value in rural areas and expansion into new agricultural markets. In many cases, renewable energy technologies and hybrid systems can provide energy services that neatly support the production process, e.g., by providing irrigation (pumps) or post harvest treatment (cooling) or processing (drying, milling, pressing). The requirements of mechanical energy in the agricultural production process are also of critical importance and include human and animal labor as well as fuels for mechanization, pumping and other activities, and indirectly the production of fertilizers and agrochemicals. However, in many cases energy consumption in agriculture tends to be inefficient because of price subsidies (Swarnkar and Singh, 2013).

Fossil fuel subsidies – public financial support for fossil fuels – come in many forms and through many methods. These subsidies can be facilitated and applied through numerous policy processes and tools, and subsidies vary in their form across different countries. Environmental degradation, air pollution and health costs stemming from extracting and burning fossil fuels are not carried by the industry but paid by society. Therefore, these 'external costs' are also considered as fossil fuel subsidies. It is believed that all forms of fossil fuel subsidies are inefficient, harmful to the environment and blocking the transition to clean energy systems (Ozturk et al., 2010). As a result many countries including Iran have developed a roadmap to gradually phase out fossil fuel subsidies in the next decades. Agriculture is central to Iranian economy accounting for about 10% of gross domestic product, 20% of employment and more than 20% of non-oil exports (CBI[1], 2015).

Iranian government has been committed to support farmers by providing cheap inputs for decades. In spite of liberalization in market for some inputs (like pesticide), energy continued to be supplied with subsidized prices until recent years. Based on World Bank data global gasoline price has increased from 0.54 to 1.13 USD per liter (109%) during the period 1998–2017, while same data for Iranian economy was 0.08 and 0.36 (350% increase). As a result average energy consumption in Iranian agriculture is three times more than world level. Official data shows that gasoil has greatest share in total fossil fuels consumed in agriculture (94%), while electricity consumption has experienced an average annual growth rate of 45% during 1988–2014 (MOE[2], 2016).

Considering all the above points, main contribution of the current study can be defined as the modeling of demand for four major Iranian agricultural energy sources, and prediction of future consumption trend which clearly depicts the outcome of energy subsidy reduction policy followed by the government.

LITERATURE REVIEW

Previous studies on energy consumption can be separated in two groups. First, those studies devoted to energy-growth association at national and sector levels. For example causality and cointegration relationship between energy consumption and economic growth is reported by some researchers Mucuk and Sugozu, 2011, Senturk and Sataf, 2015, Ozturk et al., 2010, Sharma, 2010, Stern, 2010. Mehrara (2007) found strong unidirectional causality from economic growth to energy consumption in a group of oil exporting countries. Kayhan et al. (2010) analyzed dynamic causal relationship between electricity consumption and economic growth in the Romanian economy during 2001–2010. Their results support causality relationship from electricity consumption to economic growth. Second, works trying at modeling and forecasting energy price and consumption Fu and Nguyen, 2003, Meng et al., 2011, Chang et al., 2011, González-Romera et al., 2008. Camara et al. (2016)examined the forecasting performance of SARIMA and ANN methods in the context of U.S. residential energy consumption. They conclude that the neural network model has slight superiority over SARIMA. Jankovic (2017) provided forecasts for energy consumption in Serbia using ARIMA methodology for period of 2016 to 2027. Main results revealed that, in the following years, an increase can be expected in demand for oils.

In the context of Iranian economy some economy-wide works are done. For instance, Fallahi and Montazeri (2010) by using 35 years time series data showed that energy use had a negative effect on economic growth which is against theoretical expectation. But little is conducted on energy consumption modeling especially at sector level. Assessing the current situation and forecasting the future energy demand are essential for developing sustainable energy policies that support the increasing energy demand. In this context, low oil prices and public budget limit, necessitate liberalization of fossil fuel prices which definitely imposes higher cost to farmers. In order to reduce this burden, renewable energy resources seem to be the best alternatives. So the main objective of this paper is provision of economic model and reliable insight on energy (different types) consumption in agriculture. In sum, the main recognized gap in the relevant literature goes to the lack of any work dealing with modeling and prediction of different energy sources' consumption in Iranian agriculture.

The rest of the paper is organized as follows. Section 3 briefly presents methodology applied. Section 4 discusses main results obtained from the estimated model and finally Section 5 concludes.

MATERIALS AND METHODS

Background

Forecasting demand is both a science and an art. Econometric methods of forecasting, in the context of energy demand forecasting, can be described as the science and art of specification, estimation, testing and evaluation of models of economic processes that drive the demand for fuels (Meetamehra, 2009). Energy demand forecasting is a challenging task not only in developing countries where required data and proper models are lacking, but also in industrialized countries in which these limitations are somewhat less serious.

Since the first oil crisis in the early 1970's, bulk of research has been done in order to understand the nature of energy demand in both developed and developing economies. This has led to important methodological developments, and as a result a wide variety of models became available for analyzing and prediction of energy demand (Bhattacharyya and Timilsina, 2009). In a broad classification main methods include: trend analysis, end-use method, econometric approach, time series and hybrid models. Trend analysis is a non-causal method that does not explain how the values of the variable being predicted are determined. Its main disadvantage lies in the fact that it ignores possible interaction of relevant factors. Thus it could be considered only as a way of providing preliminary estimate of the predicted value of the variable. Application of end-use method in agriculture requires detail data on each of the end-uses (tractors, pumps, etc.) which is usually unavailable in developing countries. By using econometric approach one combines economic theory with statistical methods to specify an equation (or system of equations) for forecasting energy demand. This requires a consistent dataset over a reasonably long period. Moreover, researcher should assume a specific rate of change for explanatory variables during forecasting process which might seem incorrect.

Time series methods are in fact a subset of econometric models where behavior of a variable is explained only by its own lagged values. Simplicity and no need to collect data on multiple variables are among the main advantages of this method. Finally, hybrid approaches have recently been chosen by many researchers due to their more forecasting accuracy. Considering all the pros and cons of aforementioned methods in one hand and available data for Iranian agricultural fuel consumption on the other hand, time series methods are utilized in this study. More description of the applied method is provided in the next section.

The Box–Jenkins Methodology

The BJ methodology, commonly known as ARIMA[3]model, is, in theory, the most general class of univariate time series models for forecasting a variable. In literature, selecting a true model for ARIMA is considered as an art implying that a group of ARIMAs should be estimated in order to find the best model describing underlying behavior of a variable. ARIMA forecasts future values by looking only at the past pattern of the time series without using other independent variables. It is widely recognized as the most appropriate forecasting technique in many areas and is used extensively for time series modeling in practice (Oguz, 2013). The Box–Jenkins methodology of ARIMA models has some advantages over other time series methods. It provides more information from any other time series methods while using a minimum number of parameters. It also allows for flexibility in the process of choosing the correct forecasting mode. It includes a process that allows us to examine a large variety of models in our search for the correct one. On the other hand, the only problem with ARIMA is the modeling which is difficult mathematically and requires a deep knowledge of the method. Therefore, it is not easy to build an ARIMA model without training in statistical analysis and a good knowledge of the methodology. A general ARIMA model is represented as "ARIMA (p,d,q)", where:

•p is the order of autoregressive part,

•d is the number of nonseasonal differences needed for stationarity, and

•q is the order of moving average part.

Partial autocorrelation function (PACF) and autocorrelation function (ACF) are commonly used to determine proper values of p and q, respectively. Mathematically a time series (y) can be specified as an ARIMA (p,d,q) as follows:

$$\hat{y}_t = \mu + \phi_1 y_{t-1} + \cdots + \phi_p y_{t-p} - \theta_1 e_{t-1} - \cdots - \theta_q e_{t-q} \quad (1)$$

here, "Φ's" and "θ's" are autoregressive and moving average parameters, respectively.

The Box–Jenkins methodology for modeling time series consists of five different steps, namely, differencing (to make sure that series entering into the forecasting process is stationary), identification (to determine proper value for p and q), estimation (to provide numeric value for model parameters), diagnostic checking (to decide on model adequacy) and forecasting (to get future values of the series) (Asteriou and Hall, 2011).

Accuracy of forecasts provided by estimated model is evaluated by following criteria:

$$RMSE^4 = \sqrt{\frac{1}{T}\sum_{1}^{T}\left(y_{obs} - y_{pred}\right)^2}$$

$$MAPE^5 = \frac{1}{T}\sum_{1}^{T}\frac{\left|y_{obs} - y_{pred}\right|}{y_{obs}} \times 100 \quad (2)$$

where T is the number of observations (years), Z_{obs} and Z_{pred} are observed and predicted values of the variable, respectively. The lower are RMSE and MAPE the more accurate is the estimated model. Moreover, adequacy of the estimated models is investigated by detail examination of residuals. The Ljung–Box statistic could be calculated to decide on the significance of residuals at different lags. Also, three widely used statistical and econometric tests including Jarque–Bera (for normality), Breusch–Godfrey (for serial correlation) and ARCH (for heteroscedasticity) are separately applied for all estimated models.

Data

In this study consumption of four major sources of energy in Iranian agriculture are separately modeled over the period 1988 to 2014. Then, relevant future values are predicted for the next decade. Annual data on agricultural consumption of four major sources of energy including: kerosene, gasoil, gasoline and electricity for the period 1988–2014 is gathered from CBI and MOE official documents. Description of the data is provided in the next section.

RESULTS AND DISCUSSION

Fig. 1, Fig. 2 show evolution of the data. As is clear all three fossil fuels have followed decreasing trend though, due to negligible share in total fossil fuel consumption, the line for gasoline is not clear. The reverse situation is found for electricity implying more reliance of farmers on electricity as a new source of their energy needs.

Also **s** ome statistical features of the variables are presented in Table 1. All three fossil fuels consumption followed a slight declining trend, thought the major item, gasoil, shows an insignificant change. On the other hand, electricity use in agricultural activities has risen, implying relative success of government in substituting nonrenewable fuels by renewable and clean alternatives.

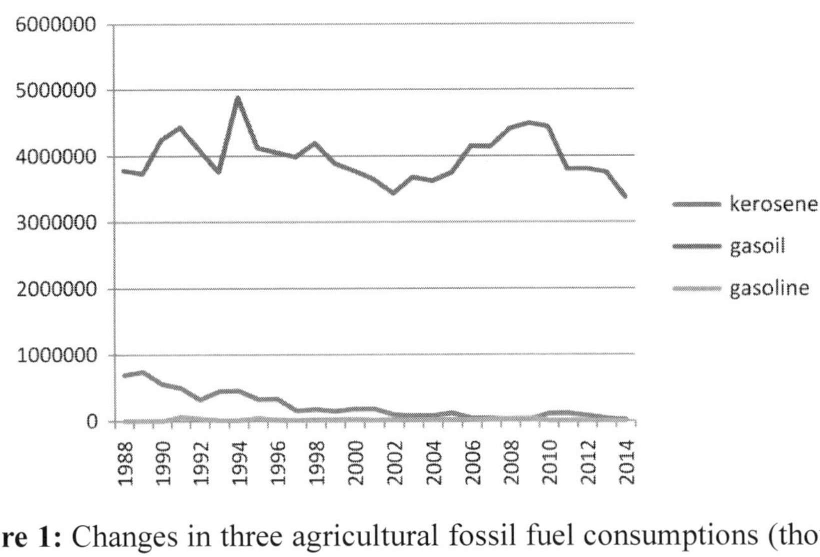

Figure 1: Changes in three agricultural fossil fuel consumptions (thousand liters).

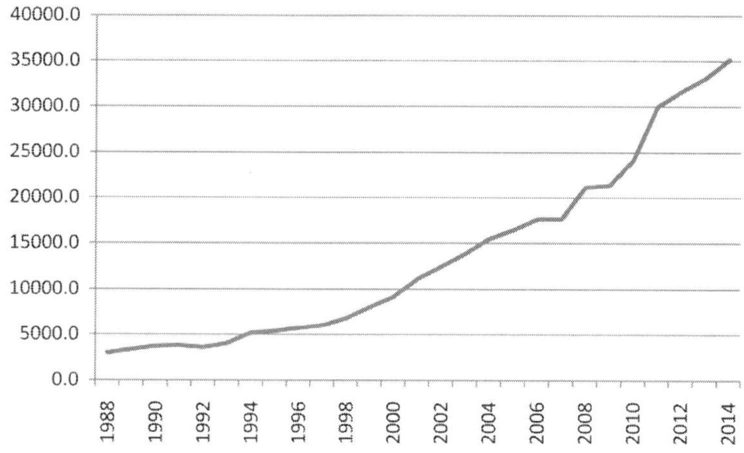

Figure 2: Changes in agricultural electricity consumptions (million kWh).

As the first stage of BJ methodology, stationarity of variables are checked. Table 2 has more.

Table 1: Descriptive statistics (1988–2014)

Variable statistic	**Kerosene**	**Gasoil**	**Gasoline**	**Electricity (million kWh)**
	(million liter)			
Mean	224.6	3978.1	16.4	13 669.9
Standard deviation	212.3	355.7	15.4	10163.1
Maximum	740.3	4882.3	62.9	35187.9
Minimum	190.1	337.1	119.1	2947.0
Growth rate (%)	−12.9	−0.4	−14.2	10.0
Share in total fossil fuel consumption (%)	5.1	94.5	0.4	–

Both applied tests reject the null hypothesis of stationarity in favor of the existence of one unit root. Thus the order of integration for all variables is determined at one. Next section documents the results of estimated ARIMA models for all energy sources.

Table 2: Stationarity tests results

Test variable	ADF		Philips-Perron		Result
	Level	First difference	Level	First difference	
Kerosene	−0.35	−7.56***	−2.07	−7.76***	I (1)
Gasoil	−2.89	−6.42***	−2.87	−6.55***	I (1)
Gasoline	−1.14	−5.80***	−2.60	−9.35***	I (1)
Electricity	−0.66	−4.59***	−0.18	−7.04***	I (1)

Note:***Denotes significance at one percent level.

Estimated Model for Kerosene Consumption

As the first choice of p and q in Eq. (1) one should refer to PACF and ACF of kerosene consumption data in first difference mode. Those lags representing a sudden cut-off could be a good starting candidates. Here, lags of 1, 2, 4, and 6 for p, and 1, 2, 3, 5, and 6 for q were considered. By estimation of the first model (1,1,1) and looking at its residuals, more technically significance of Ljung–Box statistic, the adequacy of estimated model was rejected. So in the second round the next model (1,1,2) was estimated, and again found to be unsuitable. After running different models and examining their adequacy, ARIMA (4,1,2) is chosen as the best fit to kerosene consumption data with minimized RMSE and MAPE. As Table 3 shows all slope coefficients are significant and the estimated model has acceptable goodness of fit. Furthermore adequacy of the model was confirmed by three diagnostic checking tests reported in Table 4.

Fig. 3 depicts predicted kerosene consumption trend for the period 2015–2026. The red lines boundary is 95% confidence interval. Obviously, on average, kerosene use is expected to stop in Iranian agriculture meaning a shift in energy sources.

Table 3: ARIMA (4,1,2) estimates for kerosene consumption

Variable	Coefficient	Std. Error	t-statistic
Constant	−15 900.8	10 662.2	−1.49
AR(2)	−0.62	0.20	−3.06***
AR(4)	−0.41	0.15	−2.59**
MA(2)	0.99	0.07	12.75***
R2=0.72	RMSE = 37858.8		MAPE = 129.3

Note: **Denote significance at 5% level.

***Denote significance at 1% level.

Table 4: Model adequacy tests results for kerosene consumption

Test	Jarque–Bera(normality)	BG-LM(serialcorrelation)	ARCH(heteroscedasticity)
Statistic	0.58(0.75)	1.48(0.26)	0.46(0.51)

Note: Figures in parentheses are p-values.

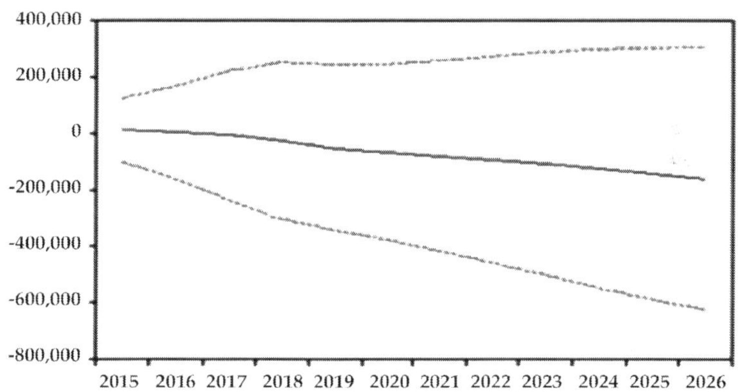

Figure 3: Kerosene consumption forecasts (2015–2026). (For interpretation of the references to colour in this figure legend, the reader is referred to the web version of this article.)

Estimated Model for Gasoline Consumption

Based on model adequacy criteria, ARIMA (0,1,4) is selected as the best model explaining gasoline consumption variations (Table 5). All five parameters are highly significant and R-squared shows an acceptable explanatory power of the model. Based on data provided in Table 6 no sign of rejection of classic regression analysis assumptions is seen implying adequacy of the estimated model.

Fig. 4 illustrates the gasoline consumption projection for the period 2015–2026. Similar to results stated for kerosene, we anticipate a declining trend and stoppage of gasoline use in next decade. This can be treated as another sign of movement to environment-friendly agriculture in the future.

Table 5: ARIMA (0,1,4) estimates for gasoline consumption

Variable	Coefficient	Std. Error	t-statistic
Constant	−591.70	274.82	−2.15**
MA(1)	−0.35	0.06	−5.39**
MA(2)	−1.03	0.10	−10.02***
MA(3)	−0.42	0.07	−6.00***
MA(4)	0.91	0.05	17.85***
R2=0.74	RMSE = 5779.6		MAPE = 242.5

Note:**Denote significance at 5% level.

***Denote significance at 1% level.

Table 6: Model adequacy tests results for gasoline consumption

Test	Jarque–Bera(normality)	BG-LM (serial-correlation)	ARCH (heteroscedas-ticity)
Statistic	2.81(0.25)	1.42(0.25)	0.33(0.57)

Note: Figures in parentheses are p-values.

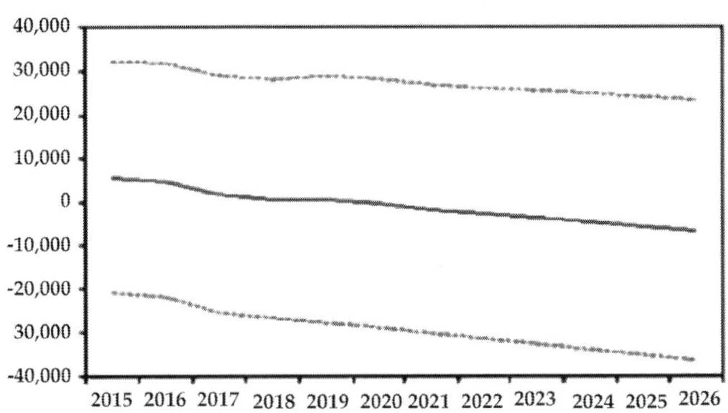

Figure 4: Gasoline consumption forecasts (2015–2026).

Estimated Model for Gasoil Consumption

Table 7, Table 8 portray the optimum ARIMA model describing gasoil consumption data and its adequacy. The order of both AR and MA parts is three. Model has passed all criteria deciding on satisfactoriness. Moreover, based on Fig. 5, dependence of agriculture on gasoil, as major fossil fuel utilized, is predicted to decline (by 13.6%) in upcoming years.

Table 7: ARIMA (3,1,3) estimates for gasoil consumption

Variable	Coefficient	Std. Error	t-statistic
Constant	−57597	43768.6	−1.31
AR(3)	0.47	0.12	3.62**
MA(3)	−0.92	0.02	−32.86***
R2=0.76	RMSE = 4525.1		MAPE = 179

Note:**Denote significance at 5% level.

***Denote significance at 1% level.

Table 8: Model adequacy tests results for gasoil consumption

Test	Jarque–Bera (normality)	BG-LM (serial-correlation)	ARCH (heteroscedasticity)
Statistic	1.02(0.60)	0.54(0.47)	0.15(0.71)

Note: Figures in parentheses are p-values.

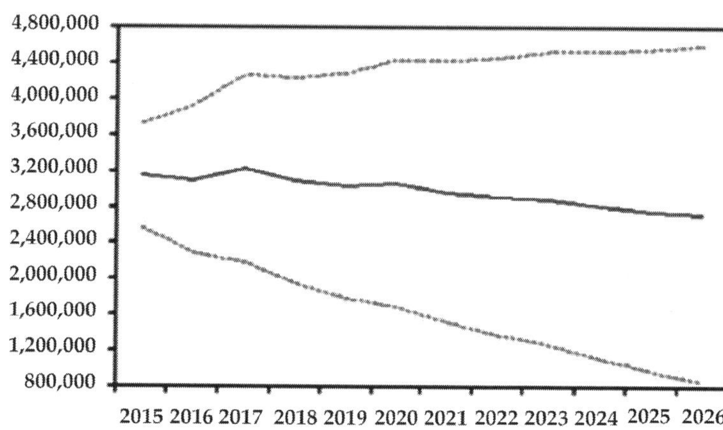

Figure 5: Gasoil consumption forecasts (2015–2026).

Estimated Model for Electricity Consumption

MOJA has tried to promote non-fossil fuels application in Iranian agriculture since mid 1990s by providing incentives like subsidized price. So it isvery important from policy making point of view to know how effective would be such policy in the future. As Table 9, Table 10 show an ARIMA (1,1,1) meeting all decision criteria, is estimated to give accurate forecasts on electricity consumption.

Forecasts obtained from the estimated model suggest an almost 65.7% increase in agricultural electricity consumption for next decade. It clearly points out to gradual substitution of electricity for other three fossil fuels (see Fig. 6).

Table 9: ARIMA (1,1,1) estimates for electricity consumption

Variable	**Coefficient**	**Std. Error**	**t-statistic**
Constant	2162.9	548.1	3.94**
AR(1)	0.88	0.08	10.77***
MA(1)	−0.99	0.10	−9.61***
R2=0.81	RMSE = 622.67		MAPE = 1.74

Note:**Denote significance at 5% level.

***Denote significance at 1% level.

Table 10: Model adequacy tests results for electricity consumption

Test	**Jarque–Bera (normality)**	**BG-LM (serial-correlation)**	**ARCH (heteroscedasticity)**
Statistic	0.94(0.72)	1.05(0.32)	0.09(0.77)

Note: Figures in parentheses are p-values.

In order to check the superiority of the estimated ARIMAs, forecasts from artificial neural network (ANN) models are produced. Table 11 compares the accuracy of forecasts from two competing methods. It is clear that for all energy sources, forecasts from ARIMAs have lower value of RMSE ad MAPE compared to those from ANNs validating the superiority of estimated ARIMAs.

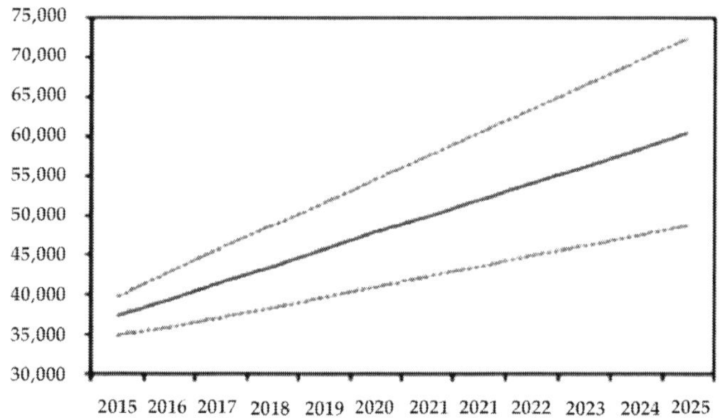

Figure 6: Electricity consumption forecasts (2015–2026).

Table 11: Forecasts' accuracy comparison

	Kerosene		Gasoline		Gasoil		Electricity	
	RMSE	MAPE	RMSE	MAPE	RMSE	MAPE	RMSE	MAPE
ARIMA	37 858.8	129.3	5779.6	242.5	4525.1	179.0	622.7	1.7
ANN	38 001.1	131.7	5829.5	253.9	4612.6	196.2	689.4	1.9

CONCLUSION

Given the fast growing population's food needs, the world's finite supply of fossil fuels and the adverse environmental impacts of using this nonrenewable resource, the existing relationship between agriculture and energy must be dramatically altered. Despite the challenges posed by the energy-intensive nature of agriculture, the careful utilization of resources and judicious application of technology has the capacity to significantly improve the long-term sustainability of food production.

Forecasting demand is both a science and an art. Econometric approaches of forecasting, in the context of energy demand modeling, can be characterized as 'the science and art of specification, estimation, testing and evaluation of models of economic processes' that drive the demand for fuels. Agriculture plays a critical role in Iranian economy in terms of production, employment, non-oil export and food security. Providing cheap fossil fuels for decades resulted in overuse of these energy sources and, of course, more pressure on environment. This has forced MOJA to revise its supportive policies with the aim of promoting application of renewable energy. Investing on the renewable energy sources can also be a way to support economic growth, and to reduce unemployment. So the Iranian government adopted targets and strategies for 2025 in order to increase the share of renewable energy sources in energy demand and to reduce the carbon emissions.

This study thus attempted to forecast the sectoral energy demand of Iranian agriculture for different energy sources as an elaboration to the existing non-source or integrated estimations. Agricultural energy consumption in Iran for four major items is modeled using BJ approach as a widespread technique. Annual data on use of kerosene, gasoline, gasoil and, electricity for the period spanning from 1998 to 2014 is employed for model estimation. Data for 2015 and 2016 is used to check prediction power of the model and finally projection is made for next decade.

Main findings reveal the upward trend in electricity demand as is projected to rise from 35 188 million kWh in 2014 to 58 764 million kWh in 2026 (almost 65.7% increase). Besides, this paper anticipates termination for kerosene and gasoline use and a 13.6% fall for gasoil consumption in forthcoming years. In sum, present study predicts that, as far as energy consumption is concerned, Iranian agriculture will move to a more sustainable situation during next decade. This definitely contributes to mitigation of environmental problems since agriculture captures nearly 20 million hectares of the country area. It could be regarded as a great achievement for government following fossil energy subsidy reduction policy.

Available policy options for more utilization of non-fossil fuels mostly include preferential loan programs and financial incentives, thought opportunities for efficient use of non-fossil energy resources tend to be rather site-specific so policy responses require some kind of outreach. Also, because of the large number of individual agricultural enterprises spread over a wide area, greater management resources are needed to implement and monitor policies.

REFERENCES

1. Asteriou, D., Hall, S.G., 2011. ARIMA models and the Box–Jenkins methodology. In: Applied Econometrics, second ed. Palgrave MacMillan.
2. Bhattacharyya, S.C., Timilsina, G.R., 2009. Energy demand models for policy formulation: a comparative study of energy demand models. working paper number 4866, The World Bank.
3. Camara, A., Feixing, W., Xiuqin, L., 2016. Energy consumption forecasting using seasonal ARIMA with Artificial Neural Networks models. Int. J. Bus. Manag. 11, 231–243.
4. Central Bank of the Islamic Republic of Iran. 2015. Economic analysis report.
5. Chang, P.C., Fan, C.Y., Lin, J.J., 2011. Monthly electricity demand forecasting based on a weighted evolving fuzzy neural network approach. Int. J. Electr. Power Energy Syst. 33, 17–27.
6. Fallahi, F., Montazeri, J., 2010. Energy consumption and economic growth: A nonlinear Approach. Iranian Econ. Res. 15, 85–109.
7. Fu, C.W., Nguyen, T.T., 2003. Models for long-term energy forecasting Power Engineering Society General Meeting, Toronto, Canada.
8. González-Romera, E., Jaramillo-Morán, M.A., Carmona-Fernández, D., 2008. Monthly electric energy demand forecasting with neural networks and Fourier series. Energy Convers. Manage. 49, 3135–3142.
9. Jankovic, R., 2017. Forecasting energy consumption in Serbia using ARIMA model, 3rd virtual In: international conference on science, technology and management in energy, Macedonia.
10. Kayhan, S., Uğur, A., Bayat, T., Lebe, F., 2010. Causality relationship between real GDP and electricity consumption in Romania (2001-2010). Rom. J. Econ. Forecast. 13, 169–183.
11. Meetamehra, , 2009. Demand forecasting for electricity. In: Body of Knowledge on Infrastructure Regulation.
12. Mehrara, M., 2007. Energy consumption and economic growth: the case of oil exporting Countries. Energy Policy 35, 2939–2945.
13. Meng, M., Niu, D., Sun, W., 2011. Forecasting monthly electric energy consumption using feature extraction. Energies 4, 1495–1507.
14. Ministry of Energy. 2016. Energy balance sheet.
15. Mucuk, M., Sugozu, I.H., 2011. Sectoral energy consumption and

economic growth nexus in Turkey. Energy Educ. Sci. Technol. Part B: Soc. Educ. Stud. 3, 441–448.

16. Oguz, M.E., 2013. Forecasting Turkey's Sectoral Energy Demand, Master Thesis in Sustainable Environment and Energy Systems Program. Middle East Technical University, Northern Cyprus Campus.
17. Ozturk, L., Aslan, A., Kalyoncu, H., 2010. Energy consumption and economic growth relationship: evidence from panel data for low and middle income countries. Energy Policy 38, 4422–4428.
18. Sathaye, J., Meyers, S., 1995. Greenhouse Gas Mitigation Assessment: A Guidebook. Springer science + business media, Dordrecht.
19. Senturk, C., Sataf, C., 2015. The Determination of panel causality analysis on the relationship between economic growth and primary energy resources consumption of Turkey and Central Asian Turkish Republics. Procedia - Soc. Behav. Sci. 195, 393–402.
20. Sharma, S.S., 2010. The relationship between energy and economic growth: empirical evidence from 66 countries. Appl. Energy 87, 3565–3574.
21. Stern, D., 2010. The role of energy in economic growth, The Australian National University, CCEP working paper 3.10.
22. Swarnkar, N., Singh, S., 2013. Analysis on electrical energy consumption of agricultural sector in Uttarakhand state. Int. J.. Emerg. Technol. Adv. Eng. 3, 344–347.
23. United Nations, 2015. World Population Prospects: Key Findings and Advance Tables, the 2015 Revisions, Department of Economic and Social Affairs, report number: ESA/P/WP.241.

CHAPTER 12

CONFRONTING TRADEOFFS BETWEEN AGRICULTURAL ECOSYSTEM SERVICES AND ADAPTATION TO CLIMATE CHANGE IN MALI

Kurt B.Waldman[a] **and Robert B.Richardson**[b]

[a]Department of Geography, Indiana University, 513 N. Park Ave., Bloomington, IN 47408, USA
[b]Department of Community Sustainability, Michigan State University, 480 Wilson Rd., East Lansing, MI 48824, USA

ABSTRACT

Changing climatic conditions present new challenges for agricultural development in sub-Saharan Africa. Sorghum has proven to be an adaptable and resilient crop despite limited funding for crop development. Recent breeding efforts target hybrid and perennial technologies that may facilitate

---

Citation: Waldman KB, Richardson RB. Confronting Tradeoffs Between Agricultural Ecosystem Services and Adaptation to Climate Change in Mali. Ecological Economics 2018; 150: 184-193.

adaptation to climate change. Advantages of perennial crops over their annual counterparts include improved soil quality and water conservation and reduced inputs and labor requirements. In contrast, hybrid crops are often bred for improved grain yield and earlier maturation to avoid variable conditions. We use discrete choice experiments to model adoption of sorghum as a function of attributes that differ between these technologies and traditional varieties in Mali. Overall, the main perceived advantage of perennial crops is agricultural ecosystem services such as soil improvement, while adoption of hybrid crops is hampered by the inability to reuse seed. Women farmers are less concerned about higher labor requirements associated with perennial crops and the ability to reuse hybrids seeds than male farmers. Farmers prefer traditional sorghum to perennial sorghum and are indifferent between traditional and hybrid sorghum. These findings have important policy implications for understanding tradeoffs that are central to farmer decision making when it comes to breeding technologies for climate adaptation.

Keywords: Agricultural ecosystem services, Choice experiments, Climate change adaptation, Gender, Mali, Sorghum

INTRODUCTION

Ensuring access to nutritious and environmentally sustainable food to all people at all times is one of the greatest challenges currently facing global society (Hall et al. 2017). Climate change is projected to impact food security in the developing world, especially in countries already facing chronic hunger and malnutrition (Lobell et al. 2008). Climate change is expected to have a direct impact on food production as changes to both temperature and rainfall patterns affect crop yields, water availability, pests and diseases, and livestock health, and smallholder farming systems in Africa (Hall et al. 2017). Countries in sub-Saharan Africa are predicted to be worst affected due to already high temperatures, high dependency on rain-fed agriculture, and economic fragility. The problem is particularly acute in West Africa, where the dual forces of population growth and climate change are likely to exacerbate the persistent challenge of food insecurity. West Africa is also subject to significant rainfall variability and drought occurrences, and this is expected to increase as climate change impacts the region (Mason et al. 2015). Sorghum is among a limited number of crops that have the resiliency to adapt to changing climate change conditions, particularly increasing drought, soil salinity and higher temperatures (ICRISAT, 2015). Despite these benefits, agricultural policies concerning subsidies and crop improvement have been inconsistent and often do not target sorghum. In the past two decades, domestic production has increased at an average rate of about 1% for sorghum, while maize has increased at an average rate of about 7% (Kelly et al. 2012), despite a steady increase in sorghum grain yield in West Africa since the 1970s (ICRISAT, 2015). Predicted increases in climate variability require strategies that improve the genetic heterogeneity of sorghum traits and improve the capacity for adaptation to increased variability (Haussmann et al. 2012).

Farmers require a diversity of crop options that can utilize different spatial niches of nutrients and water resources, and have different mechanisms for responding to annual variation in precipitation (Altieri 2002; Hall et al. 2017). Diversification is one strategy that allows farmers to spread short-term risk and create a more resilient farm system within the context of increasing climatic variation. Developing crop varieties that are resilient to variable climate conditions can improve the diversity and resilience of these systems (Morton 2007). Farmers in West Africa experience both long-term and year-to-year variability in rainfall and access to diverse crop types can improve their resilience (Mason et al. 2015).

Hybrid crops and perennial crops represent two radically different technologies which each have the potential to improve farmers› ability to adapt to variable climatic conditions. Through hybridization, desirable traits from diverse germplasm may be crossed with traditional landraces in order to create more productive varieties and varieties that combine useful traits, such as disease resistance, with regionally adapted and accepted plant morphology. Development of early maturing hybrid varieties of crops is a common way to reduce exposure to climate variability by reducing the amount of time a crop is in the field.

Cultivation of perennial crops on the other hand, is an investment in the long-term resilience of cropping systems. Perennials potentially fill an important niche for farming households and may have numerous ecological and agronomic advantages over their annual counterparts (Glover et al. 2012). While perennial crops often have drawbacks such as lower yields, many of the ecosystem services associated with the perennial nature of the crop make it more resilient and potentially an important component of climate change adaptation, particularly in water scarce areas. There has been little research to date on preferences for crop duration—or more specifically household-level tradeoffs and preferences for long duration or perennial crops versus short duration, early maturing hybrids—and the prospects for using these technologies for climate adaptation. The literature on the economics of and preferences for perennial crop production in Africa is thin, largely because the development of perennial crops is in a nascent stage. Most of the literature focuses on plant physiology or developing breeding lines that have the capacity to regrow post-harvest and produce yield over successive years (Hayes et al. 2012; Kane et al. 2016). Breeding programs in Africa target a variety of crops from maize and beans to lesser-known or more traditional crops like sorghum and millet but little attention has been paid to the tradeoffs involved with perennial version of these crops.

There has also been relatively little attention to the gender implications of crop preferences. Many rural African smallholder societies are patriarchal and men make most of the agricultural decisions for the household, even though women are often the providers of food for the family. However, in polygamous households, common in Mali, women often manage land and make agricultural decisions independent of their husbands. In addition to examining the differences in tradeoffs embodied by perennial and hybrid crops, we specifically consider the variation in preferences among attributes of hybrid and perennial varieties of sorghum between male and female farmers.

BACKGROUND

Depending on the agroecological region, sorghum, millet, and maize are the primary staple crops in Mali. Sorghum is widely cultivated across the Sudan Savannah and covers a range of agroecological zones characterized by a gradation in rainfall from the edge of the Sahara Desert to the fertile Niger River Delta. Development efforts led by the Malian government have been aimed at improving staple crop yields since the drought years that occurred in the 1970s and 1980s. Initially these efforts involved breeding with foreign genetic material but more recently has been focused on Guinea races that are traditional West African sorghum varieties (Rattunde et al. 2013).

Adaptation requirements for new sorghum varieties are specific to the agroecological zone, and no single variety predominates across all of the sorghum growing areas in Mali (Bazile et al. 2008). Yapi et al. (2000) found that farmers prefer the local sorghum landraces in Mali more than varieties derived from foreign genetic material. Local breeding programs developed both pure Guinea-race hybrids as well as Guinea-caudatum hybrid varieties in response.[1] The Guinea race of sorghum has a relatively broad geographic distribution, and research has shown that it comprises more genetic diversity than other races (Folkertsma et al. 2005). Given the wide range of agroecological zones in Mali and the highly variable climatic conditions there is a need for a diversity of plant types.

The backbone of the green revolution in Asia was the development of high yielding hybrid varieties of crops that respond well to fertilizer (Eicher 1995). Hybridized versions of crops can address specific problems that plague farmers by breeding in pest or disease resistance, drought tolerance, resistance to abiotic stress that may problematic in a given region. Hybrid breeding, particularly with maize in Africa, has focused on improving low yields, but also more recently on developing shorter duration varieties that are more suitable for shortened growing seasons in many regions. There are drawbacks with hybrid crops, notably that hybridized seed tends to be more expensive and can only be planted once before hybrid vigor is lost. Hybrid crops also tend to require greater fertilizer inputs than local varieties to achieve their yield benefits. However, research has shown that despite the higher inputs costs, hybrid crops can return more per acre because of the greater yield advantage, which is true of hybrid sorghum in Mali (Smale et al., 2016).

Perennial lines of sorghum have only recently been tested in Mali and are at this point still hypothetical to farmers. The primary tradeoff with perennial crops is yield reduction since most perennial crops yield less than their annual counterparts (Pimentel et al. 2012). For this reason, efforts to develop perennial wheat in developed countries have focused on improved grain yield, and to a lesser extent grain quality (Cox et al. 2006; Jaikumar et al. 2012; Hayes et al. 2012). There are numerous advantages to perennial crops such as lower seed inputs since they require reseeding every three to five years as opposed to every year like annual crops (Bell et al. 2008). Less tillage also translates into less soil disturbance and erosion. As a result, the farm labor costs, energy usage, and technological inputs required for yearly tillage will significantly decrease in perennial crop systems (Pimentel et al. 2012). Additionally, the large root mass helps retain soil, prevent future erosion, sequester more carbon, and hold more soil water (Glover et al. 2010). Perennial cropping systems also have natural mechanisms that make them more resilient to pests (Van Mele and Chien 2004; Cox et al. 2006). Perennial systems have more soil fauna diversity and natural below ground processes since they have more year-round vegetation (Culman et al. 2013). Perennials are also more efficient than annual crops at absorbing nutrients, so fertilizer requirements are lower and nutrient leaching is reduced.

Perennial crops present attributes and management requirements that are very different from the attributes of the hybrid varieties more commonly found in breeding programs. In many ways, perennial crops are the antithesis of the typical hybrid crop variety, which is often input-intensive, early maturing, and high yielding. Perennial crops are not necessarily high yielding, they have lower labor requirements and are longer duration, but they may be more amenable to the low-input, low-output farm management style in Africa. Perennials are adaptable to climate change through extensive root growth that permits greater access to water and nutrients during dry spells, and they are thus often associated with improved agricultural ecosystem services. Hybrid varieties on the other hand tend to be bred to fit into a new and different cropping system that is characterized by more variable rain or a shorter growing season. Early maturing hybrids can avoid changing climatic conditions and variable rainfall by through crossbreeding with varieties that are short duration. Hybrids can also be bred for tolerance to temporal midseason drought and dry spells, and are often associated with adaptation to climate change.

Hybrid and perennial crops present different tradeoffs to farmers and this paper examines farmers' preferences for these different cropping systems and the associated attributes of each using choice experiments. We employ two sets of choice experiments to look at farmers' preferences for attributes of perennial and hybrid crops using the case of sorghum in Mali. We also chose to design the study with a gender focus since the structure of households in Mali is such that most male and female farmers manage agricultural landindependently and are likely have independent preferences for crop attributes. We are most interested in the heterogeneity in preferences, with specific attention paid to differences by gender of the respondent so that we are able to characterize the relationship between crop duration and gender.

METHODS

Study Area and Data

Fig. 1 depicts the rainfall gradient and the study sites in Mali. The gradient runs from the southern border with the Ivory Coast, where the highest rainfall occurs (about 1400 mm annually) northeast to the border of the Sahara desert, where rainfall is too low to support crop cultivation (below

400 mm of rainfall per year). This rainfall gradient also extends across the rest of West Africa. There are four main agroecological zones in Mali of which the Sahelian zone and Saharan Zone are too dry to cultivate. Sorghum is cultivated in the southern half of the country in the Subhumid Zone and the Sudanian Savannah, which comprise 6% and 12% of land respectively. In the Sudanian Savannah, the soils are relatively heavy and sorghum is the dominant cereal crop, often rotated with cotton, maize, groundnuts or cowpeas.

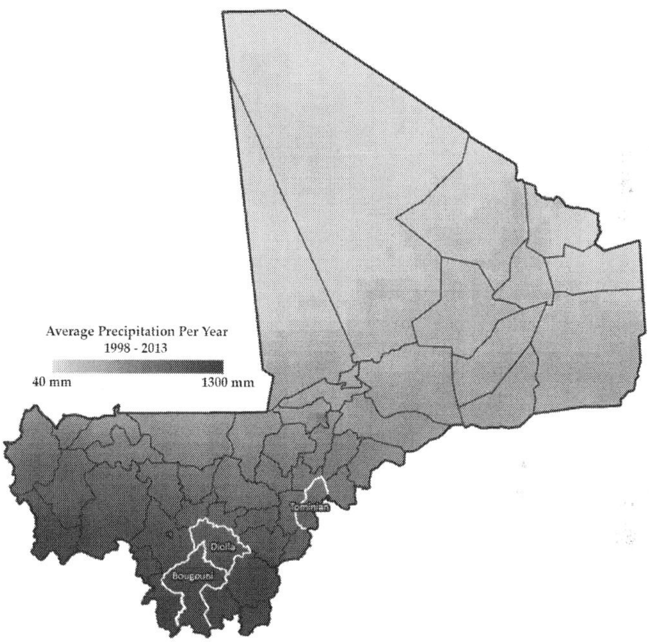

Figure 1: Precipitation map of Mali with research locations highlighted.

We use a spatially explicit clustered multistage sampling approach to capture variation along the rainfall gradient. Out of the 8 regions in Mali, we select three regions where annual sorghum is cultivated under low, medium, and high rainfall; Sikasso, Koulikoro, and Segou, respectively. Within each region we select one Cercle, which is the second level administrative unit. The selected Cercles were Bougouni, Sikasso; Dioila, Koulikoro; and Tominian, Segou (highlighted in Fig. 1) and we randomly choose six villages from a list of all villages within each Cercle. Prior to visiting households we received permission from village elders to interview 20 people in each village (approximately 10 men and 10 women) and local guides brought enumerators to houses at an increasing distance from the center of the village.

We have a total of 318 complete observations after discarding observations from two villages in Bougoni where there were some problems with initial survey enumeration quality.

The Cercle of Bougoni is at an elevation of 344 m, covers 20,028 km^2 and according to the 2009 Census, has a total population of 459,509 and a population density of 23/km^2. The average elevation of Dioïla Cercle is 331 m and it covers a total area of 12,794 km^2 with a population of 249,403 and a population density of 19/km^2. Tominian Cercle sits at an average elevation of 301 m, covers a total area of 6573 km^2 with a total population of 219,853 and a population density of 33/km^2. Annual rainfall ranges between 450 and 700 mm in Segou region and the Tomininan district is located at North in the driest part of the region. In Koulikoro region, Dioïla has a rainfall pattern which ranges between 800 and 900 mm per year. While in the Sikasso region, often referred to as the "bread basket" of Mali, the annual rainfall varies between 800 mm in the North to 1200 mm in the South-West. Bougouni is located in the wettest part of the region.

Sikasso and Koulikoro regions are the principal sorghum-producing regions, with 31% and 22% of national production respectively (Kelly et al. 2015). These areas have been priority target areas for sorghum breeding, especially for hybrid development in Mali since sorghum is the primary staple in much of this zone (ICRISAT, 2015). Rattunde et al. (2013) found average yield advantages of 380 kg/ha to 660 kg/ha with hybrids relative to an adapted landrace.

Choice Experiments

We use choice experiments to study farmers' preferences for the attributes of hybrid and perennial sorghum crops. Choice experiments measure stated preferences of participants as opposed to revealed preferences that come from observed retail transactions, yet perform similarly to market price analysis (Scarpa et al. 2003). Choice modeling is a method used to estimate the marginal value of various attributes of a good and is especially useful when observed transactions of a good do not occur (Train 2009). Choice experiments are a flexible tool useful for understanding hypothetical choices and demand for new agricultural products or technologies (Useche et al. 2009; Useche et al. 2013). To minimize hypothetical bias, increase comprehension of the choice experiment, and reduce the cognitive burden of the exercise, the choice sets were illustrated and presented to farmers on laminated cards.

Choice experiments are widely used in the agricultural and environmental economics literature and are increasingly common for understanding preferences in a development context (Bennett and Birol, 2010). They can be used to understand breeding traits in different production systems (Roessler et al. 2008), to estimate willingness to pay for a new agricultural technology (Birol et al., 2012), to preferences for cropping systems (Ortega et al. 2014), and investigate preferences for environmental adaptability and yield stability (Asrat et al. 2010). Numerous studies use choice experiments to evaluate preferences of developing country farmers concerning environmental or ecological tradeoffs involved in agricultural production. Choice experiments were used to estimate demand for perennial pigeon pea in Malawi (Waldman et al. 2017), and they have also been used to estimate Mexican farmers' preferences for biodiversity in their cropping systems (Birol et al. 2009). Choice experiments were used in the analysis of tradeoffs regarding pineappleproduction and environmental management in Costa Rica (Richardson et al. 2013), and to value incentives to households that communally manage natural resources (Pienaar et al. 2014).

Choice experiments have also been used to measure farmer adoption of improved staple crop technologies (Sánchez-Toledano et al., 2017a, Sánchez-Toledano et al., 2017b). We build on this literature by using choice experiments to understand preferences for an agricultural technology with strong ecological advantages that is a nascent stage of breeding development, and compare them with preferences for an existing technology, with an emphasis on tradeoffs between agricultural ecosystem services and adaptation to climate change.

Modeling preferences for perenniality using choice experiments

We design two choice experiments, one examines preferences for hybrid sorghum and the other examines preferences for perennial sorghum. In both, we assume a farmer chooses between perennial or hybrid and annual sorghum, and we assume that farmers seek to maximize the utility derived from their cropping decision. We use random parameters logit (RPL) model that allows us to control for the standard deviation in the choice attributes. We are interested in exploring the heterogeneity among preferences of individuals as opposed to a latent class approach, which involves identification of groups of farmers that are heterogeneous across classes and homogenous within a group. We explicitly want to understand how individual preferences differ, particularly along gender lines. An RPL model with interaction effects between the choice attributes and the gender variable is more consistent with this aim than a model such as the covarianceheterogeneity model which allows for scale to be a function of attributes.

We suppose that farmer n faces K alternatives contained in choice set ψduring situation s. We define an underlying latent variable $V_{njs}{}^*$ that denotes the value function associated with farmer n choosing option $j \in \psi$ during choice situation s. Farmer n will choose alternative j so long as $V_{njs}{}^* > V_{nks}{}^* \ \forall \ k \neq j$. Indirect utility $V_{njs}{}^*$ is not directly observed; what is observed is the actual utility maximizing choice V_{njs}, where

$$V_{njs} = \begin{cases} 1 \text{ if } V^*_{njs} = \max(V^*_{n1s}, V^*_{n2s}, \ldots, V^*_{nKs}) \\ 0 \qquad\qquad \text{Otherwise} \end{cases} \tag{1}$$

And we can therefore write farmer n's utility function as

$$V^*_{njs} = X'_{njs}\beta + \varepsilon_{njs} \tag{2}$$

where $X_{njs}{}'$ is a vector of characteristics of the cropping system for the jth alternative, β is a vector of preference parameters or vector of weights that map the attribute levels into utility, and ε_{njs} is a stochastic component of utility that is independently and identically distributed (iid) across individuals and alternative choices, and takes a predetermined (Gumbel or extreme valuetype I) distribution. This stochastic component of utility implies that predictions cannot be made with certainty and captures unobserved variations in preferences as well as errors in farmer's perceptions and optimization.

We estimate perennial and hybrid sorghum models in both preference space and willingness-to-pay (WTP) space. The coefficients on the attributes in the preference space model reflect the marginal utility of each attribute to an individual and the ratio of two marginal utilities is the marginal rate of substitution of one attribute for another. The advantage of estimating a model in WTP-space is that one can estimate the marginal rate of substitution directly (Scarpa et al. 2008). By specifying an attribute for comparison, one avoids the problem of comparing attributes from coefficients with different scales. Similar to Ortega et al. (2016), we estimate WTP for choice attributes in terms of the amount of crop yield an individual would be willing to substitute for each of the other attributes.

The most important tradeoffs and design attributes of the choice sets for annual, perennial, and hybrid sorghum production were estimated from a combination of literature review, informal discussion with breeders, and a series of focus groups with farmers in Mali in October 2014. We also consulted with local agronomists and experts in legume cultivation in Mali to design the parameters for the choice experiment. We identified the main attributes to be used in the choice experiments through informal interviews with farmers and consultation with agronomists and made a series of modeling assumptions based on these interviews.

Rather than present one choice as perennial and one as annual we decided to decompose the main attributes of a perennial crop in order to understand valuation of the attributes without the stigma that may be attached to other perennial crops or varieties of sorghum that are known to regenerate. This is consistent with Lancaster's theory of consumer behavior, which states that consumers are not seeking the goods themselves but the characteristics they embody (Lancaster 1966). Some farmers were aware of the concept of hybrid and perennial sorghum and we wanted to avoid associating their choices with their perceptions of these crops so we avoided labeling the choices as such. We model sorghum as a sole crop that is part of a crop rotation as opposed to intercropped, although various forms of intercrop are common in Mali.

The yield reduction from a perennial crop was identified as a key factor in farmers' decision-making processes so we include this as a "price variable" in the choice experiment. Given the prevalence and importance of sorghum in Mali, this attribute serves as a substitute for a cost or price variable when evaluating tradeoffs with the other attributes. This indirect measure is preferred over a direct monetary variable, as yield is a more intuitive measure of value to farmers given the subsistence nature of agriculture in the region (Birol et al. 2009). This is also consistent with the notion of staple grains such as sorghum serving as a form of currency for smallholders (Smale et al., 2016). We chose to frame sorghum yield as a percentage of an average year's harvest in order to avoid having to specify the plot size being harvested, or control for variation between villages and farms due to environmental factors.

Other factors identified as critical tradeoffs concerning annual versus perennial sorghum was the reduction in seed and labor requirements resulting from a single planting with multiple harvests. We also model the benefits of perennial sorghum, which theoretically include soil improvementfrom more extensive perennial root growth, and higher levels of biomass/forage, since less energy is required for root development. We assume that breeders can reach a point where perennial sorghum can reliably survive the dry season and regenerate but that on average the yield of perennial sorghum is lower than traditional sorghum. With hybrid sorghum, we assume less seed is required, the price of the seed is higher than traditional sorghum seed, and the seeds cannot be reused in subsequent years. Furthermore, we assume that there are multiple plant durations of hybrid sorghum, and the grain yield of hybrid sorghum is higher on average than traditional sorghum.

Parameterization of the choice models

Smale et al. (2016) calculate the unit variable cost of traditional (annual) sorghum to be 125 CFA/kg producing an average of 950 kg/ha, improved sorghum to be 105 CFA/kg producing an average of 1500 kg/ha and hybrid sorghum the lowest at 65 CFA/kg or 2500 kg/ha.[2] Based on these figures we model the choice between a traditional and perennial crop (or hybrid crop) as a reduction (or increase) in yield based roughly on the variation in current versus previous yield of traditional varieties and the percentage range observed for perennial or hybrid to traditional. The values of the parameters were wide enough to capture both the high and low ends of the spectrum.

Cropping system attributes used in the perennial choice experiment included seed requirements, labor requirements, forage quantity, soil improvement, and sorghum yield. Cropping system attributes and their levels are described below in Table 1. We modeled the variation in seed requirements as ranging from the same requirements as their current varieties to 15% and 30% reductions to capture the lower amounts of seeds required for perennial cultivation. We treated labor as a binary variable (low and high) since there is such wide variation in labor availability across farms and regions. Perennial crops tend to produce much higher levels of above-ground biomass since they have to devote less energy to root production after establishment, and so we have also modeled this variable as a binary choice to avoid misleading amounts, since we do not have accurate quantification of levels of biomass from perennial sorghum. Similarly, we modeled soil improvement as a binary variable since one of the often-stated benefits of perennial crops is long-term soil improvement from the additional carbon resulting from the larger root mass, although we do not have empirical evidence of the magnitude. Modeling attributes as binary that are continuous is a significant limitation of the study but in the absence of detailed empirical estimates this approach allows us to evaluate how important this attribute is to farmers.

Table 1: Attributes used in the perennial sorghum choice experiment

Attribute	Levels	Description
Seed requirements	Average, −15%, −30%	Amount of seed required compared to current sorghum varieties

Labor requirements	Low, high	Amount of labor required with the sorghum crop/variety
Forage quantity	Low, high	Level of residue produced over the planting period
Soil improvement	Low, high	Extent that the soil quality is improved from a crop
Sorghum yield	Same, −20%, −40%, −60%	Percent of previous year sorghum grain yield

The main attributes identified as important in the decision to plant hybrid versus traditional sorghum varieties were the seed requirements, the plant duration, the ability to reuse the seeds, the cost of the seeds and the grain yield (see Table 2). We included two attributes of seeds that differ between traditional and hybrid varieties—the amount of seed required (hybrid seeding rates recommended by seed companies are roughly half that of local varieties) and the cost of the seed (approximately 600 kg for hybrid, 100 kg for traditional) (Smale et al., 2016).

Table 2: Attributes used in the hybrid sorghum choice experiment

Attribute	Levels	Description
Seed requirements	5, 10 kg/ha	Amount of seed required compared to current sorghum varieties
Plant duration	Early, Medium, Late	The days to maturity of the plant (early is about 4 months while late is >6 months)
Ability to reuse seeds	No, Yes	Whether the seeds can be harvested and reused in the following season
Cost of seed	100 CFA, 600 CFA	The cost of 1 kg of the seed
Sorghum yield	120%, 140%, 160%, 180%	Percent of previous year sorghum grain yield

We created an orthogonal experimental design using these attributes and levels using the software Ngene 5.0. The experiment was divided into 6 blocks with 4 choice sets in each block. An option to opt out or select "neither" choice was also presented each time, labeled option C. We created illustrated booklets of the choice sets in each of the 6 blocks. A sample choice set is presented in Fig. 2.

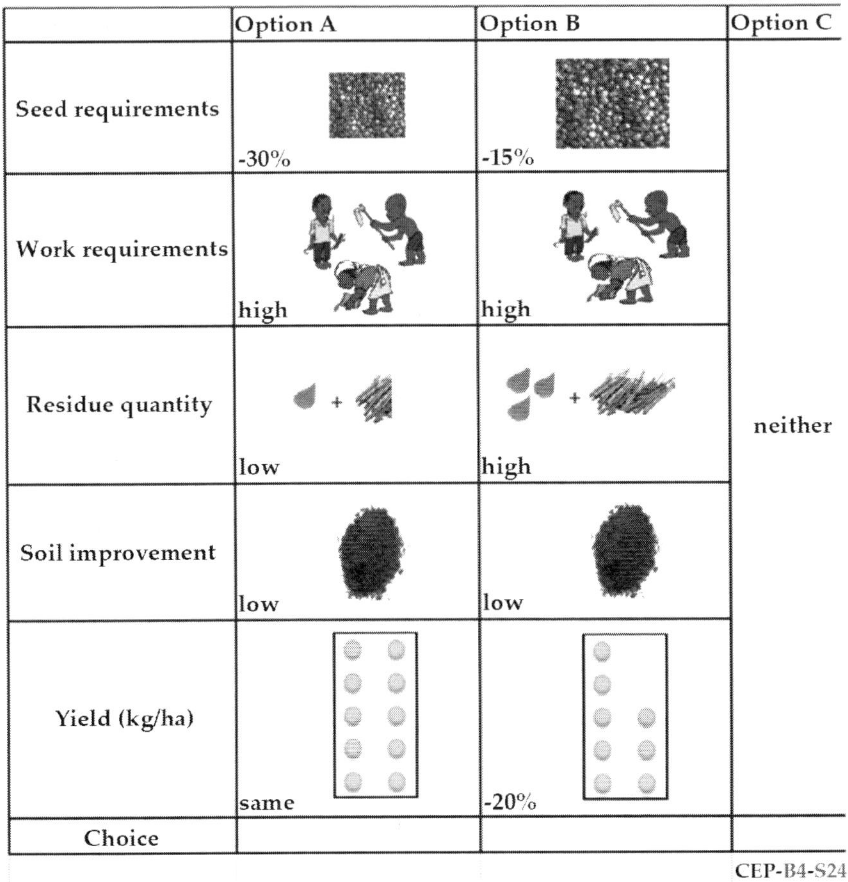

Figure 2: Sample choice set for perennial sorghum experiment.

RESULTS

Descriptive Statistics

We present basic summary statistics for farmers in Table 3 below. The sample is 65% male, with an average age of about 42, and with only 3 years of formal education. The average unit of agricultural production (*Unité de Production Agricole*, or UPA) is approximately 18 people including 8 children, 4 students and one person over 70 years old. A UPA is an agricultural production unit made up of farmers, mostly members of the same family group, whether living or not in the same household (Diallo 2011). The average land holdings per UPA are 9 ha, 42 km from a main road, and 18 km from a market. They receive on average 5.5 extension visits and 86% of the sample belong to an agricultural association. The UPA size is significantly larger in Bougouni than the other locations, and the average landholding is largest in Dioila.

Table 3: Summary statistics of farmers in sample

Variable	Bougoni		Dioila		Tominian		Total	
	Mean	Std.	Mean	Std.	Mean	Std.	Mean	Std.
Gender of respondent (male = 1)	0.62	0.49	0.73	0.45	0.61	0.49	0.65	0.48
Age of respondent (years)	43.86	13.85	41.45	12.24	40.52	13.05	41.92	13.09
Education (years)	3.06	1.42	3.07	1.44	2.97	1.42	3.03	1.42
Number of people in UPA	24.70	17.07	17.72	10.81	13.31	9.18	18.50	13.56
Children under 16 in UPA	10.36	7.56	8.56	6.26	6.24	3.56	8.36	6.22
Students in UPA	5.27	4.14	4.19	3.17	2.85	2.45	4.09	3.45
Older than 70 in UPA	0.76	0.91	1.41	6.77	0.64	0.87	0.93	3.98
UPA landholdings (ha)	9.94	7.42	10.41	6.83	7.43	4.32	9.27	6.46
Personally farmed (ha)	1.74	3.32	1.96	4.07	2.06	3.53	1.93	3.65
Years farming this plot	10.04	10.45	7.45	13.63	11.27	8.85	9.19	11.71
Distance to road (km)	7.57	12.31	87.46	185.85	31.23	116.79	42.34	131.33
Distance to market (km)	28.78	96.44	5.20	2.88	22.35	57.68	18.70	65.23
Extension visits	8.48	8.52	6.00	7.47	2.07	2.59	5.50	7.18
Association membership	0.81	0.67	0.76	0.45	0.49	0.50	0.68	0.56
N	104		106		108		318	

We also compare the differences in income and assets between the various sites. Participants in Bougouni have the highest non-agricultural income of the three villages while Tominian has the least opportunities for off farm income. The majority of income in Dioïla comes from small businessactivities. Participants in Dioïla take more loans than participants in the other villages on average, particularly in Tominian. On average

Bougouni appears to be a wealthier community than Dioïla and Tominian and have more agricultural and non-agricultural assets. Large ruminants (cattle) are more common in Bougouni and least common in Tominian but small ruminants (sheep and goats) and pigs are most common in Tominian.

We asked farmers a series of questions related to sorghum cultivation and use (see Table 4). In our sample, average rainfall and percentage of farmers growing sorghum is inversely related— the lower the rainfall the higher the percentage of farmers producing sorghum. However, the largest mean land area planted to sorghum was in Dioïla (3.69 ha). There was wide variation in reported yield per hectare with the highest figures in the highest rainfall areas but the standard deviations were large and so these figures are not statistically different from one another. The total quantity consumed and sold was higher in the higher rainfall areas but the percent of the harvest consumed was smaller when more sorghum was harvested (86%, 94%, and 95% in Bougoni, Dioïla, and Tominian respectively). There is a relatively high level of intercropping with sorghum (about 50%) with the most in Dioïla and least in Tominian. The vast majority of the sorghum that was intercropped was intercropped with cowpea (89%).

Table 4: Sorghum production and mean descriptive statistics for sorghum production

	HH/UPAs producing sorghum	Area (Ha^{-1})	Quantity harvested (kg/Ha^{-1})	Quantity consumed (kg/Ha^{-1})	Quantity sold (kg/Ha^{-1})	Inter-cropping
Bougouni	59%	1.34	836 (1018)	717 (979)	649 (824)	51%
Dioila	87%	3.69	715 (401)	669 (407)	288 (443)	68%
Tominian	90%	2.29	519 (521)	495 (482)	303 (275)	22%

Note: standard deviations are in parenthesis.

It is common that women in Mali will manage land separate from the main agricultural fields, producing small amounts of cereal crops for household consumption or sale. These cereals are often intercropped with women›s primary crops, such as cowpea or groundnut. Since males head the UPA, when we interviewed them we asked them to report all of the sorghum they managed under the UPA while women were asked only to report the land that they managed personally. In some UPAs, there was sorghum cultivation taking place that was not managed by the women and of which they were not aware, so the figures presented above roughly illustrate the differences between the three sites but may not be precise estimates of total sorghum production. This is a function of the polygamous household structure and a major hurdle to collecting "household" level production data in Mali. Since we are interested in factors that impact individual decisions regarding sorghum production we thought it was more appropriate to ask respondents only about the land that they personally manage rather than interview married husband and wife pairs.

Sixty-three percent of farmers are aware of the concept of perennial sorghum but only 1% of the sample had tried it. Only 17% of the sample has tried "improved" varieties. Eighty-nine percent of the sample use sorghum residue as forage, 60% collect the residue and transport it to the animals, while 40% leave the residue in the field. Almost all of the farmers sampled (92%) source seed from their own production. Anecdotally, women do not have access to draft power when they needed it because the men use it for the main fields. However, in our sample, 80% of women reported having access to draft animals (and 90% of men) with slightly lower reporting not having access to draft animals when they needed it (78% of women and 85% of men).

Preferences for Attributes of Perennial Sorghum

Estimations of preferences for perennial sorghum are presented in Table 5. The first column displays the results of a RPL, column two includes the same estimation with gender interactions, and the third column estimates the same model in WTP-space using sorghum yield as the scale factor. Differences in preferences by gender are estimated by interacting the gender of the respondent with the choice attributes.

Table 5: Random parameters logit model for perennial sorghum

Variable	RPL		RPL w/ gender interactions		WTP-space w/ gender interactions	
	Coefficient	Error	Coefficient	Error	Coefficient	Error
Random parameter means						
Seed requirements	0.037***	0.008	0.036***	0.009	0.512***	0.098
Labor requirements	−0.107	0.096	−0.203*	0.106	−0.732	1.273
Forage	0.314***	0.088	0.288***	0.096	3.798***	1.035
Soil improvement	1.369***	0.139	1.149***	0.114	12.84***	1.034
Yield	0.103***	0.009	0.089***	0.007		
Non-random parameter means						
Opt out dummy	−4.719***	0.395	−4.113***	0.286	−5.203***	0.367

Female * seed requirements			−0.007	0.013	−0.014	0.014
Female * labor requirements			0.304*	0.180	0.175	0.186
Female * forage			−0.013	0.154	−0.115	0.164
Female * soil improvement			0.063	0.162	0.074	0.185
Female * yield			−0.001	0.006	0.000	0.009
Random parameter standard deviations						
Seed requirements	0.081***	0.011	0.070***	0.010	0.593***	0.100
Labor requirements	0.958***	0.172	0.894***	0.155	8.467***	1.327
Forage	0.563***	0.195	0.524***	0.178	4.048*	2.104
Soil improvement	0.658***	0.176	0.216	0.287	4.044**	1.962
Yield	0.033***	0.006	0.026***	0.006		
N	1272		1272		1272	
K	11		16		16	
Log-likelihood	−877.81		−876.84		−857.13	
Pseudo R^2	0.372		0.373		0.387	
AIC	1777.6		1785.7		1746.3	

Note: *** indicates significance at the 1% level, ** indicates significance at the 5% level, and * indicates significance at the 10% level, respectively. The sorghum yield coefficient estimate in WTP-space model is omitted since this is the scale factor.

We use the yield variable to calculate the marginal rate of substitution, and this is framed as a percentage reduction in the level of a respondent's sorghum harvest from the previous growing season (−60%, −40%, −20%, the same). A positive valuation of yield can be understood as how much of a yield reduction a participant would be willing to tradeoff (or accept) to have higher levels of an attribute (e.g., soil improvement). Since seed requirements are also framed as a reduction in level (average, 15% less or 30% less), a positive sign on this variable indicates a positive valuation of a reduction in seed required. Labor requirements, forage biomass, and soil improvement are all binary variables so a positive sign would indicate the difference in valuation from the low to high level. Coefficients in the WTP-space model should be interpreted as tradeoff values relative to other attributes in terms of sorghum yield. Note that the coefficient listed for yield in WTP-space is the associated preference space estimate.

The value of the coefficient on seed requirements is small, about 0.5 CFA (or about $0.005) per 1% of sorghum yield, suggesting that this attribute is a minor concern relative to the other attributes involved in the choice of the sorghum variety. Labor requirements are negative but not significant, indicating that labor requirements are not necessarily an important determinant of the choice between an annual and perennial sorghum variety overall. The random parameter standard deviation on labor is significant and relatively large however, indicating that there is heterogeneity in labor valuation. In the RPL with interactions model the value of labor to women is positive and marginally significant, based on the coefficient of the interaction between the gender and labor variables, indicating that women are less concerned than men about labor savings. This effect however washes out in the WTP model when coefficients are scaled.

As forage increases from low to high there is a positive and significant valuation but the random parameter estimate for this is not significant suggesting that this is not important to all farmers. The tradeoff value for soil improvement indicates that this attribute is roughly three times more important to farmers than forage/biomass production. In the RPL model yield is positive and significant as is the random parameter standard deviation and there is some heterogeneity in yield valuation. The opt out dummy is negative and very large and significant for perennial sorghum, indicating that many respondents prefer traditional sorghum over perennial.

Preferences for Attributes of Hybrid Sorghum

Model results of preferences for hybrid sorghum are presented in Table 6. The same three models are presented: an RPL choice model, an RPL model with gender interactions, and a model with gender interactions estimated in WTP space. One important difference in the way the choice sets are constructed is that the hybrid sorghum yield variable is framed as an increase in yield compared to traditional sorghum whereas perennial was framed as a yield reduction. In the perennial model a positive valuation of yield can be interpreted as how much of a yield reduction a participant would be willing to accept while the interpretation of yield in the hybrid sorghum model is how much the respondent values a percentage increase in yield. Also in the perennial model the seed requirements variable is framed as a percentage of seed required (since less seed is required with perennial but it is not clear exactly how much) whereas in the hybrid model we specifically capture the seeding rate difference using experimental data (hybrid sorghum has a higher germination rate and so only one half the amount of seed is required).

Table 6: Random parameters logit model for hybrid sorghum

Variable	RPL		RPL w/ gender interactions		WTP-space w/ gender interactions	
	Coefficient	Error	Coefficient	Error	Coefficient	Error
Parameter means						
Seed requirements	−0.007	0.034	−0.031	0.044	−0.809	0.705
Plant duration	−0.437***	0.064	−0.445***	0.080	−6.133***	1.083
Ability to reuse	2.059***	0.259	2.321***	0.309	33.300***	4.171
Cost of seed	0.000	0.000	0.000	−0.280	−0.001	0.005
Yield	0.071***	0.007	0.073***	0.008		
Non-random parameter means						
Opt out dummy	1.073	0.872	1.139	0.872	3.898***	0.610
Female * seed requirements			0.063	0.067	0.091	0.073
Female * plant duration			0.004	0.121	−0.054	0.137
Female * ability to reuse			−0.758**	0.345	−0.354	0.435
Female * cost of seed			−0.001	0.001	−0.001	0.001
Female * yield			−0.004	0.008	−0.005	0.006
Random parameter standard deviations						
Seed requirements	0.134*	0.076	0.132*	0.079	1.468	1.710
Plant duration	0.017	0.133	0.025	0.133	0.463	15.248
Ability to reuse	1.845***	0.274	1.829***	0.282	30.929***	3.934
Cost of seed	0.002***	0.000	0.001***	0.000	0.033***	0.006
Yield	0.041***	0.006	0.041***	0.006		
N	1272		1272		1272	
No. of parameters	11		16		16	
Log-likelihood	−744.984		−740.203		−765.51482	
Pseudo R²	0.466		0.470		0.451769	
AIC	1512		1512.4		1563	

Note: *** indicates significance at the 1% level, ** indicates significance at the 5% level, and * indicates significance at the 10% level, respectively. The sorghum yield coefficient estimate in WTP-space model is omitted since this is the scale factor.

The respondent is presented with a choice that is not explicitly hybrid versus non-hybrid sorghum but rather two choices that differ in attributes that characterize the main differences between hybridized and non-hybridized sorghum crops. Seed requirements are not an important component of the tradeoff to farmers and there is little variation in this preference. Sorghum maturity is important, indicated by the significant tradeoff value and there is a significant and homogenous preference for shorter duration sorghum. The ability to reuse the seed is the most important attribute to participants in the choice, approximately 6 times more important than earlier maturity. On average women put slightly less emphasis on the ability to reuse seed than their male counterparts (seen in the RPL model), but there is significant variation within this group and this effect washes out in the WTP-space model.

Sorghum yield is positive, relatively large and also significant indicating the importance of yield for hybrid seeds. However, there is heterogeneity in the valuation of sorghum yield. The coefficient on the opt out dummy in the WTP-space model is positive and significant, indicating an overall preference for hybrids sorghum over traditional sorghum.

DISCUSSION

Perennial and hybrid crops present fundamentally different technological options to farmers and there is evidence of demand for attributes of each technology and evidence that they might fit into men and women›s cropping systems differently. The two statistics that are the most prominent from these estimations is that the most important attribute of perennial sorghum is soil improvement and the most important attribute of hybrid sorghum is whether the seeds can be reused. We address each of these attributes in more detail.

Farmers' main interest appears to be improvement in soil quality from the perennial plant, which is consistent with pigeon pea farmers in Malawi (Waldman et al. 2017). Soil improvement is the most important determinant of farmer choice, approximately three to four times as important as biomass production. The benefits of perenniality in terms of reduced seed and labor requirements appear to be negligible to farmers. While previous research has found that perennial crops improve soil (Culman et al. 2013) and soil structure(Kell 2011), there is no evidence of direct soil fertility improvement from perennials since the improvement happens over such a long time horizon. Developing perennial sorghum varieties that are primarily soil enhancing with more extensive root systems and higher above ground biomass is an area where further breeding efforts could increase adoption by farmers. Efforts to improve soil fertility should aim to keep yields within relatively similar to annual sorghum yields in subsequent years of production to remain appealing to farmers. However, since labor and seed requirements are not as important to farmers and distinguishing attributes of a perennial crop there will likely be extensive behavioral barriers to adoption of perennial sorghum in Mali.

Hybrid crops will likely face resistance because of the long tradition of saving and reusing seeds among traditional cultures in Mali and elsewhere. The traditional practice of hand selecting varieties with desirable attributes and replanting them in subsequent seasons (and seed recycling) is ingrained in farmers. Hand selecting crops over time allows them to adapt to local variation in climate and to localized pest and disease outbreaks. Manual seed selection and reusing seed gives farmers control over their own seed supply and buffers them fluctuating seed prices. There is also a great deal of misinformation associated with hybrid crops, and associations with genetic modification of crops that carry a negative stigma in rural parts of Mali (as well as across Africa). The process of hybridization is thus a barrier to the proliferation of the technology. However, because of the yield advantages from hybrids some farmers are likely to incorporate hybrid sorghum varieties into their existing cropping systems alongside traditional or even perennial varieties.

The duration of the plant is important to farmers and there is a strong and homogenous preference for shorter duration crops. This appears to extend to perennial sorghum which had high and significant opt out values among farmers. Seed requirements and labor requirements are the core of the differences between an annual and a sorghum crop and both of these attributes were valued very little by farmers. This suggests that unless perennial crops can make significant ecosystem service contributions, particularly in terms of soil improvements they are not likely to be adapted simply by the different management requirements of the cropping system.

Anecdotally, sorghum is not traditionally the woman's domain in Mali, although some women grow small plots of sorghum for income generation. Women might not have much decision-making power over the land or the inputs used and so increasing forage and soil through perennial crops could be more appealing to female farmers. Given the prevalence of polygamous households in Mali, women generally have less access to cash and assets and often farm smaller plots of land with fewer resources. For some women, labor may not be a limiting constraint, likely because the plots of land they are farming are so small that they can personally supply enough labor. However, for women farmers with a labor shortage a perennial crop might be an advantage to them since fewer plowings are required. Regardless there is heterogeneous demand for labor savings, so a perennial crop is potentially an important technology for resource-poor women and may provide an important incentive for adoption of sorghum.

CONCLUSIONS

Attributes of both perennial and hybrid varieties are positively valued by farmers and the technologies could be considered as complementary and part of a risk diversification strategy for smallholder farmers in Africa. There are constraints to widespread adoption of either technology, which highlights the importance of understanding the tradeoffs between agricultural ecosystem services and climate change adaptation. These findings have important policy implications for adaptation to climate change. This is particularly true of perennial grain crops that have not been well developed in the region or on the continent. Hybrids also face numerous behavioral barriers and in particular the hybridization process runs contrary to cultural drivers of seed choice, specifically the prevalence of seed reuse or recycling. To be consistent with farmers' preferences, breeding efforts in Africa should continue to focus on crop diversity and diversification into crops other than maize to provide risk averse farmers with more options which fit their agro climatic conditions and individual farming systems.

Breeders should be mindful of the two major findings of these experiments. The first is that perennial crops have little appeal to farmers unless they can provide substantial ecological benefits. The ability to grow in drier climates is a potentially very powerful aspect of perennial crops in a future characterized by rising temperatures and greater variability in precipitation. Increased forage biomass and improved soil fertility from perennial crops is promising to farmers, but only at a relatively small yield reduction. The second major finding is that acceptability of hybrid crops is largely hampered by the ability to recycle the seeds. Farmers value shorter duration crops, but this is far outweighed by their aversion to seeds they cannot recycle.

Developing perennial lines of sorghum that can survive the dry seasons in Mali could be a very important technology for climate change adaptation. Enhancing and documenting the ecosystem services and other ancillary benefits provided by perennial crops will be key to farmer adoption. Early maturing hybrids are also a promising technology to reduce exposure to climate risk and adapt to changing climatic conditions. Working with farmers to produce hybrid seed at the local level is one potential way to both create a local market for hybrids and to enable farmers to create a sustainable and appealing seed system that involves hybrids.

ACKNOWLEDGEMENTS

We would like to thank Alpha Kergna from Institut d'Economie Rurale who helped organize and lead the fieldwork for this project. The research design benefited from discussions with Dr. Eva Weltzein at the International Crops Research Institute for the Semi-Arid Tropics (ICRISAT) in Bamako and Dr. Sieglinde Snapp at Michigan State University. We extend thanks to Dr. Krista Isaacs at ICRISAT for helpful comments on an early draft. We also would like to express our gratitude to the Malian farmers who graciously welcomed us and were willing to speak with us.

REFERENCES

1. Altieri, M.A., 2002. Agroecology: the science of natural resource management for poor farmers in marginal environments. Agric. Ecosyst. Environ. 93, 1–24. http://dx.doi. org/10.1016/S0167-8809(02)00085-3.
2. Asrat, S., Yesuf, M., Carlsson, F., Wale, E., 2010. Farmers' preferences for crop variety traits: lessons for on-farm conservation and technology adoption. Ecol. Econ. 69 (12), 2394–2401. http://dx.doi.org/10.1016/j.ecolecon.2010.07.006.
3. Bazile, D., Dembélé, S., Soumaré, M., Dembélé, D., 2008. Utilisation de la diversité varietale du sorgho pour valoriser la diversité des sols du Mali. Cah. Agric. 17, 86–94. http://agritrop.cirad.fr/544653/1/document_544653.pdf.
4. Bell, L.W., Byrne, F., Ewing, M.A., Wade, L.J., 2008. A preliminary whole-farm economic analysis of perennial wheat in an Australian dryland farming system. Agric. Syst. 96 (1–3), 166–174. http://dx.doi.org/10.1016/j.agsy.2007.07.007.
5. Bennett, J., Birol, E., 2010. Choice Experiments in Developing Countries: Implementation, Challenges and Policy Implications. Edward Elgar, Northampton, MA.
6. Birol, E., Smale, M., Yorobe Jr., J.M., 2012. Bi-modal preferences for Bt maize in the Philippines: a latent class model. AgBioforum 15 (2), 175–190. http://www. agbioforum.org/v15n2/v15n2a06-birol.htm.
7. Birol, E., Villalba, E.R., Smale, M., 2009. Farmer preferences for milpa diversity and genetically modified maize in Mexico: a latent class approach. Environ. Dev. Econ. 14, 521. http://dx.doi.org/10.1017/S1355770X08004944.
8. Cox, T.S., Glover, J.D., Tassel, D.L.V., Cox, C.M., DeHaan, L.R., 2006. Prospects for developing perennial grain crops. Bioscience 56, 649–659. http://dx.doi.org/10.1641/ 0006-3568(2006)56[649:PFDPGC]2.0.CO;2.
9. Culman, S.W., Snapp, S.S., Ollenburger, M., Basso, B., DeHaan, L.R., 2013. Soil and water quality rapidly responds to the perennial grain kernza wheatgrass. Agron. J. 105, 735–744. http://dx.doi.org/10.2134/agronj2012.0273.
10. Smale, M., Kergna, A., Diakité, L., 2016. Impact Assessment Report No. 2. An Economic Assessment of Sorghum Improvement in Mali.

International Crops Research Institute for the Semi-Arid Tropics, Patancheru 502 324, Telangana, India (52 pp).

11. Diallo, A., 2011. An Analysis of the Recent Evolution of Mali's Maize Subsector (Master's Thesis, Michigan State University, Department of Agricultural, Food, and Resource Economics). Michigan State University, East Lansing, MI.
12. Eicher, C.K., 1995. Zimbabwe's maize-based green revolution: preconditions for replication. World Dev. 23, 805–818. http://dx.doi.org/10.1016/0305-750X(95)93983-R.
13. Folkertsma, R.T., Rattunde, H.F.W., Chandra, S., Raju, G.S., Hash, C.T., 2005. The pattern of genetic diversity of Guinea-race Sorghum bicolor (L.) Moench landraces as revealed with SSR markers. Theor. Appl. Genet. 111, 399–409. http://dx.doi.org/10.1007/ s00122-005-1949-0.
14. Glover, J.D., Reganold, J.P., Bell, L.W., Borevitz, J., Brummer, E.C., Buckler, E.S., Cox, C.M., Cox, T.S., Crews, T.E., Culman, S.W., DeHaan, L.R., Eriksson, D., Gill, B.S., Holland, J., Hu, F., Hulke, B.S., Ibrahim, A.M.H., Jackson, W., Jones, S.S., Murray, S.C., Paterson, A.H., Ploschuk, E., Sacks, E.J., Snapp, S., Tao, D., Van Tassel, D.L., Wade, L.J., Wysem, D.L., Xu, Y., 2010. Increased food and ecosystem security via perennial grains. Science 328, 1638–1639. http://dx.doi.org/10.1126/science. 1188761.
15. Glover, J.D., Reganold, J.P., Cox, C.M., 2012. Agriculture: plant perennials to save Africa's soils. Nature 489, 359–361. http://dx.doi.org/10.1038/489359a.
16. Hall, C., Dawson, T.P., Macdiarmid, J.I., Matthews, R.B., Smith, P., 2017. The impact of population growth and climate change on food security in Africa: looking ahead to 2050. Int. J. Agric. Sustain. 15, 124–135. http://dx.doi.org/10.1080/14735903. 2017.1293929.
17. Haussmann, B.I.G., Rattunde, H.F., Weltzien-Rattunde, E., Traoré, P.S.C., vom Brocke, K., Parzies, H.K., 2012. Breeding strategies for adaptation of pearl millet and sorghum to climate variability and change in West Africa. J. Agron. Crop Sci. 198, 327–339. http://dx.doi.org/10.1111/j.1439-037X.2012.00526.x.
18. Hayes, R.C., Newell, M.T., DeHaan, L.R., Murphy, K.M., Crane, S., Norton, M.R., 2012. Perennial cereal crops: an initial assessment of wheat derivatives. Field Crop Res. 133, 68–89. http://dx.doi.org/10.1016/j.fcr.2012.03.014.

19. Institute for the Semi-Arid Tropics (ICRISAT), 2015. Explore it at ICRISAT. Institute for the Semi-Arid Tropics, Sorghum. Bamako, Mali Retrieved from. http://exploreit. icrisat.org/profile/Sorghum/193.

20. Jaikumar, N.S., Snapp, S.S., Murphy, K., Jones, S.S., 2012. Agronomic assessment of perennial wheat and perennial rye as cereal crops. Agron. J. 104, 1716–1726. http:// dx.doi.org/10.2134/agronj2012.0291.

21. Kane, D.A., Rogé, P., Snapp, S.S., 2016. A systematic review of perennial staple crops literature using topic modeling and bibliometric analysis. PLoS One 11, e0155788. http://dx.doi.org/10.1371/journal.pone.0155788.

22. Kell, D.B., 2011. Breeding crop plants with deep roots: their role in sustainable carbon, nutrient and water sequestration. Ann. Bot. 108, 407–418. http://dx.doi.org/10. 1093/aob/mcr175.

23. Kelly, V., Murekezi, A., Me-Nsope, N., Perakis, S., Mather, D., 2012. Cereal market dynamics: the Malian experience from the 1990s to present. In: MSU International Development Working Paper 128. Michigan State University, East Lansing, Michigan Retrieved from. http://ageconsearch.umn.edu/record/146935/files/idwp128.pdf.

24. Kelly, V., Diakité, L., Teme, B., 2015. Sorghum Productivity in Mali: Past, Present, and Future.

25. Lancaster, K., 1966. A new approach to consumer theory. J. Polit. Econ. 74, 132–157.

26. Lobell, D.B., Burke, M.B., Tebaldi, C., Mastrandrea, M.D., Falcon, W.P., Naylor, R.L., 2008. Prioritizing climate change adaptation needs for food security in 2030. Science 319, 607–610. http://dx.doi.org/10.1126/science.1152339.

27. Mason, S.C., Ouattara, K., Taonda, S.J.-B., Palé, S., Sohoro, A., Kaboré, D., 2015. Soil and cropping system research in semi-arid West Africa as related to the potential for conservation agriculture. Int. J. Agric. Sustain. 13, 120–134. http://dx.doi.org/10. 1080/14735903.2014.945319.

28. Morton, J.F., 2007. The impact of climate change on smallholder and subsistence agriculture. Proc. Natl. Acad. Sci. 104, 19680–19685. http://dx.doi.org/10.1073/pnas. 0701855104.

29. Ortega, D.L., Wang, H.H., Olynk Widmar, N.J., Wu, L., 2014. Reprint of: "Chinese producer behavior: aquaculture farmers in southern China". China Econ. Rev. 30, 540–547. http://dx.doi.org/10.1016/j.chieco.2014.05.007.

30. Ortega, D.L., Waldman, K.B., Richardson, R.B., Clay, D.C., Snapp, S., 2016. Sustainable intensification and farmer preferences for crop system attributes: evidence from Malawi's central and southern regions. World Dev. 87, 139–151. http://dx.doi.org/ 10.1016/j.worlddev.2016.06.007.
31. Pienaar, E.F., Jarvis, L.S., Larson, D.M., 2014. Using a choice experiment framework to value conservation-contingent development programs: an application to Botswana. Ecol. Econ. 98, 39–48. http://dx.doi.org/10.1016/j.ecolecon.2013.12.015.
32. Pimentel, D., Cerasale, D., Stanley, R.C., Perlman, R., Newman, E.M., Brent, L.C., Mullan, A., Chang, D.T.-I., 2012. Annual vs. perennial grain production. Agric. Ecosyst. Environ. 161, 1–9. http://dx.doi.org/10.1016/j.agee.2012.05.025.
33. Rattunde, H.F.W., Weltzien, E., Diallo, B., Diallo, A.G., Sidibe, M., Touré, A.O., Rathore, A., Das, R.R., Leiser, W.L., Touré, A., 2013. Yield of photoperiod-sensitive sorghum hybrids based on Guinea-race germplasm under farmers' field conditions in Mali. Crop Sci. 53 (6), 2454. http://dx.doi.org/10.2135/cropsci2013.03.0182.
34. Richardson, R.B., Kellon, D., Leon, R., Arvai, J.L., 2013. Using choice experiments to understand household tradeoffs regarding pineapple production and environmental management in Costa Rica. J. Environ. Manag. 127, 308–316. http://dx.doi.org/10. 1016/j.jenvman.2013.05.043.
35. Roessler, R., Drucker, A.G., Scarpa, R., Markemann, A., Lemke, U., Thuy, L.T., Valle Zárate, A., 2008. Using choice experiments to assess smallholder farmers' preferences for pig breeding traits in different production systems in North-West Vietnam. Ecol. Econ. 66, 184–192. http://dx.doi.org/10.1016/j.ecolecon.2007.08.023.
36. Sánchez-Toledano, B.I., Kallas, Z., Gil, J.M., 2017a. The Importance of Farmers' Social, Environmental and Economic Objectives in Improved Corn Seeds Adoption in Chiapas, México. Revista de la Facultad de Ciencias Agrarias, pp. 20.
37. Sánchez-Toledano, B.I., Kallas, Z., Gil-Roig, J.M., 2017b. Farmer preference for improved corn seeds in Chiapas, Mexico: a choice experiment approach. Span. J. Agric. Res. 15, e0116.
38. Scarpa, R., Ruto, E.S.K., Kristjanson, P., Radeny, M., Drucker, A.G., Rege, J.E.O., 2003. Valuing indigenous cattle breeds in Kenya: an empirical comparison of stated and revealed preference value estimates. Ecol. Econ. 45 (3), 409–426. http://dx.doi.org/ 10.1016/

S0921-8009(03)00094-6.

39. Scarpa, R., Thiene, M., Train, K., 2008. Utility in willingness to pay space: a tool to address confounding random scale effects in destination choice to the alps. Am. J. Agric. Econ. 90, 994–1010. http://dx.doi.org/10.1111/j.1467-8276.2008.01155.x.
40. Train, K., 2009. Discrete Choice Methods With Simulation. Cambridge University Press, New York.
41. Useche, P., Barham, B.L., Foltz, J.D., 2009. Integrating technology traits and producer heterogeneity: a mixed-multinomial model of genetically modified corn adoption. Am. J. Agric. Econ. 91, 444–461. http://dx.doi.org/10.1111/j.1467-8276.2008. 01236.x.
42. Useche, P., Barham, B.L., Foltz, J.D., 2013. Trait-based adoption models using ex-ante and ex-post approaches. Am. J. Agric. Econ. 95, 332–338. http://dx.doi.org/10.1093/ ajae/aas044.
43. Van Mele, P., Chien, H.V., 2004. Farmers, biodiversity and plant protection: developing a learning environment for sustainable tree cropping systems. Int. J. Agric. Sustain. 2, 67–76. http://dx.doi.org/10.1080/14735903.2004.9684568.
44. Waldman, K.B., Ortega, D.L., Richardson, R.B., Snapp, S.S., 2017. Estimating demand for perennial pigeon pea in Malawi using choice experiments. Ecol. Econ. 131, 222–230. http://dx.doi.org/10.1016/j.ecolecon.2016.09.006.
45. Yapi, A.M., Kergna, A.O., Debrah, S.K., Sidibe, A., Sanogo, O., 2000. Analysis of the Economic Impact of Sorghum and Millet Research in Mali. ICRISAT Impact Series No. 8 International Crops Research Institute for the Semi-Arid Tropics, Andhra Pradesh, India.

CHAPTER 13

DO GREENHOUSE GAS EMISSION CALCULATIONS FROM ENERGY CROP CULTIVATION REFLECT ACTUAL AGRICULTURAL MANAGEMENT PRACTICES? – A REVIEW OF CARBON FOOTPRINT CALCULATORS

Christiane Peter [a,b] **, Katharina Helming** [c,d] **, Claas Nendel** [a]

[a] Leibniz Centre for Agricultural Landscape Research, Institute of Landscape Systems Analysis, Eberswalder Straße 84, 15374 Müncheberg, Germany

[b] Justus Liebig University Gießen, Institute of Farm and Agribusiness Management, Senckenbergstraße 3, 35390 Gießen, Germany

[c] Leibniz Centre for Agricultural Landscape Research, Institute of Land Use Systems, Eberswalder Straße 84, 15374 Müncheberg, Germany

[d] Eberswalde University for Sustainable Development, Faculty of Landscape Management and Nature Conservation, Schicklerstraße 5, 16225 Eberswalde, Germany

ABSTRACT

A wide range of calculators have been developed to assess the greenhouse gas(GHG) emissions of agricultural products, including biomass for

Citation: Peter C, Helming K, Nendel C. Do greenhouse gas emission calculations from energy crop cultivation reflect actual agricultural management practices? – A review of carbon footprint calculators. Renewable and Sustainable Energy Reviews 2017; 67: 461-476.

bioenergyproduction. However, these calculators often fail in their ability to take into account the differences in pedoclimatic conditions, agricultural management practices and characteristics of perennial crops and crop rotations. As a result, the predictions of GHG emissions by these calculators are characterized by a high level of uncertainty, and calculators may fail in their ability to detect mitigation options along the production chain. The aim of this study was to analyze the available calculators for calculating GHG emissions from energy crop cultivation based on Carbon Footprint (CFP) approaches according to the goal and scope of the calculator, the methodology used to account for GHG emissions from energy crop cultivation, energy crop cultivation management practices and the ability to model crop rotation. Out of 44 environmental assessment calculators for agricultural products, we identified 18 calculators which are capable of assessing GHG emissions from energy crop cultivation. These calculators differ in their goal and scope and which farming operations related to crop management are taken into account; this makes it difficult to compare and interpret the results from these CFP assessments. Only seven calculators out of 18 can calculate GHG emissions from energy crop rotations. At the moment, none of these calculators are able to consider actual effects from energy crops in rotation in the context of nutrient shifts, reductions in the use of agricultural operating needs, or the sequence and composition of crop rotations. However, by expanding the system boundaries of the CFP study, by taking the whole energy crop rotation and local agricultural management practices into account, the opportunity to identify more GHG mitigation options increases.

Keywords: Crop rotation, Carbon Footprint, Crop management, Energy crops, Life Cycle Assessment, Greenhouse gas emissions

INTRODUCTION

Human influence on climate change was again confirmed by the latest report from the Intergovernmental Panel on Climate Change (IPCC) [1]. Anthropogenic greenhouse gas (GHG) emissions from fossil fuel combustionand industrial processes contributed about 78% to the total increase of GHGs in the atmosphere over the last 40 years [1]. Furthermore, the Agriculture, Forestry and Other Land Use sector (AFOLU) accounted for about a quarter of anthropogenic GHG emissions [1]. In response to this, a growing number of governments have begun introducing renewable energy policies in an effort to reduce GHG emissions by replacing non-renewable fossil fuels with renewable energy sources. The European Commission has committed itself to increase the proportion of renewable energy to 20% of the overall share of the energy consumption and to 10% of transportation-related energy consumption by 2020 [2]. In 2008, 12.9% of the total global primary energysupply had already originated from renewable energy sources, of which bioenergy contributed the dominant share (80%) [3]. This implies that the production and use of biomass to generate power, heat and fuel has significantly increased in recent years [4].

Biomass for the supply of energy is traditionally obtained from fuelwood. However, in the last decade, the use of crop residues and dedicated energy crops delivering the demanded biomass increased. Energy crops are agricultural crops solely cultivated for energy-related use. Several food crops (e.g. maize or sugar beet) can also be grown as energy crops if they have high yields and, preferably, a low demand for agrochemical inputs [5].

Energy generation from energy crops has an almost-closed CO_2 cycle (in which the combustion of biomass releases the same amount of CO_2 as was captured by the crop during growth). However, it is not carbon neutral over its whole production chain, since GHG emission occurs during the production stage, e.g. through production of fertilizer, pesticides, farming machinery or fuel combustion from machinery used [5]. Agricultural management practices have a considerable effect on the amount of GHG emissions from energy crop production and, correspondingly, on the entire biomass energy production chain [6]. Consequently, agriculture, including energy crop cultivation, holds significant potential for reducing GHG emissions [7].

However, appropriate assessment tools are required to identify the GHG emission benefit of bioenergy compared to its fossil alternatives. The most widely used approach is the Life Cycle Assessment (LCA) defined by ISO Standards 14040 [8] and 14044 [9], [10].

LCA is defined as a method for compiling and evaluating all inputs, outputs and the potential environmental impact of a production system throughout its life cycle. It enables the user to measure and quantify the environmental impacts of a product. Furthermore, it helps to identify hot spots where the most significant impacts occur, giving the user the opportunity to develop strategies for improving the product's environmental performance [8].

In addition to the LCA guidelines, the Carbon Footprint (CFP) defined by ISO Standard 14067 [11] provides requirements and guidelines for the quantification and communication of GHG emissions in a production chain. The CFP is a specific method within the LCA approach and summarizes all GHG emissions and removals occurring within the established product system boundaries, expressed as CO_2 equivalents. There are a considerable number of tools working with the CFP approach for calculating the GHG emissions from agricultural products [12], [13]. An overview of currently available tools for quantifying GHG emissions at landscape scale from AFOLU was provided by Denef et al. [13]. They divided those tools into three categories: (1) calculators, (2) protocols and guidelines, and (3) process-based models. Based on these results a review of these tools was conducted by Colomb et al. [14,12] to evaluate the methodological differences between these tools, to promote transparency and to provide guidance for the user to choose the most appropriate tool. As distinct from Colomb et al. [14], our review focuses only on calculators, including web-based and software-based calculation tools, which are able to quantify GHG emissions from energy crop cultivation at farm scale. For this subset we provide an extended analysis of the complex crop cultivation system, including an evaluation of the calculators for their ability to take energy crop production specific characteristics, crop rotation effects and farm specific management practices into account. CFP calculators are used by farmers, agricultural suppliers and scientists to identify the potential for GHG mitigation in their local agricultural production chains [15]. In order to be able to detect these GHG emission mitigation potentials, however, calculators should account for local agricultural management practices on the farm and especially for energy crop specifications by taking into account differences in pedoclimatic conditions, farming practices, farming technologies [16], the characteristics of perennial crops [17], and crop rotations (sequence and composition of crops) [18]. Diversification of crop rotation patterns is one option for GHG emission reduction in cropping systems [19], but CFP studies from crop cultivation typically only take into account one vegetation period of one single crop [18]. Accordingly, as agriculture systems are highly complex, not all underlying material flows can be quantified when the assessment is limited to such a short time period. As result, calculation systems leave out crop rotation effects, including all interactions between the previous crop and the assessed crop, such as nutrient shifts, reduction in the use of agricultural operating needs, different intensity and the timing of farming activities [18]. Furthermore, CFP studies frequently fail to adequately consider the specifics of energy crop cultivation, such as differences in the timing of sowing and harvesting dates, the allocation of byproducts (e.g. the production of digestate and its reuse as fertilizer), and cultivation management (e.g. increased fuel use for the whole plant harvest, tillage frequency, and fertilizer quantities) [5], [20].

There are various case studies that use the CFP approach to assess the GHG emissions of biomass energy production. Cherubini and Strømman [20]presented an overview of these case studies and an assessment of the key methodological issues. They pointed out that there are wide ranges and uncertainties in bioenergy CFP case studies due to differences in methodological assumptions (e.g. different reference systems, the database used, functional units, and allocation procedures) and the many variables involved in this calculation (e.g. selection of system boundaries, including land use change and accounting for field emissions from different fertilizer types and crop residues). Furthermore, some of these key parameters regarding agricultural processes are still not well understood and depend heavily on local and climate conditions [21].

The aim of this paper is to review currently available calculators for their ability to quantify GHG emissions from energy crop cultivation by taking into account the specific features of energy crop production and local management practices (as explained above). Following Buytaert et al. [22], who note that LCA is the most suitable assessment tool to assess emissions from bioenergy production systems, we focused our review on calculators that are based on the specifications of the LCA approach for GHG emission assessment, the CFP. Additionally, for CFP studies focusing on agricultural processes, the system boundary can be restricted to "cradle to farm gate" instead of "cradle to grave" to avoid complications of a full CFP study [23]. Following the recommendation of Audsley et al. [23], we set the system boundaries of our study at the farm gate ending with crop harvest, but including byproducts such as organic fertilizers (e.g. digestate, manure, slurry). Our analysis of the calculators is based on four criteria: (1) the goal and scope of the calculator, (2) the methodology used to account for GHG emissions from energy crop cultivation, (3) energy crop cultivation management and (4) the ability to model crop rotation.

MATERIALS AND METHODS

The review process was performed in three steps: first, we identified calculators which account for GHG emissions of agricultural products. Out of these, in the second step, we identified the calculators which could account for GHG emissions from energy crop cultivation. In the third step, we analyzed and compared the resulting calculators regarding to the four criteria described in Table 2.

Material: Identifying GHG Calculators for Energy Crop Cultivation

Our search for calculators was carried out in English and German and covered only published information, which includes peer-reviewed literature, reports, calculator descriptions and websites. A systematic database search of peer-reviewed articles was conducted using the electronic Web of Science. All analyses were conducted between January and November 2014. The following thematic search terms were used: energy crops AND review, LCA AND agriculture, environmental impact AND bioenergy, Carbon Footprint AND energy crops, LCA AND modeling AND bioenergy. The composite terms were placed inside quotation marks, and an asterisk was used at the end of each term to capture all possible extensions and variations of a particular word. The documents were considered relevant if they matched at least one of the topical search terms in their titles, abstracts or keywords and were published in the last 25 years. After identifying the relevant papers, we used references and citations from these papers to search for cited reports, websites and models. In two cases we consulted the software developer directly for further information. Calculators that were developed exclusively for internal use by companies, consultancies or scientists for a very specific product were not included in our review. In the end, we identified 44 environmental assessmentcalculators for agricultural products. An overview of these 44 calculators, including their specific properties (user interface, method, GHG indicator, availability, target user group and literature source) is provided in Table 1.

Fig. 1 depicts our selection process. Methodologies for governmental certifications of bioenergy sustainability often use CFP methodology [24], [25]and the guidelines of the IPCC for AFOLU [26] to assess GHG emissions from biomass production [24], [25]. Therefore, we selected all calculators out of the 44 earlier identified calculators which were able to calculate GHG emissions from crop cultivation (focusing on all processes occurring from "cradle to farm gate") following this methodology and these guidelines. The result is that all GHG emissions occurring during the production process are aggregated into one single impact category of "climate change" by using the category indicator the Global Warming Potential (GWP) or CML 2001 [7], [27]. Calculators following other environmental assessment methods as described by Payraudeau and van der Werf [28] were excluded, as well as calculators designed to model detailed soil-plant-atmosphere processes on farms. Adopting these criteria in our selecting process, we identified 34 calculators matching these criteria.

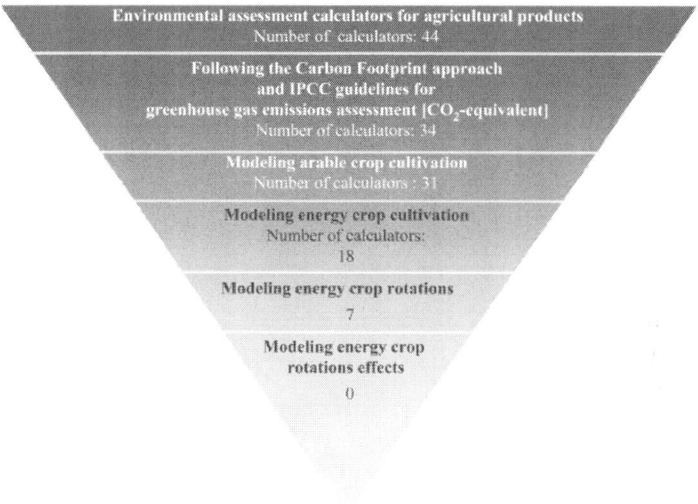

Figure 1: Selection process for GHG emission calculators for energy crop cultivation.

From the remaining 34, we selected all calculators that were able to model GHG emissions from arable crop cultivation. Calculators modeling only horticultural crops and calculators working only with fixed datasets for crop cultivation, without the possibility to modify the inputs and outputs or cultivation processes, were excluded. This resulted in 31 suitable calculators.

Finally, from the remaining 31 we selected all calculators that were capable of calculating GHG emissions from energy crop cultivation. Since several arable crops for food, feed or fiber production can also be grown as energy crops, our analysis included calculators designed for GHG emission assessment from arable crop cultivation; while these are not exclusively designed for energy crops, they nevertheless are capable of assessing GHG emissions from energy crop cultivation as well. Calculators modeling crops without any specification of crop type or only with broad crop categories (e.g. general cropping system, rice fields) without a category for energy crops were excluded. Eighteen calculators were identified that fulfill the requirements (Table 1).

Table 1: Overview of the 18 selected calculators for GHG emissions from energy crop cultivation

Name	Full title	Developed by	Reference
Agri-LCI models	Agricultural Life Cycle Inventory models	Cranfield University, UK	http://www.cranfield.ac.uk/about/people-and-resources/schools-and-departments/school-of-applied-sciences/groups-institutes-and-centres/cwsi-software/CWSI-AgriLCA-download.html? ref=161050, accessed: 25.09.2014
BioGrace	Biofuel Greenhouse Gas Emissions in Europe	Agency NL, IFEU, BIO IS	http://www.biograce.net/content/ghgcalculationtools/recognisedtool, accessed: 29.10.2014
CAPRI	Common Agricultural Policy Regionalized Impact analysis	University of Bonn	http://www.capri-model.org, accessed:11.11.2014
CFF	Farm Carbon Calculator	Farm Carbon Cutting Toolkit	http://www.cffcarboncalculator.org.uk, accessed: 29.10.2014
COMET FARM	Whole Farm And Ranch Carbon And Greenhouse Gas Accounting System.	United States Department of Agriculture, Colorado State University	http://www.cometvr.colostate.edu, accessed: 26.09.2014
CFT	Cool Farm Tool	Cool Farm Alliance	http://www.coolfarmtool.org/CftExcel, accessed: 15.01.2015;[54]
C-Plan	Carbon Footprint Calculator	CPLAN 2014 (Scottish Farmer)	http://www2.cplan.org.uk, accessed: 12.11.2014
FarmGAS	FarmGAS Calculator And Financial Tool	Australian Farm Institute	http://calculator.farminstitute.org.au, accessed: 12.11.2014
FSGGEC	Farm System Greenhouse Gas Emissions Calculator	Michigan State University	http://surf.kbs.msu.edu, accessed: 13.11.2014
GaBi	Product Sustainability Solution	PE International	http://www.gabi-software.com, accessed: 22.01.2014

GEMIS	Global Emission Model Integrated Systems	International Institute for Sustainability Analysis and Strategy	http://www.iinas.org/news-de.html, accessed: 21.02.2014
HGCA 1	Biofuel Greenhouse Gas Calculator	Agriculture and Horticulture Development Board	http://www.hgca.com/tools/biofuel-greenhouse-gas-calculator.aspx, accessed: 14.08.2014
HGCA 2	Carbon Footprinting Decision Support Tool	Agriculture and Horticulture Development Board	http://www.hgca.com/tools/carbon-footprinting-decision-support-tool.aspx, accessed: 14.08.2014
IFSC	Illinois Farm Sustainability Calculator	University of Illinois	https://ideals.illinois.edu/handle/2142/13458, accessed: 15.08.2014
openLCA	Open Source LCA Software	GreenDelta	http://www.openlca.org, accessed: 14.09.2014
SALCA	Swiss Agricultural Life Cycle Assessment	AGROSCOPE	http://www.agroscope.admin.ch/oekobilanzen/01199/index.html?lang=de, accessed: 22.01.2014
SimaPro	Sustainable Performances Of Products And Services	Pré	http://www.pre-sustainability.com/software, accessed: 21.02.2014
Umberto	LCA And Environmental Product Declaration Software	ifu Hamburg	http://www.umberto.de/en, accessed: 21.02.2014

Thirteen of the calculators are freely available and can be downloaded directly from their website or by contacting the developer. We tested these calculators to determine their features, inputs and outputs, functionality and operability. For all 18 calculators, their methodological and practical aspects were obtained from published information, including peer-reviewed literature, reports and calculator descriptions on the websites, plus the results from our calculator tests. Before analyzing them in the third step of our review, we screened them for their ability to assess GHG emissions from energy crop rotations and their effects. Seven of these 18 calculators were capable of modeling energy crops in rotation, but none of these calculators could assess energy-crop rotation effects like interactions between crops such as nutrient management or green manuring.

Method: Analytical Framework for Analyzing GHG Calculators for Energy Crop Cultivation

The CFP methodology defines four phases to assess GHG emissions along the production chain: (1) the goal and scope definition phase, (2) the inventory analysis phase, (3) the impact assessment phase and (4) the interpretation phase [11]. The first phase defines the general framework of the CFP study. The data collection for each process is carried out in the second phase, and this data is summarized into one CFP result in the third phase. In the final CFP phase, the results from the first three phases are evaluated in light of their completeness and sensitivity; on this basis, researchers form their conclusions, including any limitations of the study and finally give recommendations. The adoption of these phases in GHG emission calculators is essential for the calculated result and therefore for the applicability and utilization of the results. As the first three CFP phases are particularly relevant for the design and development of the calculator and for the results of the CFP study, we focused on the first three phases in our study.

Table 2: Assignment of the four criteria for analyzing the 18 selected GHG emissions calculators from energy crop cultivation to the CFP phases and their related indicators and variables

Criteria	CFP phase	Indicator	Variables
Goal and scope of the calculator	Goal and scope definition	Goal of the calculator	-Raising awareness
			-Reporting
			-Project evaluation
			-Product assessment
		System boundaries	-Process definition: “cradle to grave” or “cradle to farm gate”
			-Calculation scale: global, national or farm level
			-Time horizon: one year or multiple years
		Allocation method	-Expanding and substituting other products
			-Specific indicator
			-Avoid allocation
		Functional unit	-Per unit area
			-Per unit product
			-Per emission category
			-Per farm

Methodology used to account for GHG emissions from energy crop cultivation	Goal and scope definition	Data requirements, assumptions and quality requirements	-Calculation pathway: Tier 1, 2 or 3
			-Country specific (calculation method)
	Inventory analysis	Kind of database used	-E.g. Ecoinvent, RED, IPCC, Agri-Footprint LCI database
	Impact assessment	Impact category "Climate Change"	-Global warming potential for 20, 100 or 500 years
Energy crop cultivation management	Inventory analysis	Indirect GHG emissions (caused by the manufacture of agricultural farming operating needs)	-Including: fertilizer, pesticides, building materials, seeding material, energy, fuel or machinery
			-Distinguish among pesticides
			-Distinguish among fertilizer
		Direct GHG emissions (induced by farming processes)	-Distinguish among: mineral fertilizer types
			-Including: organic fertilizer (digestate)
			-Distinguish among: organic fertilizer types
			-Including: crop residues
			-Including: fuel combustion
			-Distinguish among: tillage types
			-Including: land use change
		Calibrated energy crops	-Energy crop species
			-Including: perennial crop
			-Including: undersowing
Ability to model crop rotations	Inventory analysis	Crop rotation effect	-Including: catch crops or green manure
			-Including interaction between previous crop and assessed crop: nutrient management, timing of farming activities

We chose the following four criteria for the comparison of the selected calculators: (1) the goal and scope of the calculator, (2) the methodology used to account for GHG emissions from energy crop cultivation, (3) energy crop cultivation management and (4) the ability to model crop rotation. These criteria were assigned to the first three CFP phases, and indicators and variables related to these CFP phases were identified (Table 2). The importance and relevance of each criterion (including CFP phases, indicators and variables) are described in detail in the following paragraphs.

Goal and Scope

The following indicators should be considered and clearly described in the first CFP phase: the goal of the study, the system boundary, the allocation method and the functional unit [11]. By defining these indicators, the limits of the processes included in each calculator as well as the working plan of the entire CFP study can be defined.

When characterizing the goal of CFP studies, the intended application and audience as well as the reason for the study has to be defined [8]. GHG emission assessment of crop production can be undertaken for various reasons. Colomb et al. [14] divided these reasons into four categories: raising awareness, reporting, project evaluation and product assessment. Calculators whose goal is *raising awareness* often have an educational purpose by giving information about climate change in crop cultivation and are often used by farmers and farming consultants. Calculators in the second category, *reporting,* assess GHG emissions at the farm level (used by farmers) or landscape level (used by policy-makers) to compare results with other farms or countries and to help propose GHG mitigation options. The third category, *project evaluation,* includes calculators assessing the GHG emissions of a project or a policy, often used by policy-makers, NGOs, technicians or consultants comparing different projects (e.g. different management systems, agricultural innovations). The fourth category, *product assessment,* covers calculators used by private businesses for assessing GHG emissions from agricultural production chains to compare different production systems and to provide GHG reduction plans.

The system boundary defines processes, inputs and outputs of the production system to be included in the inventory analysis [8]. The CFP study may be performed for the complete production chain "from cradle to grave" to the end product – e.g. biodiesel from oilseed rape, or just for the first product in the production chain, "from cradle to farm gate" – e.g. rapeseed cultivation. If the objective of the study is to evaluate GHG emissions from the cultivation process, the post farm gate processes can be neglected in the assessment. However, for determining the global impact up to consumption, all processes "from cradle to grave" (including post-farm gate processes) should be considered in the assessment [17]. Depending on the goal of the calculator, the scale of the assessment can range from the global level, to the national, regional or individual farm level, or even down to individual farming processes.

The time scale is another important factor in the system boundary consideration. CFP can be carried out for the whole life cycle of one crop, which could be less than one year for annual crops or more than one year for perennial crops. It is essential to define the time scale for each CFP and describe the findings in the CFP report in order to make it comparable to other CFP studies. Annual crops are typically assessed for one vegetation period from seedbed preparation to harvesting. The influence of the previous crop on the assessed crop is often outside the system boundary of typical CFP studies. For perennial crops, the system boundaries can be set either to one single production year or to the entire life cycle, from crop establishment to the final harvesting period. Further reflections on this complex issue of modeling crop rotations and the effects it has on single crops, will be discussed in Section 2.2.3.

Allocation issues occur when a single process delivers more than one product or service (multifunctional process). Energy crop cultivation and processing of biomass can lead to multiple outputs, e.g. oil and oilseed meal from oilseed crushing, or biogas and digestate from anaerobic biomass digestion. There are three different methods available to allocate the processes emissions to different products [29]. The first allocation method expands the system boundary (until the use of the byproduct is included) and then applies the substitution method. The second method divides the emissions of the entire system among the different byproducts by using a specific indicator (either a physical indicator, e.g. weight or energy content, or a socioeconomic indicator, e.g. market value). The third allocation method ignores the allocation process and allocates all emissions to the main product or avoids allocations by using a suitable fictional unit. The chosen allocation method is extremely important for bioenergy systems, due to its large impact on the final CFP result [21].

The functional unit should be consistent with the goal and scope of the CFP study and provide a reference unit for all life cycle flows and indicators, allowing the comparison between systems [9]. The results from CFP studies from energy crop production can be expressed as kg of CO_2 equivalent per unit area; per unit product; per emission category or per farm.

Methodology used to Account for GHG Emissions from Energy Crop Cultivation

Defining the data requirements, assumptions and quality requirements of the data is part of the first CFP phase and is influenced by the goal of the calculator and the goal of the CFP study correspondingly. The IPCC provides three calculation pathways, called Tiers, in the AFOLU guidelines to account for land-based GHG emissions [26]. The Tiers differ in their degree of complexity: Tier 1 is the least accurate methodology, though the simplest to use, as it provides equations and global default values; Tier 2 may use the same methodological approaches as Tier 1, but requires specific regional data and emission factors, while Tier 3 level methodologies are based on actual measurements or model simulations. Using a higher Tier generally improves the accuracy of the inventory analysis and reduces uncertainty, but requires a higher amount and quality of input data. Making sure that the chosen GHG emissions calculation pathway (Tier) corresponds to the geographical coverage of the calculator is very important [14]. Global or national calculators use the Tier 1 approach, in which only a small amount of input data is required and global or country average emission factors are used. Calculators using the Tier 2 approach often focus on a regional application, and pedoclimatic and management data is needed. Using the Tier 3 approach for assessing the GHG emissions enables the calculator to obtain farm-specific results in different timeframes (day, month and/or year). However, this requires specific measurements or complex pedoclimatic and management input data, which is often too time-consuming to obtain. Furthermore, calculators using the Tier 3 approach are locally restricted or focus on a specific product or emission processes; this could be unfeasible for most CFP studies. The results of the GHG calculator and the integrated calculation pathway can only be as precise and reliable as the input data used to compute these results.

Various LCA databases are available, providing datasets from agriculture, energy supply, transportation, biofuels and biomaterials, bulk and special chemicals, construction and packaging materials, basic and precious metals, and metal processing, as well as waste treatment. These datasets integrated in the calculators enable users to calculate their production chain by simply combining the single production steps which are provided in a kit of modules from the chosen database. All datasets are representative of previously completed LCA study results. The result of a CFP study largely depends on which database is used.

In the third CFP phase, the impact categories and category indicators are selected consistent with the goal of the study. The collected emission data from the inventory analysis are assigned to the selected impact category [9]. In our review, we focused on calculators following the IPCC guidelines [26] and using the impact category "climate change" with the category indicator GWP [7], [27]. The GWP can be calculated over a specific time interval: 20, 100 or 500 years, and aggregates all emitted GHG into one unit (kg of CO_2 equivalent per functional unit), which makes it easier to compare GHG emissions from different products.

Energy Crop Cultivation Management

The production of biomass from energy crops requires multiple steps: tillage, sowing, fertilization, use of pesticides, and harvest. GHGs are emitted from each farming operation. However, agricultural croplands that are intensively managed offer many opportunities for reducing GHG emissions through changes in agronomic practices [7]. The IPCC [26] recommends taking into account all indirect and direct emissions caused by farming operations when calculating the CFP of crop cultivation. Indirect emissions occurring during the production of all inputs (agricultural operating needs such as seeds, fertilizer, pesticides, agricultural machinery, fuel, building materials and energy) and have often been considered by using combined emission factors expressed in CO_2 equivalent from available databases. Indirect emissions from the production of agricultural operating needs can have a significant impact on the CFP results [23]; fertilizer production in particular is responsible for high GHG emissions [30]. Therefore, distinguishing between fertilizer types used in agricultural production can have a great impact on the CFP results [30]. The production of pesticides is less GHG emission intensive, but the distinction between different types can affect the CFP result as well.

Direct GHG emissions occur on the field through the application of crop residues and fertilizer (organic and mineral). According to the IPCC guidelines for AFOLU [26], CO_2, N_2O and CH_4 should be considered for direct emissions and NH_3 and NO_x for indirect emissions when estimating anthropogenic GHG released during crop cultivation. CO_2 emissions result from liming and urea application. Nitrous oxide (N_2O) emissions from managed soils arise from anthropogenic nitrogen input, such as mineral and organic fertilizer and crop residues, including N_2O through two indirect emissions pathways. The first indirect N_2O pathway is nitrogen volatilization, which occurs for example when NH_3 and NO_x are deposited onto soil and water. Leaching and runoff of nitrogen from fertilizer application is the second pathway for indirect N_2O emissions. Both indirect emissions pathways lead to further processes in which N_2O emissions occur. The type of fertilizer applied on the field affects the direct and indirect GHG emissions caused by the fertilizer [31]. This can be seen in a simple form, by distinguishing between mineral and organic fertilizer in general, but also in a more advanced distinction as seen between mineral and organic fertilizer types. The use of digestate as an organic fertilizer to substitute for mineral fertilizer is one option for reducing GHG emissions [32] by eliminating the GHG emissions from mineral fertilizer production. However, digestate application increases diesel consumption and correspondingly GHG emissions [33]. Consequently, the possibility of distinguishing between fertilizer types could be beneficial for the CFP of energy crop cultivation.

From crop residues remaining on the field, GHG emissions occur through the process of nitrification and denitrification, and should be included in the CFP calculations. The amount of nitrogen added to the field annually through crop residues (above-ground and below-ground) including nitrogen-fixing crops, is related to crop type, yield, residual nitrogen content, ratio of below-ground and above-ground biomass and the crop management system (what is left on the field, e.g. straw, stubble) [26].

CO_2, CH_4 and N_2O and other air pollutants (e.g. CO, NO_x) are emitted during fuel combustion [34]. The amount of fuel used for each cropping system as a function of machinery operation for tilling, drilling, seeding and harvest is related to machinery performance (technical standard) and the type of machinery used, soil type, and harvest yield, as well as crop management (e.g. the tillage system, the amount and type of fertilizer and pesticides applied) [35].

CO_2 emissions can occur through changes in soil organic carbon (SOC) stock changes caused by changes in the land use and management regime, called a Land Use Change (LUC). According to ISO 14067 [11], GHG emissions through LUC should be integrated, but documented separately in CFP studies.

The choice of crop can have a high impact on GHG emissions from the whole production system as well as on N_2O and NO emissions from fertilized fields [31]. Therefore, a parameter addressing the type of energy crop should be included in GHG emission calculators. A wide range of species can be used as energy crops, but the intensity of crop management depends on the species selected. Energy crop production management is in many ways similar to conventional food crop management. Crops with rapid growth, a high yield of usable biomass, an ability to grow under adverse weather and poor soil conditions, and with a high resistance against pests and diseases are favored as energy crops [36]. Sometimes, energy crops can have different crop management requirements than food crops [36], especially if the selected species has not traditionally been grown in the area (e.g. *Sorghum* in Central Europe) or if perennial crops are used instead of annuals (e.g. *Miscanthus sinensis, Silphium perfoliatum*). If food crops are grown as energy crops, alternative genotypes less suited for food production but with lower input requirements may be used [37]. However, biomass yield still depends on climate and soil conditions, fertilizer supply, and the timing of sowing and harvesting [38].

Perennial crops can have several benefits compared to annual crops. The inputs of a perennial cropping system are lower because the crop only has to be established once and the long-living roots can interact with the ecosystem, which can be beneficial to the nutrient balance of the soil [36]. When describing the crop management of perennial crops, the whole life cycle should be taken into account, since the agricultural performance of the crop correlates with the age of the plants. During crop establishment and at the end of the crop cycle, productivity is lower than in the years between these two phases. Consequently, the CFP of perennial crops may be underestimated when assessing only a single productive cultivation year and ignoring the other cultivation stages. Hence, the inclusion of detailed inventories of agricultural management at each stage of perennial crop cultivation would improve CFP calculation and the reliability of the assessment results [17]. Integrating undersowing crops (sowing a secondary crop underneath the main crop) into crop management may also have a positive influence on the CFP result [39].

Ability to Model Crop Rotation

Energy crops can be included in traditional food crop rotations or can be grown in self-contained rotations. In general, crop rotation improves soil fertility (by enhancing soil structure, reducing soil erosion and maintaining sufficient content of soil organic matter), nutrient use efficiency (reduced and demand-oriented fertilizer use), and biodiversity (improved crop diversity). Crop rotation also tends to reduce the input of crop protection agents and increase crop yields [37]. The system boundaries in agricultural CFP are typically set at one vegetation period of one single crop [18]. However, as agriculture systems are highly complex, often not all underlying material flows can be quantified when the assessment is limited to such a short time period. Including all interactions (crop rotation effects) between the previous crop and the assessed crop in the CFP was recommended as a possible solution by Brankatschk and Finkbeiner [18]. When looking at only one vegetation period, it can be difficult to consider the exact nutrient supply, since each crop uses different amounts of nutrients and leaves different residue nutrients in the field. Another effect of crop rotations can be the improvement of phytosanitary conditions by reducing the pressure of disease and infestation by parasites. Therefore, the previous crop can affect nutrient and pesticide management for subsequent crops. By switching crops in a crop rotation, the intensity and timing of farming activities can be influenced, since the soil structure and texture are influenced. Crop residues remaining on the field or the introduction of green manure crops or catch crops in the crop rotation can have a major impact on the subsequent crop and on the crop rotation as a whole by affecting the soil properties and fertility, and correspondingly the achievable yield [18], [19]. Today's CFP studies typically assess each crop independently if crop rotations are assessed, but by doing so they lose the ability to reflect the effects of crop rotation itself [18]. For this reason, the whole crop rotation should be assessed in CFP of energy crop cultivation in order to assess all effects related to crop rotation. This includes the consideration of all shifts of inputs in the crop rotation from one crop to the subsequent crop; otherwise the previous crops within the crop rotation carry the GHG emission burden from the following crops.

RESULTS

Goal and Scope of the GHG Assessment Calculators

Not all 18 identified calculators have assessment of GHG emissions from energy crop cultivation as their only goal, with a target audience of farmers or private companies. However, all of them included a possibility for GHG assessment from energy crop cultivation in their goal and scope definition. Nevertheless, the objective of each calculator and user group varies (Table 3). C-Plan and Farm GAS (both web-based) were mainly designed to raise awareness among farmers, consultants, students and land managers. They are both focused on giving an initial overview of farm-related GHG emissions and the impact of farm management decisions on GHG emissions.

Table 3: Comparison of goal and scope from the 18 selected GHG emissions calculators from energy crop cultivation based on the LCA approach, including the indicators: goal, system boundaries, allocation method and functional unit

Name	Goal	Cradle to	Scale	Time horizon	Allocation method	Functional unit [per unit]
Agri-LCI models	Product assessment (agricultural products)	Farm gate	Farm	Single year	Indicator: economic value	Product
BioGrace	Reporting; CFP of biofuels	Farm gate and grave	Farm	Single year	Indicator: energy content	Product and emission category
CAPRI	Project evaluation; decision support tool for policies applied within the agricultural sector	Farm gate	National and regional	Single year	Indicator: economic value and physical value	Product
CFF	Reporting; CFP of a farm	Farm gate	Farm	Single year	No allocation	Emission category on farm
COMET FARM	Reporting; CFP of a farm	Farm gate	Farm	Single year or multiple years	No allocation	Product and area
CFT	Product assessment; CFP of a farm	Farm gate	Farm	Single year	Indicator: economic value	Product, area and emission category
C-Plan	Raising awareness; CFP of a farm	Farm gate	Farm	Single year	Avoid allocation (outside system)	Farm and emission category
Farm GAS	Raising awareness; CFP and economics of a farm	Farm gate	Farm	Single year	Indicators: provided by Australian National GHG report or user defined **Avoid allocation**	Area and emission category

FSGGEC	Reporting; CFP of a farm	Farm gate	Farm	Single year or multiple years	No allocation	Area and emission category
GaBi	Product assessment (agricultural and other industrial products)	Farm gate and grave	Global, and national	Single year or multiple years	User specific	Product, area and emission category
GEMIS	Product assessment (energy production and transport systems)	Farm gate and grave	National	Single year	User specific	Product and emission category
HGCA 1	Reporting; CFP of biofuel production	Farm gate and grave	Farm	Single year	Expanding and substituting other products	Product and emission category
HGCA 2	Product assessment; CFP of a farm	Farm gate	Farm	Single year	Expanding and substituting other products	Product and emission category
IFSC	Reporting; farm sustainability	Farm gate	Farm	Single year	No allocation	Area and emission category
openLCA	Product assessment (agricultural and industrial products)	Farm gate and grave	Global, national, and farm	Single year or multiple years	User specific, according to database implemented	Product, area and emission category
SALCA	Product assessment (agricultural products)	Farm gate and grave	Global, national, and farm	Single year or multiple years	Indicator: economic, area, arable area	Product and emission category
SimaPro	Product assessment (agricultural and industrial products)	Farm gate and grave	Global, national, and farm	Single year or multiple years	Indicator: physical value and economic value	Product, area and emission category
Umberto	Product assessment (agricultural and industrial products)	Farm gate and grave	Global, national, and farm	Single year or multiple years	User specific	Product, area and emission category

BioGrace, CFF, COMET-FARM, FSGGEC, HGCA 1 and IFSC have the purpose of reporting accurate GHG emissions for subsequent comparisons between farms or countries. BioGrace and HGCA 1 are both based on Microsoft Excel, and were developed for the purpose of calculating the entire CFP ("from cradle to grave") of biofuel production from biomass. Both calculators were designed for politicians or consultants to support decision-making and to design national GHG reduction programs, and for farmers to see how changes in management practices could affect the overall GHG emissions of the resulting biofuel production. CFF, COMET-FARM and FSGGEC are web-based calculators developed to support farmers to estimate the CFP of their farm or single products with a focus on carbon sequestration and crop management. IFSC addresses the same topic, but is Excel-based. All four calculators have the goal of identifying mitigation options of GHG emissions on the farm and to report the CFP results voluntarily to national GHG emission reports or for CFP labeling and for comparing CFPs from similar products.

Only one of the 18 calculators under review, CAPRI, was developed for the purpose of project evaluation. CAPRI is a multi-purpose modeling system software for EU agriculture developed for policy-makers and scientists to analyze research questions in relation to specific agricultural policies [40].

Nine out of 18 calculators were developed for product assessment. Four of them (GaBi, openLCA, SimaPro and Umberto) are software solutions and were originally developed to assess the life cycle from industrial products. These tools were designed to be used by scientists, companies or policy-makers to assess the potential environmental burdens of a product in its production, use and disposal, and to detect mitigation options in the production chain. Agri-LCI models, CFT and HGCA 2 (all Excel-based) and SALCA (software solution) were developed for farmers, companies and policy-makers to assess the LCA from agricultural products and different management systems and to derive recommendations from these results for GHG reduction. GEMIS is a life cycle calculating software program developed for companies and policy-makers to model energy, material and transportation flows.

All 18 calculators under study include the assessment of GHG emission from crop cultivation "from cradle to farm gate" in their system boundaries, but only eight of them (BioGrace, GaBi, GEMIS, HGCA 1, openLCA, SALCA, SimaPro and Umberto) are able to extend the system boundary to the end of the life cycle of the assessed production chain (Table 3). Despite the similar system boundary of "cradle to farm gate", each calculator includes different direct and indirect GHG emissions sources related to crop cultivation (Table 5, Table 6).

OpenLCA, SimaPro, SALCA and Umberto offer assessment of GHG emissions at the global, national and farm scale, GaBi on the global and the national scale, GEMIS at the national level, and CAPRI at the national and regional scale. The other 11 calculators were only developed to assess the GHG emissions at the farm scale.

In 11 of 18 calculators, the time horizon for GHG emission assessment was limited to one crop vegetation period (one year) only. However, seven calculators can extend the assessment to multiple crop cultivation periods.

Different allocation methods are used by the 18 calculators to allocate the GHG emissions to co-products. HGCA 1 and 2 expand the system boundaries and substitute the byproducts with other products already included in the CFP study. Six calculators share the system›s emissions among byproducts by using specific indicators. Three of them use only one indicator for allocation: Agri-LCI models and CFT use economic indicators; energy content is used by BioGrace. Sometimes in a multifunctional production process it is impossible to find one appropriate indicator which works for all byproducts. As a result, two indicators (physical and economic, as used by CAPRI and SimaPro) or three indicators (economic information, area and arable area as SALCA offers) can be used for allocation. CFF, COMET FARM, C-PLAN, FSGGEC and IFSC avoid using an allocation method and either allocate all emissions to the main product or choose a suitable functional unit by which the byproduct is outside the system boundary. The other calculators provide all allocation methods and the user can choose a suitable method according to the goals defined for the particular CFP study.

Table 3 provides an overview of the functional units used by the calculators to report the CFP results. COMET FARM, CFT, GaBi, openLCA, SimaPro and Umberto provide results per unit product and per unit area. All calculators except the Agri-LCI models, CAPRI and COMET FARM provide the GHG emissions separately for each emission category in addition to the total result. IFSC and COMET FARM provide the results in imperial units, whereas all the other calculators are metric.

Methodology used to Account for GHG Emissions from Energy Crop Cultivation

The amount of data required by each calculator depends on the processes, activities and sources included in the calculator and on the GHG emission calculation pathway (Tier) used. The calculators are classified into different Tiers in Table 5 in order to distinguish their degree of complexity of integrated methodology to account for land-based GHG emissions. In seven out of 18 calculators, all three Tiers were combined into one approach, e.g. COMET FARM and FSGGEC. In GaBi, openLCA, SALCA, SimaPro and Umberto, own more detailed data can be integrated as well. Seventeen out of 18 calculators not only use the IPCC guidelines [26], but mix different assessment methodologies. However, the assessment methodologies chosen are always on the same complexity level as the Tier methodology applied in the calculator. BioGrace, CFF and C-Plan use the Tier 1 approach with global or national default values. Seven of the 18 calculators use Tier 1 and Tier 2 approaches for GHG emission calculation and can be modified with country-specific emission values. Calculators including the Tier 2 approach focus on regional application, and pedoclimatic and management data are required. COMET FARM and FSGGEC provide a country map, with climate and soil data, where the user can locate their farm and run the dynamic process-based crop–soil–atmosphere models. The Tier 3 approach requires measurements or high-resolution input data for model simulation, and is locally restricted. Seven calculators use Tier 3 approach methods, but only to calculate some processes included in the CFP that are crucial for the result of the CFP; they calculate other processes using approaches of lower complexity.

When choosing a GHG emission calculator, it is important to know for which region or country it was developed, and consequently which GHG emission default values were implemented. Only CFT, GaBi, openLCA, SimaPro and Umberto can be used worldwide. All the other reviewed calculators were calibrated for specific countries.

Six out of the 18 calculators under review use the Ecoinvent Database [41] and other agricultural and product databases (Table 4). The Ecoinvent Database provides around 10,000 datasets from agriculture, energy supply, transportation, biofuels, biomaterials and other industrial processes [41]. The datasets integrated in the models enable users to calculate their production chain by simply combining the individual production steps, which are provided in a kit of modules from the Ecoinvent Database. If a specific crop or production step is not available, it is possible to modify an existing dataset or create a dataset from scratch (e.g. new energy crops). For most modules, global, European or national mean values are available.

Table 4: Comparison of methodology used from the 18 selected calculators to account for GHG emissions from energy crop cultivation, including the indicators: calculation pathway (Tier), country specialization, database used and time horizon for the GWP assessment

Name	Tier	Country	Database, data source	GWP [years]
Agri-LCI models	1,2	England, Wales	Ecoinvent, UK Inventory Report, DEFRA and MAFF publications, farm production databases, IPCC	100
BioGrace	1	EU	RED[a]; JEC[b] consortium, IPCC	100
CAPRI	1,2	EU, Norway, Western Balkans and Turkey	Data from: EUROSTAT, FAOSTAT, OECD, FADN	100
CFF	1	UK	UK DEFRA, IPCC	100
COMET FARM	1,2,3	USA	DAYCENT, IPCC	100
CFT	1,2	Global	Ecoinvent, ASABE, IPCC	100
C-Plan	1	UK	IPCC, UK National Inventory	100
Farm GAS	1,2	Australia	Australian Methodology for the Estimation of Greenhouse Gas Emissions and Sinks	100
FSGGEC	1,2,3	USA	IPCC, SOCRATES soil carbon change model	100
GaBi	1,2,3	Global	GaBi Database, Ecoinvent, US LCI	20, 100, 500
GEMIS	1	Germany	RED[a], IPCC	100
HGCA 1	1,2	UK	IPCC, DEFRA, UK specific emission factors	100

HGCA 2	1,2	UK	IPCC, DEFRA, UK specific emission factors	100
IFSC	1,2	Illinois, USA	IPCC, Literature, COMET-VR soil carbon model	100
openLCA	1,2,3	Global	Ecoinvent, ELCD, GaBi Databases, LCA Food, NEEDS, ProBAS	20, 100, 500
SALCA	1,2,3	EU, Switzerland	SALCA LCI Database	20, 100, 500
SimaPro	1,2,3	Global	Ecoinvent, ELCD, LCA Food DK, US LCI, Agri-Footprint LCI database, US Input Output library, Swiss Input Output Database	20, 100, 500
Umberto	1,2,3	Global	Ecoinvent	20, 100, 500

[a]Renewable Energy Directive.

[b]European Commission Joint Research Center.

SALCA uses its own GHG assessment concept for agriculture. This covers LCA methods adapted to the agri-food chain, such as GHG emission calculators, and the SALCA life cycle inventory database, based on the Ecoinvent Database [42]. OpenLCA is an open-source software program into which freely and commercially available databases can be integrated. The methodological approach of this calculator is equivalent to GaBi, SimaPro and Umberto, and the same databases can be integrated. In a similar way to other software, it works like a kit, in which the individual production steps are provided as freely combinable modules. The general methodology of the other calculators is based on global, European or national guidelines (e.g. IPCC and RED) which provide GHG emission default factors for different cultivation-related GHG sources. Furthermore, datasets from literature are used to assess specific indirect and direct agricultural GHG emissions. COMET FARM integrated the dynamic agro-ecosystem model DayCent (the official U.S. National Greenhouse Gas Inventory model) to estimate emissions on the field and through LUC. IFSC uses the COMET-VR soil carbon model from COMET FARM.

Regarding GWP, the calculators GaBi, openLCA, SALCA, SimaPro and Umberto can determine GWP for 20, 100 and 500 years, and the user can choose the preferred indicator. All other calculators only provide GWP for 100 years (Table 4).

Energy Crop Cultivation Management

As mentioned above, the selected system boundaries can significantly affect the processes, activities and sources included in each calculator as well as the amount and quality of the input data required. Table 5 gives an overview of the included indirect emissions arising from farming processes for each calculator. In two calculators (COMET FARM, FarmGas), indirect emissions from the production of operating resources (e.g. machinery, pesticides and fertilizer) are omitted. In contrast, in eight out of 18 calculators, GHG emissions for the production of all agriculturally related inputs are embedded in the assessment (CAPRI, CFF, GaBI, GEMIS, openLCA, SALCA, SimaPro and Umberto). This group consists of calculators in which a large amount of datasets is provided, allowing the user to decide which indirect emissions should be included in the CFP. In GEMIS and CFF, indirect emissions from the production of agricultural inputs are included in the calculators; in CFF, the user can influence the values by modifying the amount of inputs. The remaining calculators provide indirect emissions from production for only a limited number of farming inputs – sometimes only one. Twelve out of 18 calculators include indirect emissions from manufactured pesticides. However, only seven can distinguish among types of pesticides by dividing pesticides into categories: herbicides, insecticides, fungicides and lubricants. Furthermore, with GaBi, openLCA, SALCA, SimaPro and Umberto, the user can calculate the CFP of the individual pesticides by aggregating the GHG emissions of the pesticide ingredients provided by the database included in these calculators. The other five calculators include pesticide production by aggregating emissions from pesticide use in one category; they do not distinguish between types of pesticides. Fourteen out of 18 calculators include indirect emissions from manufactured mineral fertilizer, but only 12 out of these 14 can distinguish among types of mineral fertilizers. They all provide a different number of fertilizer types. BioGrace, for example, only distinguishes among mineral fertilizer ingredients (N, CaO, K_2O and P_2O_5) while CFT provides 35 different types of mineral fertilizer, and the user can add new types or edit existing ones.

Table 5: Comparison of energy crop cultivation management related indirect GHG emissions from the 18 selected calculators regarding indirect emissions from the manufacture of agricultural operating needs, and the possibility of distinguishing among different types of pesticides and fertilizer (+=yes; −= no)

Name	Included operating needs (emissions from manufacturing)	Distinguish among types of	
		Pesticides	Fertilizer
Agri-LCI models	Fertilizer, building materials, fuel	−	−
BioGrace	Fertilizer, pesticides, seeding materials, energy	−	+
CAPRI	All agricultural related inputs	+	+
CFF	All agricultural related inputs	+	+
COMET FARM	No indirect emissions from input production	−	−
CFT	Fertilizer, pesticides, energy, fuel	−	+
C-Plan	Energy	−	−
Farm GAS	No indirect emissions from input production	−	−
FSGGEC	Fertilizer	−	−
GaBi	All agricultural related inputs	+	+
GEMIS	All agricultural related inputs	−	+
HGCA 1	Fertilizer, pesticides	−	+
HGCA 2	Fertilizer, pesticides	−	+
IFSC	Energy	−	−
openLCA	All agricultural related inputs	+	+
SALCA	All agricultural related inputs	+	+
SimaPro	All agricultural related inputs	+	+
Umberto	All agricultural related inputs	+	+

Table 6 gives an overview of the included direct GHG emissions arising from farming processes for each calculator. C-Plan and FSGGEC do not distinguish among mineral fertilizer types; they only take the amount of N fertilizer (sum of N in kg) applied on the field into account. The other 16 calculators can distinguish among mineral fertilizers to a different degree of accuracy. With the exception of FSGGEC, all calculators included organic fertilizer in their assessment; 13 calculators can even distinguish among different types of organic fertilizers. Regarding digestate, only nine calculators take this particular organic fertilizer into account. However, they cannot distinguish among application methods (manure chisel plow, drag shoe, drag hose or incorporated in one hour after application) to account for the GHG emission arising from different digestate applications. Other than Agri-LCI models, COMET FARM and C-Plan, all calculators of the 18 include direct GHG emissions from crop residues applied on the field.

Table 6: Comparison of energy crop cultivation management related direct GHG emissions from the 18 selected calculators regarding the included emissions arising from the application of organic fertilizer, crop residues, fuel combustion from machinery use, and emission occurring after land use change. Furthermore, if the calculator distinguishes among different types of mineral and organic fertilizer use and tillage (+=yes; −= no)

Name	Distinguish among types of			Including			
	Mineral fertilizer	Organic fertilizer	Tillage	Organic fertilizer (digestate)	Crop residues	LUC	Fuel combustion
Agri-LCI models	−	−	+	+ (−)	−	+	+
BioGrace	+	−	−	+ (−)	+	+	+
CAPRI	+	+	+	+ (+)	+	+	+
CFF	+	+	−	+ (−)	+	+	+
COMET FARM	+	+	+	+ (−)	−	−	−
CFT	+	+	+	+ (−)	+	+	+
C-Plan	−	−	−	+ (−)	−	+	+
Farm GAS	−	−	−	+ (−)	+	−	−
FSGGEC	−	−	+	− (−)	+	+	+
GaBi	+	+	+	+ (+)	+	+	+
GEMIS	+	+	+	+ (+)	+	−	+
HGCA 1	+	+	+	+ (+)	+	−	+
HGCA 2	+	+	+	+ (+)	+	−	+
IFSC	+	+	+	+ (-)	+	+	+
openLCA	+	+	+	+ (+)	+	+	+
SALCA	+	+	+	+ (+)	+	+	+
SimaPro	+	+	+	+ (+)	+	+	+
Umberto	+	+	+	+ (+)	+	+	+

Through the use of machinery during energy crop cultivation, GHG emissions arise from fuel combustion. All calculators take this into account except COMET FARM and Farm GAS. Fourteen calculators can even distinguish among different types of tillage. Agri-LCI models and COMET FARM use categories (e.g. reduced tillage, plow-based, direct drilling) to account for different crop management systems and the amount of diesel used, respectively. In the other 12 calculators, it is possible to calculate the amount of diesel used by selecting each crop management step within the calculator (e.g. CFT provides a list of management steps) or by adding the actual amount of diesel used (e.g. BioGrace).

Thirteen out of the 18 calculators account for GHG emissions from LUC, but not all calculators document these results separately and some just account for emissions through land use change and not through management change. However, CAPRI, COMET FARM, FSGGEC GaBi, IFSC, open LCA, SALCA, SimaPro and Umberto all feature integrated process-dynamic models to determine emissions from soil carbon change through management changes.

Only BioGrace, HGCA 1 and GEMIS were originally designed to calculate GHG emissions from energy crop cultivation, but the calibration is limited to traditional energy crops for bioenergy production. However, the other 15 calculators provide datasets and calibrations for energy crops in addition to food crops, and also provide the possibility to modify or add crops. An overview of the calibrated energy crops in the calculators under study is given in Table 7. If the category "other" is selected on the calculator, other energy crops can be calculated without a specific calibration for this crop.

Table 7: Overview of calibrated energy crops in 18 calculators of GHG emissions from energy crop cultivation

C. Peter et al. / Renewable and Sustainable Energy Reviews 67 (2017) 461–476

Overview of calibrated energy crops in 18 calculators of GHG emissions from energy crop cultivation.

Name	**Alfalfa** *(Medicago sativa)*	**Barley** *(Hordeum vulgare)*	**Grass** *(Poaceae)*	**Legumes** *(Fabaceae)*	**Maize** *(Zea mays)*	**Millet**	**Miscanthus** *(Miscanthus sinensis)*	**Oil Seed Rape** *(Brassica napus)*	**Palm** *(Arecaceae)*	**Perennial Grass**	**Rey** *(Secale cereale)*	**Sorghum**	**Soya bean** *(Glycine max)*	**Sugar cane** *(Saccharum officinarum)*	**Sugar beet** *(Beta vulgaris)*	**Sunflower** *(Helianthus annuus)*	**Switchgrass** *(Panicum virgatum)*	**Triticale** *(X triticosecale)*	**Wheat** *(Triticum aestivum L)*	**Other**
Agri-LCI models		+			+			+					+					+	+	
BioGrace					+			+	+				+	+	+	+			+	
CAPRI								+ +	+				+			+			+	+
CFF		+		+	+			+			+		+		+			+	+	
COMET FARM	+	+	+	+	+	+							+	+	+	+			+	
CFT	+	+		+	+					+	+	+							+	+
C-Plan		+			+			+			+				+			+		+
Farm GAS		+			+	+		+				+						+	+	+
FSGGEC					+								+				+		+	+
GaBi			+		+			+	+	+			+			+			+	+
GEMIS			+		+			+	+					+					+	
HGCA 1								+											+	
HGCA 2								+			+							+	+	
IFSC		+		+	+		+	+			+	+	+			+		+	+	
openLCA			+		+			+	+				+			+				+
SALCA			+		+			+	+	+			+			+			+	+
SimaPro			+		+			+	+	+			+			+			+	+
Umberto			+		+			+	+	+			+			+			+	+

Datasets from energy crop cultivation are included in CAPRI, GaBi, openLCA, SALCA, SimaPro and Umberto, but they are limited to the traditional energy crops as shown in Table 7. Previously unconsidered energy crop species, such as *Silphium perfoliatum*, could be added by users by modifying existing datasets or creating their own.

Perennial crops are omitted in most calculators or integrated as an annual average whenever it is impossible to distinguish among the different stages of cultivation. CFT, GaBi, openLCA, SALCA, SimaPro and Umberto can calculate the GHG emissions from perennial crop cultivation. However, the user has to check if the full life cycle (from the establishment to the end of the crop productivity) of the perennial crop is considered. Undersowing crops were not addressed in any of the 18 calculators under review.

Ability to Model Crop Rotation

Seven (COMET FARM, FSGGEC, GaBi, openLCA, SALCA, SimaPro and Umberto) of the 18 identified GHG emissions calculators for energy crop cultivation based on the CFP approach can calculate energy crop rotations. For crop rotation modeling with GaBi, openLCA, SALCA, SimaPro and Umberto, the existing modules (datasets) from crop cultivation can be combined, e.g. three years of maize cultivation can be calculated by using the same maize cultivation module three times. Within the single modules, the management system can be changed by the user. With COMET FARM, it is possible to calculate GHG emissions on a farm for a longer period. The user can enter management data on an annual basis, which can cause problems if the cultivation period spans over two calendar years (e.g. winter crops). FSGGEC offers a simple type of crop rotation calculation to the user: for each year, a single crop can be cultivated and calculated at Tier 1 or 2 level. The result is a very simple CFP where only a few GHG emission sources are taken into account. COMET FARM is the only calculator which has catch crops integrated.

It is difficult in all seven calculators to assess crop rotation effects, such as shift of nutrients or reduced farming activities and inputs. Most of these calculators generate their crop modules as single annual crops, which makes it difficult to display and to determine the effects of the crops on each other.

DISCUSSION

Goals of GHG Assessment Calculators

The most important stakeholders for biomass cultivation in bioenergyproduction are farmers, energy industries, politicians and NGOs. All of them require information about GHG emissions and calculators to assess this information for their own purposes. None of the calculators discussed here can meet the needs of all target groups, but many calculators are available with varying levels of complexity and target different goals and user groups. Raising awareness is the goal of C-Plan and Farm GAS. These calculators require little time and knowledge of GHG emissions and climate change. They can be used without training and need only small amounts of input data to estimate GHG emissions. The results are displayed as simple graphics and guide the user toward identifying mitigation opportunities. However, they are not usually designed to assess changes in management and to take into account alternative and more sustainable management practices.

Results from calculators designed for reporting can be used as the reporting basis for the certification of sustainable biofuel production and for the verification of compliance with sustainability criteria for biofuels of the Renewable Energy Directive and the Fuel Directive [2]. BioGrace, for example, was developed to harmonize the different European calculators and calculation methods for GHG emissions from biofuel production, which is necessary to comply with the Renewable Energy Directive and Fuel Quality Directive [43]. The calculation scheme (calculation rules, default values) of BioGrace is often used in combination with other national calculators (national default values and legal frameworks) for reporting the national specific GHG emissions from biofuel production, e.g. $ENZO_2$ in Germany. The calculators in this group are either available as an Excel document, in which case calculation functions, emission values and intermediate results can easily be reproduced, or have a web-based user interface where modification can only be rendered manually via input data.

Calculators designed for product assessment are well suited for revealing the relationship between different production levels. The software-based calculators are in general more time consuming and require a basic knowledge of agronomy and basic computer skills. Standard values for energy crops are available and different scenarios can be calculated by the user with only a small amount of input data. However, these standard modules only contain global or national mean values, and have to be modified by the user for regional calculations. In order to model new energy crops, datasets from farm operations, machinery, and mineral and organic fertilizers (including digestate) are available in the integrated databases. However, the user must pay particular attention to the inclusion of field emissions and also to which Tier is used to calculate these emissions.

The goals of a CFP study should always correspond to the goals of the chosen calculator and to the defined target user groups, otherwise the results of the study could be misinterpreted. At the very least, the calculator chosen should be in the same purpose category as defined by Colomb et al. [14]. The user should bear in mind that it is difficult to draw a meaningful comparison of results across similar production chain studies using different calculators with different goals, as these goals affect the system boundaries and the calculation approaches used.

All 18 investigated calculators can calculate the GHG emissions from "cradle to farm gate", and these results can be integrated in further CFP of bioenergy production chains. The defined system boundary affects the processes, activities and sources included in each calculator. System boundaries in CFP studies from agricultural production systems vary greatly within and among the same production chains [17]. Significant differences in GHG emission results can occur from the same dataset of one bioenergy production chain, depending on the calculator used [55]. The results show that it is crucial which farming processes are integrated in the calculator, which calculation pathway and allocation method is used, and if the whole cropping cycle (e.g. perennial crops) or crop rotation is included [44].

Various crop cultivation CFP studies have been based on secondary data from statistics or literature. Input data based on global or national statistics can be used to assess the GHG emissions from typical cropping systems at the global or national level, but not to assess the influence of regional pedoclimatic conditions and specific management practices on GHG emissions [17]. Therefore, the user should identify the type of available input data and the assessment goal and scale for the CFP study before choosing the calculator and the calculation scale, respectively.

The allocation of GHG emissions among the individual byproducts of energy crop cultivation, as well as the subsequent use of the byproduct›s burdens in other production cycles, are major methodological challenges. The inaccuracy of the CFP results can increase with each allocation step performed in one LCA, and the results are fundamentally affected by the choice of allocation method [45]. Six of the calculators in the analysis share the emissions of the system among the different co-products by using a specific indicator as recommended by ISO Standard 14067 [11]. Physical indicators (e.g. weight or energy content) appear to be most scientifically accurate, as they use physical principles instead of societal values, but economic indicators reflect the driver of the process through product demand. However, market prices can differ among countries and can lead to different CFP results [20]. Expansion of the system boundary can help to foresee the effects on GHG emissions through changes induced by substituting products. The integrated allocation method should always be transparent for the user of the calculator and in compliance with the intended purpose of the CFP study.

According to the scope of the study, the user should select the functional unitcarefully, because different functional units can lead to contradictory interpretations of the results. Calculators can provide the results in two ways: for the assessed process, or as result of a comparison of two scenarios (baseline vs. end of project). Calculators providing the GHG emissions separately for each emission category have a greater potential for identifying mitigation options. Users should pay attention if the GHG emissions are reported in CO_2 equivalents or as individual GHGs, e.g. N_2O, CH_4, and CO_2, or as N_2O-N, CH_4-C and CO_2-C. The simplest reporting unit for energy crop GHG emissions assessment is by area. However, this unit is not suitable for reporting GHG emissions in the context of renewable energy sustainability, and cannot be included in the calculation pathways for biofuels or bioenergy. In CFP from bioenergy, connecting the GHG emissions to the product is more appropriate. However, this includes several units which are associated with production (kg product, MJ product) and several outputs (main product e.g. kg grain and byproduct e.g. kg straw). For bioenergy-oriented crop cultivation, results should always be related in some way to the next production phase in the production chain.

Methodology used to Account for GHG Emissions from Energy Crop Cultivation and Management

GHG emissions from crop cultivation depend on local conditions [23]. Therefore, the results from the CFP can be improved by using one of the 14 calculators which take into account national or regional climate and soil conditions. Datasets from different databases representing the same process can result in different emission factors (emission assessments) affecting the comparison of CFP studies with datasets from different databases. Consequently, for similar inputs (e.g. fertilizer production), emission factors from the same database should be used to quantify different types of inputs (e.g. fertilizer types).

In addition to the GWP based on the IPCC guidelines [26], GaBi, SALCA, SimaPro and Umberto are able to calculate the GHG emissions for other impact assessment methodologies as well (e.g. ReCiPe, Impact 2002+, Eco-Indicator 99, CML, TRACI and IPCC). Furthermore, these calculators offer the possibility of extending the LCA with other impact categories provided, such as acidification, eutrophication, aquatic and terrestrial ecotoxicity, human toxicity, land use and/or ozone depletion. Regardless of the calculator chosen, the user should bear in mind that it is difficult to draw a meaningful comparison of results across similar production chain studies using different time horizons in terms of the GWP for 20, 100 or 500 years or using different impact assessment methodologies to translate life cycle flows to the same impact category impact.

Cultivation of energy crops differs from that of conventional food crops in some aspects which may significantly influence the GHG emissions and their estimation. LCA methodologies have been recently adopted for agricultural products to account better for location characteristics and differences in farming practices, focusing on annual crops [17]. The amount of GHG emissions from energy crop cultivation can be controlled by the choice of crop type, fertilizer, pesticides and machine management and by the design of crop rotations [46].

Indirect emissions from on-farm operations (e.g. machinery use) have a significant impact on the CFP results [23]. However, emissions from production of the agricultural operating needs (e.g. seeds, pesticides, fertilizer, machinery, fuel) are sometimes ignored or only partly addressed by the calculators. Since each calculator accounts for different GHG emission sources, potential users of these calculators need to check which key sources (e.g. production of fertilizer, pesticides, machinery or seeding materials) are covered by the calculator in order to derive mitigation options for their investigated production chain from the results. Furthermore, differences in crop cultivation management can be better detected if the calculator distinguishes among type of fertilizer and pesticides used for crop cultivation, since it has a significant impact on the whole CFP – especially the amount and form of nitrogen (e.g. NH_4-N, NO_3-N, urea-N) of the fertilizer used [30]. Optimizing crop management and nutrient use efficiency by adjusting the use of nitrogen fertilizer according to the crop's needs can directly reduce GHG emissions on the field and also indirectly through reduced fertilizer manufacture [46].

Direct GHG emissions, especially N_2O emissions from managed soils, have a significant impact on the CFP result [47]. However, the calculation of land-based GHG emissions from energy crop cultivation is the stage where the calculators differ most, since different calculation pathways are applied. The methodology used to calculate N_2O emissions from N-fertilization is the main contributor to the derivation among calculator results [44]. In most CFPs, the Tier 1 approach [26] is used to calculate N_2O emissions from managed crops. This approach uses a default emission factor of 1% of nitrogen added to the soil assuming that 1% from mineral and organic fertilizer and crop residues is lost as N_2O to the atmosphere. Using this method, no distinction is made between fertilizer types, crop types or soil characteristics. Calculators following the Tier 2 approach [26] often use the approach from Bouwman et al. [31]. This approach takes into account different regional conditions as well as different crops and fertilizer types (mineral and organic). COMET FARM is the only calculator that adopts the Tier 3 approach and uses the dynamic agro-ecosystem model DayCent to calculate field emissions. Fifteen out of the 18 investigated calculators take into account GHG emissions from crop residues, calculating these GHG emissions according to the Tier 1 approach (as described above). The accuracy of this calculation method can be improved by including crop residues management in the calculation such as the amount of straw left on the field (e.g. HGCA 1 and HGCA2), the quantity of crop residues or the amount of stubble burnt (e.g. Farm GAS) and by using the real nitrogen content of the above-ground biomass (grain and straw) to calculate the nitrogen content of the above-ground and below-ground biomass (this can be integrated by the user in GaBi, open LCA, SALCA, SimaPro and Umberto). Consequently, before choosing one of these calculators, the user should check which nitrogen sources are considered in the calculator and which Tier approach is used, and decide if these are sufficient for their specific goal.

The amount of NH_3 emissions induced by organic fertilizers (i.e. slurry and manure, digestate, poultry manure) depends on the fertilizer type, the fertilizer application rate and method, the daily temperature and a binary variable indicating whether the fertilizer was incorporated within one hour [48]. Seventeen of the investigated calculators can distinguish among types of organic fertilizer. Calculators using the Tier 2 approach of Bouwman et al. [31]for calculating GHG emissions from organic fertilizer application take into account different application methods (e.g. CFT). However, none of these calculators take into account the daily temperature or incorporation time. The properties of digestate are different from conventional organic fertilizer (slurry ore manure) and are affected by the anaerobic, microbial fermentation process and by the substances used in the process [49]. During the production, storage and application of digestate, CH_4, NH_3 and CO_2 emissions can occur [50]. Through organic fertilizer production and storage management as well as the application method, the amount of GHG emissions can be influenced and should be included in the consideration of CFP calculations.

LUC should be included in the CFP assessment, but should be reported separately in the results (ISO 14067). However, some calculators exclude LUC for practical reasons since the methodology used to detect LUC is very complex. Models like RothC [51] (Tier 3 level) can calculate the SOC change on a monthly and regional basis, but also require a lot of input data. Using the Tier 1 approach [26] is less complex, because global emission factors (CO_2emissions occurring over a period of 20 years) and reference native soil carbon content, depending on soil type and climate region, are provided. It is very important to consider the period of time over which emissions occur, since calculators that do not account for time are unable to calculate LUC-induced emissions [14]. Generally, with a longer time horizon, the yearly rate of SOC change decreases, since SOC change is always faster during the first years after disturbance. This aspect has already been highlighted in Petersen et al. [52], where the authors suggested using a 100-year time horizon when simulating SOC change for CFP studies, based on a 100-year GWP calculation. However, a 100-year time horizon conflicts with the confidence time of many other factors characterizing the agricultural sector (e.g. land use, cropping systems, management regimes) as their defining framework conditions (e.g. consumer demand, economic trends, societal transformation, public policy) are highly volatile and it is difficult to elaborate predictions in the longer term. For agricultural land use decision-making, 20 years should be considered as a more reasonable time horizon, which is why it has been used to include SOC change into CFP according to the Tier 1 approach. However, when changing the cultivation system each year, the effect of management change on the SOC content is not stable and may be disregarded when calculating the CFP from annual crops [26]. In CFP calculation, different tillage systems are accounted for through the different amount of resources used [35]. New technologies and crop cultivation methods have been shown to reduce the direct fossil fuel (diesel) consumption. Diesel consumption is either modeled by the calculator or the user can include the real amount of diesel used. Using mean values for diesel consumption estimates can overestimate the amount of diesel consumption by 47% [35]. Taking real diesel consumption data from the farm is always the most precise way for GHG emission calculations. However, if this data is not available, using diesel consumption models which distinguish among farming operations (tillage, seeding, fertilizing and harvesting) and considering the soil characteristic (e.g. CFT) may be a good alternative to simply dividing the results among tillage systems in general (categories as, e.g. reduced tillage, no tillage) or making no distinction whatsoever.

Not all calculators in this study were designed for specific energy crop calculations. Most calculators are calibrated for a small number of crops and it is not possible to integrate new ones. Furthermore, characteristics related to energy crop cultivation, e.g. digestate application on the field and whole plant harvest, are often ignored or insufficiently considered.

New cropping management systems, such as undersowing, were not considered in any of the calculators under review. Not only does undersowing offer benefits for reducing GHG emissions by minimizing the farming operations required, thus saving fuel, weeds may be replaced by the undersowing crop and the second crop will be further ahead than if it were sown after the primary crop was harvested [39].

The GHG assessment of perennial cropping systems is complex, since it is sometimes impossible to gather data for the whole cropping cycle [53]. Perennial cropping systems are insufficiently considered in the available GHG assessment methods and calculators. As previously mentioned, the crop type is a driving factor for N_2O emissions, but in most approaches perennial crops are not represented and can only be classified as "other crops" or "grass" [54]or representative data for proper calibration of the models is lacking [17]. Hence, more research on perennial cropping systems and their field emissions is needed. The whole cropping cycle and detailed inventories of agricultural management at each stage of perennial crop cultivation should be included in order to improve the CFP calculation and the reliability of the assessment results [17]. Including the specific characteristics of energy crop type, cultivation management and new cropping management, e.g. undersowing, in GHG emission accounting calculators can reduce the uncertainty in GHG emission assessment and can help users to detect GHG mitigation options in the cultivation process. But to carry out this concept, a high amount of input data with high quality requirements and specific high Tier level GHG emission calculation pathways are necessary.

Ability to Model Crop Rotation

Seven of the 18 investigated calculators are able to calculate energy crop rotations, but none of these cover the consequences of optimizing the management, sequence and composition of crop rotations. Most of these calculators generate their crop modules as single annual crops, which makes it difficult to display and to determine the effects of the crops on each other. For this reason, it seems challenging for the user to have to evaluate new energy crops and their effect at a specific position in the crop rotation and to model crop rotation effects, such as savings in operating resources (e.g. fertilizer, machinery use) and effects on yield. Neglecting nutrient shifts from one crop to the subsequent crop leads to free-rider situations for crops that consume nutrients left by preceding crops [18]. Consequently, the amount of GHG emissions of the subsequent crop decreases, since the crop does not get charged for its true nutrient and fertilizer consumption. This points out the need to include the effects of crop rotation in CFP. Diverse crop rotations (including the use of catch crops or green manure) can help to reduce the CFP [19]. Expanding the systems' boundaries to consider the whole crop rotation could improve the CFP calculations, because in this way all crops (and thus the effects between them) are included in the CFP. However, most energy crop cultivation CFP studies are performed for one single crop and therefore for a specific product. Hence, the effort for including the whole crop rotation is often too high for users. For this reason, new LCA approaches to account for crop rotation effects in single crop cultivation assessment should be developed, such as the agricultural allocation approach developed by Brankatschk and Finkbeiner [18], and integrated in the existing calculator.

CONCLUSION

In this paper, we identified 18 calculators for GHG emissions for energy crop cultivation that followed the CFP guidelines [11] and adopted the IPCC approaches [26] for calculating emissions from managed soils. However, using the same calculation guidelines does not guarantee the same accuracy of results across all calculators.

Each calculator addresses different goals and user groups, and consequently has individual advantages and disadvantages. This is why users have to work out for themselves the balance between efficiency (time and input data) and accuracy (desired output) when deciding which calculator to use.

The integrated methodology and default emission factors given by the calculator as well as the amount of farming processes included in an assessment correspond to the level of input data required. The main limitations in the assessment of energy crop cultivation management are the failure to account for LUC and to distinguish among fertilizer types including digestate, the lack of distinction among tillage types, and the lack of parametrization of many energy crops in the calculators. Furthermore, the impact on the CFP result by using regional GHG emission assessment methodologies is often overlooked. The ability of the calculators to detect GHG mitigation options through improvements in cultivation management is therefore limited. The methodologies used and the farming operations included in any study have a significant impact on the CFP results, thus emphasizing why CFP results should be carefully interpreted. Differences in integrated methodology and accuracy in energy crop cultivation management accounting make any comparison of results from current calculators virtually impossible.

Only seven calculators are capable of calculating GHG emissions from perennial crops and from energy crops in rotation. This may be due to both a lack of methodological guidance to account for crop rotations (or an entire life cycle of a perennial crop, respectively) and a lack of focus on the agronomical specifics of crop rotations systems. Expanding the system boundaries of a CFP by taking into account the whole energy crop rotation increases the likelihood of identifying GHG mitigation options. However, currently, no reviewed calculator can process the effects from energy crops in rotation as nutrient shifts, reduction in use of agricultural operating needs, sequence and composition of crop rotations as well as integration of catch crops of green manuring. To overcome this shortcoming, existing calculators should be extended by integrating energy crop rotations, or new calculators and methods need to be created.

ACKNOWLEDGMENTS

The authors wish to express their gratitude to the members of the project "Development and comparison of optimized cropping systems for the agricultural production of energy crops" (FKZ 22013008), funded by the German Federal Ministry of Food, Agriculture and Consumer Protectionthrough the Agency for Renewable Resources (FNR).

REFERENCES

1. IPCC. Summary for Policymakers. In: Edenhofer O, Pichs-Madruga R, Sokona Y, Farahani E, Kadner S, Seyboth K, et al., editors. Climate Change 2014: Mitigation of Climate Change Contribution of Working Group III to the Fifth Assessment Report of the Intergovernmental Panel on Climate Change. Cambridge, United Kingdom and New York, NY, USA: Cambridge University Press; 2014. p. 1–32.
2. European Commission. Directive 2009/28/EC of the European Parlament and of the Council of 23 April 2009 on the promotion of the use of energy from renewable sources and amending and subsequently repealing Directives 2001/ 77/EC and 2003/30/EC. In: Union OJotE, editor. Brussels: The European Parliament and the Council of the European Union; 2009.
3. Moomow W, Yamba F, Kamimoto M, Maurice L, Nyboer J, Urama K, et al. Introduction. In: Edenhofer O, Pichs-Madruga R, Sokona Y, Seyboth K, Matschoss P, Kadner S, et al., editors. IPCC special report on renewable energy sources and climate change mitigation. United Kingdom and New York, NY, USA: Cambridge University Press; 2011.
4. Allen B., Kretschmer B., Baldock D., Menadue H., Nanni S., Tucker G. Space for energy crops – assessing the potential contribution to Europe's energy future. London: Report produced for BirdLife Europe, European Environmental Bureau and Transport & Environment. IEEP; 2014. p. 1-69.
5. Cherubini F, Bird ND, Cowie A, Jungmeier G, Schlamadinger B, Woess-Gallasch S. Energy- and greenhouse gas-based LCA of biofuel and bioenergy systems: Key issues, ranges and recommendations. Resour Conserv Recycl 2009;53:434–47.
6. Blengini GA, Brizio E, Cibrario M, Genon G. LCA of bioenergy chains in Piedmont (Italy): a case study to support public decision makers towards sustainability. Resour Conserv Recycl 2011;57:36–47.
7. Smith P., Martino D., Cai Z., Gwary D., Janzen H., Kumar P., et al. Agriculture. In: Metz B, Davidson OR, Bosch PR, Dave R, Meyer LA, editors. Climate Change 2007: Mitigation Contribution of Working Group III to the Fourth Assessment Report of the Intergovernmental Panel on Climate Change. Cambridge, United Kingdom and New York, NY, USA: Cambridge University Press; 2007. p. 498– 540.
8. ISO 14040. Environmental management - Life cycle assessment -

Principles and framework. ISO 14040. Geneva: International Standard Organisation; 2006.

9. ISO 14044. Environmental management - Life cycle assessment - Requirements and guidelines. ISO EN 14044. Geneva: International Standard Organisation; 2006.
10. Buratti C, Fantozzi F. Life cycle assessment of biomass production: Development of a methodology to improve the environmental indicators and testing with fiber sorghum energy crop. Biomass Bioenerg 2010;34:1513–22.
11. ISO 14067. Carbon footprint of products — Requirements and guidelines for quantification and communication. ISO EN 14067. Geneva: International Standard Organisation; 2013.
12. Colomb V, Touchemoulin O, Bockel L, Chotte J-L, Martin S, Tinlot M, et al. Selection of appropriate calculators for landscape-scale greenhouse gas assessment for agriculture and forestry. Environ Res Lett 2013;8:015029.
13. Denef K., Paustian K., Archibeque S., Biggar S., Pape D. Report of Greenhouse Gas Accounting Tools for Agriculture and Forestry Sectors. Interim report to USDA under Contract No. GS23F8182H; 2012. p. 1-135.
14. Colomb V, Bernoux M, Bockel L, Chotte J-L, Martin S, Martin-Phipps C, et al. Review of GHG calculations in agricultural and forestry sectors - a guideline for appropriate choice and use of landscape based tools. 2 ed. Angers Cedex, Fance: ADEME (French Environment & Energy Managmnetn Agency), IRD (Istitut de recherche pour le développment) and FAO (Food and agriculture Organization); 2012. p. 43.
15. Hillier J, Hawes C, Squire G, Hilton A, Wale S, Smith P. The carbon footprints of food crop production. Int J Agric Sustain 2009;7:107–18.
16. Bessou C, Lehuger S, Gabrielle B, Mary B. Using a crop model to account for the effects of local factors on the LCA of sugar beet ethanol in Picardy region, France. Int J LCA 2013;18:24–36.
17. Bessou C, Basset-Mens C, Tran T, Benoist A. LCA applied to perennial cropping systems: a review focused on the farm stage. Int J Life Cycle Assess 2013;18:340–61.
18. Brankatschk G, Finkbeiner M. Modeling crop rotation in agricultural LCAs — Challenges and potential solutions. Agric Syst 2015;138:66–76.

19. Nemecek T, Hayer F, Bonnin E, Carrouée B, Schneider A, Vivier C. Designing eco-efficient crop rotations using life cycle assessment of crop combinations. Eur J Agron 2015;65:40–51.
20. Rehl T, Lansche J, Müller J. Life cycle assessment of energy generation from biogas—Attributional vs. consequential approach. Renew Sustain Energy Rev 2012;16:3766–75.
21. Cherubini F, Strømman AH. Life cycle assessment of bioenergy systems: State of the art and future challenges. Bioresour Technol 2011;102:437–51.
22. Buytaert V, Muys B, Devriendt N, Pelkmans L, Kretzschmar JG, Samson R. Towards integrated sustainability assessment for energetic use of biomass: a state of the art evaluation of assessment tools. Renew Sustain Energy Rev 2011;15:3918–33.
23. Audsley E, Albert S, Clift R, Cowell S, Crettaz P, Gaillard G, et al. Harmonisation of environmental life cycle assessment for agriculture. Final Rport Concerted Action AIR3-CT94-2028.Brussels, Belgium: European Commission DG VI; 1997.
24. Scarlat N, Dallemand J-F. Recent developments of biofuels/bioenergy sustainability certification: a global overview. Energ Policy 2011;39:1630–46.
25. Whittaker C, McManus MC, Hammond GP. Greenhouse gas reporting for biofuels: a comparison between the RED, RTFO and PAS2050 methodologies. Energ Policy 2011;39:5950–60.
26. IPCC. IPCC Guidelines for National Greenhouse Gas Inventories. In: Agriculture, forestry and other land use, 4. Hayama, Japan: Prepared by the National Greenhouse Gas Inventories Programme; 2006.
27. Myhre G, Shindell D, Bré on F-M, Collins W, Fuglestvedt J, Huang J, et al. Anthropogenic and Natural Radiative Forcing. In: Stocker TF, Qin D, Plattner G-K, Tignor M, Allen SK, Boschung J, et al., editors. Climate Change 2013: The Physical Science Basis Contribution of Working Group I to the Fifth Assessment Report of the Intergovernmental Panel on Climate Change. Cambridge, United Kingdom and New York, NY, USA: Cambridge University Press; 2013. p. 659–740.
28. Payraudeau S, van der Werf HMG. Environmental impact assessment for a farming region: a review of methods. Agric Ecosyst Environ 2005;107:1–19.
29. Benoist A, Dron D, Zoughaib A. Origins of the debate on the life-cycle

greenhouse gas emissions and energy consumption of first-generation biofuels - A sensitivity analysis approach. Biomass- Bioenerg 2012;40:133–42.

30. Hasler K, Bröring S, Omta SWF, Olfs HW. Life cycle assessment (LCA) of different fertilizer product types. Eur J Agron 2015;69:41–51.
31. Bouwman AF, Boumans LJM, Batjes NH. Modeling global annual N2O and NO emissions from fertilized fields. Glob Biogeochem Cycles 2002;16:1080.
32. Jury C, Benetto E, Koster D, Schmitt B, Welfring J. Life Cycle Assessment of biogas production by monofermentation of energy crops and injection into the natural gas grid. Biomass- Bioenergy 2010;34:54–66.
33. Gissén C, Prade T, Kreuger E, Nges IA, Rosenqvist H, Svensson S-E, et al. Comparing energy crops for biogas production – Yields, energy input and costs in cultivation using digestate and mineral fertilisation. Biomass Bioenergy 2014;64:199–210.
34. IPCC. IPCC Guidelines for National Greenhouse Gas Inventories Chapter 3. Energy, 2. Hayama, Japan: Mobile Combustion; 2006. p. 78.
35. Sorensen CG, Halberg N, Oudshoorn FW, Petersen BM, Dalgaard R. Energy inputs and GHG emissions of tillage systems. Biosyst Eng 2014;120:2–14.
36. López-Bellido L, Wery J, López-Bellido RJ. Energy crops: prospects in the context of sustainable agriculture. Eur J Agron 2014;60:1–12.
37. Zegada-Lizarazu W, Monti A. Energy crops in rotation. A review. Biomass Bioenergy 2011;35:12–25.
38. Molinuevo-Salces B, Larsen SU, Ahring BK, Uellendahl H. Biogas production from catch crops: Evaluation of biomass yield and methane potential of catch crops in organic crop rotations. Biomass Bioenergy 2013;59:285–92.
39. Merker A, Eriksson D, Bertholdsson N-O. Barley yield increases with undersown Lepidium campestre. Acta Agriculturae Scandinavica, Section B — Soil &. Plant Sci 2010;60:269–73.
40. Weiss F, Leip A. Greenhouse gas emissions from the EU livestock sector: A life cycle assessment carried out with the CAPRI model. Agric Ecosyst Environ 2012;149:124–34.
41. Weidema BP, Bauer C, Hischier R, Mutel C, Nemecek T, Reinhard J,

et al. The ecoinvent database: overview and methodology Data Qual Guidel ecoinvent Database Version 3; 2013.

42. G. Gaillard, T. Nemecek. Swiss Agricultural Life Cycle Assessment (SALCA): an integrated environmental assessment concept for agriculture; 2009.
43. BioGrace. ⟨http://www.biograce.net/⟩ accessed on 13.08.2015.
44. Hennecke AM, Faist M, Reinhardt J, Junquera V, Neeft J, Fehrenbach H. Biofuel greenhouse gas calculations under the European Renewable Energy Directive – A comparison of the BioGrace tool vs. the tool of the Roundtable on Sustainable Biofuels. Appl Energy 2013;102:55–62.
45. Brankatschk G, Finkbeiner M. Application of the Cereal Unit in a new allocation procedure for agricultural life cycle assessments. J Clean Prod 2014;73:72–9.
46. Fitton N, Ejerenwa CP, Bhogal A, Edgington P, Black H, Lilly A, et al. Greenhouse gas mitigation potential of agricultural land in Great Britain. Soil Use Manag 2011;27:491–501.
47. Walter K, Don A, Fuß R, Kern J, Drewer J, Flessa H. Direct nitrous oxide emissions from oilseed rape cropping – a meta-analysis. GCB Bioenergy 2014 (n/a-n/a).
48. KTBL. Faustzahlen für die Landwirtschaft. Darmstadt: Kuratorium für Technik und Bauwesen in der Landwirtschaft; 2009.
49. Koblenz B, Tischer S, Rücknagel J, Christen O. Influence of biogas digestate on density, biomass and community composition of earthworms. Amsterdam, The Netherlands: Industrial Crops and Products; 2015. p. 206–9.
50. Zeshan Visvanathan C. Evaluation of anaerobic digestate for greenhouse gas emissions at various stages of its management. Int Biodeterior Biodegrad 2014;95:167–75.
51. Coleman K, Jenkinson DS. RothC-26.3. A model for the turnover of carbon in soil: model description and user's guide. Harpenden, UK: Lawes Agricultural Trust; 1999.
52. Petersen BM, Knudsen MT, Hermansen JE, Halberg N. An approach to include soil carbon changes in life cycle assessments. J Clean Prod 2013;52:217–24.
53. Bessou C, Ferchaud F, Gabrielle B, Mary B. Biofuels, greenhouse gases and climate change. A review. Agron Sustain Dev 2011;31:1–79.
54. Bouwman AF, Boumans LJM, Batjes NH. Emissions of N2O and NO

from fertilizes fields: Summary of available measurements data. Glob Biogeochem Cycles 2002;16:1058.

55. Christiane Peter, Angela Fiore, Ulrike Hagemann, Claas Nendel, Cristos Xiloyannis. Improving the accounting of field emissions in the carbon footprint of agricultural products: a comparison of default IPCC methods with readily available medium-effort modeling approaches. The International Journal of Life Cycle Assessment 2016;21(6):791–805.

INDEX

D

E

F

G

H

I

K

L

M

N

O

P